(1) QUEEN ELIZABETH I - engraving attributed to Remigius Hogenburg printed in Christopher Saxton's *Atlas of the Counties of England and Wales* (1579), depicting Elizabeth enthroned between the pillars of Wisdom and Intelligence.

(2) QUEEN ELIZABETH I - engraving by George Virtue, from drawing by Isaac Oliver, depicting the phoenix rising from the flames of its pyre.

THE FRANCIS BACON RESEARCH TRUST JOURNAL
Series I, Volume 5:
Festival of Unification

ARCADIA

by

Peter Dawkins

The Ancient Egyptian Mysteries

Arcadia and the Arcadian Academy

(The Life and Times of Francis Bacon, 1579-1585)

STUDIES IN ANCIENT WISDOM

First published by The Francis Bacon Research Trust in 1988
Printed in England by Coventry Printers

(Series 1) ISSN: 0262 8228
(Vol.4) ISBN: 0 86293 008 1

ACKNOWLEDGEMENTS

The author wishes to acknowledge with gratitude the help given him by the Marquess of Northampton, the Earl and Countess of Verulam, Mr. Thomas Bokenham, Mr. Noel Fermor, Mrs. Elizabeth Kirsten, Mr. Paul Lucas, Mr. David Millward, Mrs. Donna Willetts, Mrs. Elizabeth Wrigley, The Francis Bacon Society (UK), The Francis Bacon Library (USA), The Rosicrucian Order (AMORC); the staff and trustees of the various Museums, Art Galleries and Art Collections who have allowed us to reproduce their paintings, drawings and other works of art; the authors of the books referred to or quoted from; and all those many others who have helped, contributed to or supported in one way or another the work of the Francis Bacon Research Trust.

The Francis Bacon Research Trust
is a Registered Charitable Trust
(No. 280616) in the UK.

All information regarding the work
and membership of the FBRT is available from:

The Secretary, FBRT,
Tl … er,

FRANCIS BACON RESEARCH TRUST
Roses Farmhouse, Epwell Road
Upper Tysoe, Warwick
CV35 0TN, England
Tel: 0044 (0)1295 688185
Fax: 0044 (0)1295 680770

CONTENTS

Contents Page 5
List of Illustrations 7
The Francis Bacon Research Trust Journals 12

PREFACE 15

I. THE ANCIENT EGYPTIAN MYSTERIES
INTRODUCTION
The May Festival of Unification and the 'Air' Initiation 21
Egyptian Pre-History 27
The Master Plan and High Council of Ancient Egypt 32
The Temple of Egypt 36
The Country of Egypt 37
The Seven Great Centres of Egypt 39
Incarnation and Resurrection 44
The A-B-C of Life 51

THE HERMETIC ART OF ALCHEMY
The Scarab 54
The Three Works of Alchemy 54
The Thirteen Steps of Alchemy 55

THE EGYPTIAN MYSTERIES OF INITIATION
Introduction 57
Preparation for Initiation 58
The Lesser Mysteries 58
The Greater Mysteries 60
The Offices 62

II. THE ARCADIAN ACADEMY
(THE LIFE AND TIMES OF FRANCIS BACON, 1579-1585)
Return to England 71
John D: the Magus 87
The English Areopagus and Arcadia 100
Immerito 110
Spenser 129
Euphues 143
Athena's Veil 158
The 'AA' Signature of Light 181
A Rare and Unaccustomed Suite 207
Secret Intelligence and the Grand Tour 214
The Greatest Birth of Time 227
The Catholic Menace 245

APPENDIX
Dr. John Dee's *General and Rare Memorials* 251

NOTES AND BIBLIOGRAPHY
Notes: Part I 258
Notes: Part II 271
Select Bibliography 296

ILLUSTRATIONS

Page	*Ref.*	**Diagrams:**
22	A.	The Cycle of Seasons
24	B.	The Cycle of Initiation
30	C.	The Tree of Life
31	D.	Egyptian Tree of Life
41	E.	Map of Egyptian Landscape Temple
43	F.	The Chakras in Man
45	G.	Egyptian Cosmology
63	H.	Ladder of Initiation (Egyptian)
64	I.	The Tree of Life (with officers)
81	J.	Celestial Map
223	K.	Map of Europe

Page	*Ref.*	**Portraits:**
1	1.	Queen Elizabeth I - engraving attrib. to Remigius Hogenberg from *Atlas of the Counties of England and Wales* (1579), by Christopher Saxton. *Reproduced by courtesy of the Trustees of the British Museum. (Ref. 1934-6-41).*
2	2.	Queen Elizabeth I - engraving by George Vertue, after Isaac Oliver. *Reproduced by courtesy of the Trustees of the British Museum. (Ref. 105670-964-Y-5.12.6).*
70	6.	Francis Bacon, aged 18 years (1578/9) - miniature by Nicholas Hillard. *Reproduced from Baconiana, Vol. L. No. 167, Journal of the Francis Bacon Society.*
73	7.	Queen Elizabeth I - Charter authorising Sir Walter Mildmay to found Emmanuel College, Cambridge, 11th January 1584 - illuminated manuscript by Nicholas Hilliard and assistants. *Reproduced by courtesy of Emmanuel College, Cambridge.*
74	8.	Robert Dudley, Earl of Leicester - painting (Parham Park). *Reproduced by courtesy of Mrs. P.A. Tritton, Parham Park, W. Sussex.*
78	10.	Ambrose Dudley, Earl of Warwick (1528?-1590) - painting entitled "Sir Henry Sidney". *Reproduced by courtesy of Mrs. P. A. Tritton, Parham Park, W. Sussex.*
79	11.	Sir Henry Sidney (1573) - painting, artist unknown. *Reproduced by courtesy of The National Trust, photography by the Courtauld Institute of Art.*
82	12.	William Cecil, 1st Baron Burghley (*c*.. 1585) - painting attributed to Marcus Gheeraerts, junior. *Reproduced by courtesy of the National Portrait Gallery (NPG 362).*
83	13.	Robert Cecil, 1st Earl of Salisbury (1602) - painting by John de Critz . *Reproduced by courtesy of the National Portrait Gallery (NPG 107).*
84	14.	Sir Francis Walsingham - line drawing by Simon Passe for Holland's *Herwologia* (1620). *Reproduced by courtesy of the Trustees of the British Museum. (Ref. C210b2 P.82).*
86	15.	John Dee - painting, artist unknown. *Reproduced by courtesy of the Ashmolean Museum, Oxford.*
95	18.	Queen Elizabeth I as 'Astraea' - engraving by F. Delarum after N. Hilliard. *Reproduced by courtesy of the Trustees of the British Museum. (Ref. 1849-3-15-1).*
99	19.	Sir Philip Sidney (*c.* 1577) - engraving, artist unknown, from painting (see illustration no. 43).
131	35.	Sir Walter Raleigh (*c.*1585) - miniature by Nicholas Hilliard. *Reproduced by courtesy of the National Portrait Gallery (NPG 4106).*
136	38.	Robert Devereux, 2nd Earl of Essex (*c.*1588) - "Portrait of a Nobleman", miniature by Nicholas Hilliard. *Reproduced by kind permission of the Metropolitan Museum of Art, Fletcher Fund, 1935 (Ref. 35.89.4).*
140	40.	Mary Sidney, Countess of Pembroke (1618) - engraving by Simon van de Passe. *Reproduced by courtesy of the National Portrait Gallery (NPG 8445).*

Page	*Ref.*	
141	41.	Henry Herbert, 2nd Earl of Pembroke - line drawing by Simon Passe for Holland's *Herwologia* (1620). *Reproduced by courtesy of the Trustees of the British Museum. (Ref. C210b2 P115).*
142	42.	Sir Philip Sidney (*c.* 1577)- painting by unknown artist. *Reproduced by permission of the Marquess of Bath, Longleat House, Warminster, Wiltshire.*
154	47.	Queen Elizabeth I as 'Diana' (*c.* 1560) - painting attributed to Frans Floris. *Reproduced by courtesy of The Marquess of Salisbury, Hatfield House, photography by the Courtauld Institute of Art.*
206	76.	Anthony Bacon, aged 30 years (1588) - miniature by Nicholas Hilliard. *Reproduced by courtesy of the Board of Trustees of the Victoria and Albert Museum (P21-1942).*
217	78.	Sir Thomas Bodley (1598) - miniature by Nicholas Hilliard. *Reproduced by courtesy of the Bodleian library (Pr. 10950).*
238	94.	Frances Walsingham - miniature by Isaac Oliver. *Reproduced by courtesy of the Board of Trustees of the Victoria and Albert Museum (GH 1576).*
242	95.	Sir Francis Drake (1581) - miniature by Nicholas Hilliard. *Reproduced by courtesy of the National Portrait Gallery (NPG 4851).*
244	96.	Mary Queen of Scots (1578) - painting after Nicholas Hilliard. *Reproduced by courtesy of the National Portrait Gallery (NPG 429.).*

Page	*Ref.*	**Illustrations Generally:**
14	3.	'Caduceus' illustration from *Das Buchdruckerzeichen des Johannes Froben* (Basel, 1523), by Hans Holbein. *Reproduced by courtesy of Offtenliche Kunstammlung Basel, Kunstmuseum.*
19	4.	"Sapientia Pacem: Pax Opulentiam" ("Wisdom brings Peace: Peace brings Prosperity") - emblem from titlepage to *Florilegium Epigrammatum Græcorum, eorumque Latino Versu à varijs redditorum* (London, 1629) - a collection of Greek epigrams (including a poem by Francis Bacon parodying the Latin version of the *Seven Ages of Man)* collected, translated and published by Thomas Farnaby.
69	5.	"Anchora Spei" ("Anchor of Hope") - emblem from titlepage to *The Art of English Poesie* (London, 1589).
76	9.	Map of London by Brown and Hogenberg, 1572. *Reproduced by courtesy of the Museum of London.*
88	16.	John Dee's Coat of Arms, as depicted at the end of his *Letter Apologeticall* (1599).
92	17.	Titlepage, *Monas Hieroglyphica* (Antwerp, 1564) by Dr. John Dee. *Reproduced by permission of the British Library (Ref. 718g.6. TP 8770168).*
102	20.	First page of the letter to Gabriel Harvey by E.K., foreword to *The Shepheards Calender* (London, 1611) by Edm. Spenser.
105	21.	'Pallas Athena' emblem from titlepage to *Nova Atlantis* (1643) by Francis Bacon.
105	22.	'Pallas Athena' emblem from titlepage to *Scripta et universali Philosophia* (1653) by Francis Bacon.
107	23.	Titlepage, *Hymne to Elizabeth I* (1586) by George de la Mothe - manuscript. *Reproduced by courtesy of the Bodleian Library, Oxford (Ms. Fr. e.1, fol. 7r).*
108	24.	Poem to Anthony Bacon in French - anonymous. *Reproduced by courtesy of the Lambeth Palace Library (Ms. 653 Item 159).*
113	25.	"Tempore patet occulta Veritas" ("Hidden Truth brought forth by Time") - emblem from titlepage to *New Atlantis* (publ. with *Sylva Sylvarum,* 1626) by Francis Bacon.
113	26.	'Father Time' emblem from the *Countess of Pembroke's Arcadia* (1625) by Sir Philip Sydney.
115	27.	Page 1, 'Januarie', *The Shepheards Calender* (London, 1611) by Edm. Spenser.
117	28.	Titlepage, *The Shepheards Calender* (London, 1611) by Edm. Spenser.
119	29.	Titlepage, *The Shepheards Calender* (London, 1617) by Edm. Spenser.
121	30.	"To His Booke", preface to *The Shepheards Calender* (London, 1611) by Edm. Spenser.
123	31.	"To His Booke", preface to *The Shepheards Calender* (London, 1617) by Edm. Spenser.
124	32.	"Sub Umbra Alarum Tuarum" ("Beneath the Shadow of Thy Wings") - emblem assigned to Sir Francis Bacon from *The Mirror of Modestie* (1618).

Page	*Ref.*	
127	33.	Memorial (restored 1778) to Edm. Spenser, Westminster Abbey.
128	34.	Engraving of original memorial to Edm. Spenser, Westminster Abbey, from the 1679 edition of *The Faerie Queene.*
133	36.	Titlepage, *Complaints* (London, 1591) by Edm. Spenser.
134	37.	Titlepage, *Colin Clouts Come Home Again* (London,1595) by Edm. Spenser.
139	39.	St. George and the Dragon - woodcut from *The Faerie Queene* (1590) by Edm. Spenser.
145	43.	Dedication page, *Virgils Gnat* (1617), Spenser Folio.
146	44.	Titlepage, *The Countesse of Pembrokes Arcadia* (London, 1593) by Sir Philip Sidney. *Reproduced by permission of the British Library (Ref. C.21.D.21).*
147	45.	Titlepage, Part I, *The Faerie Queene* (London, 1611) by Edm. Spenser.
153	46.	Titlepage, *Rosalynde, Euphues golden legacie* (London, 1590) by T.L..
157	48.	Titlepage, *Orlando Furioso* (1591) translated by Sir John Harington.
158	49.	'Pallas Athena' frontispiece, *La Sagesse Mysterieuse des Anciens* (1641) by Francis Bacon.
161	50.	Dedication page, *Attourney's Academy (1630) by Thomas Powell.*
163	51.	Titlepage, *Minerva Britanna* (London, 1612) by Henry Peacham.
164	52.	'Shake-spear' emblem, page 33, *Minerva Britanna* (London, 1612) by Henry Peacham.
165	53.	'The Shepherd Knight' emblem, page 34, *Minerva Britanna* (London, 1612), by Henry Peacham.
167	54.	Titlepage, *Memoria Honoratissimi Domini Francisci, Baronis de Verulamio, vice-Comitis Sancti Albani Sacrum* - otherwise known as the *Manes Verulamiani* (London, 1626).
171	55.	Titlepage, *The Great Assises holden in Parnassus* (London, 1645) by "Apollo and his Assessours".
172	56.	Page III, *The Great Assises holden in Parnassus* (London, 1645) by "Apollo and his Assessours".
172	57.	Page IV, *The Great Assises holden in Parnassus* (London, 1645) by "Apollo and his Assessours".
175	58.	'Bacon' engraving from *The Mirrour of State & Eloquence* (1656). *Reproduced by permission of the British Library (Ref. BL E891 (10) FP 8671275).*
176	59.	Frontispiece, *Of the Advancement and Proficience of Learning* (Oxford, 1640) by Francis Bacon.
180	60.	Titlepage, *De Furtivis Literarum Notis, Vulgo De Ziferis* (Naples, 1563) by Ioan Baptista Porta.
183	61.	"In Dies Meliora" emblem from *Emblemata* (1546) by Alciat. *Reproduced by permission of the British Library (Ref. 8674345 245 C.7 page 33).*
184	62.	"In Dies Meliora" emblem from *Emblemata* (1564) by Alciat. *Reproduced by courtesy of the Francis Bacon Library, California, U.S.A.*
185	63.	"In Dies Meliora" emblem XLV from *Emblemata* (1577) by Alciat. *Reproduced by permission of the British Library (Ref. 8674345 12305-bbb 36 p204).*
186	64.	Francis Bacon's 'Boar' crest, from the binding of a presentation copy of his book, *Novum Organum* (1620).
187	65.	Francis Bacon's Achievement of Arms - frontispiece to Vol. II of Bacon's Works (1740).
189	66.	"Mediolanum" emblem from *Emblemata* (1546) by Alciat. *Reproduced by permission of the British Library (Ref. 8674345 245.c.7 page 45).*
190	67.	"Mediolanum" emblem from *Emblemata* (1564) by Alciat. *Reproduced by courtesy of the Francis Bacon Library, California, U.S.A.*
191	68.	"Mediolanum" emblem II from *Emblemata* (1584) by Alciat. *Reproduced by permission of the British Library (Ref. 8674345 .C.48 a18p19).*
192	69.	"Mediolanum" emblem from *Emblemata* (1589) by Alciat. *Reproduced by courtesy of the Francis Bacon Library, California, U.S.A.*
196	70.	The Egyptian hand plough - emblem of the rulers of the earth and shepherds of the people.
199	71.	Titlepage, *De Rep. Anglorum Instauranda Libri* by Thomas Chalonero (London, 1579). *Reproduced by permission of the British Library (Ref. 1070 M. 31 T.P. 8675139).*
200	72.	Titlepage, *The Essayes of A Prentise in the Divine Art of Poesie* (printed in Edinburgh by Thomas Vautrollier, 1584). *Reproduced by permission of the British Library (Ref. 8675139 G11237 TP).*
201	73.	Titlepage, *The Arte of English Poesie* (printed in London by Richard Field, 1589). *Reproduced by permission of the British Library (Ref. C71 c16 TP 777740).*
202	74.	Titlepage, *Venus and Adonis* (1593 Quarto) by Shakespeare. *Reproduced by permission of the British Library (Ref. G11180 TP 8474471).*

Page	Ref.	
203	75.	Dedication page, *Venus and Adonis* (1593 Quarto) by Shakespeare. *Reproduced by permission of the British Library (Ref. 8675139 G8 1180 DP).*
210	77.	Dedicatory page, *Hymne to Elizabeth I* (1586) by George de la Mothe. *Reproduced by courtesy of the Bodleian Library, Oxford (Ms. Fr. e.1, fol. 7 r).*
221	79.	Titlepage, *An Apologie* (London, 1564), translated by Anne Bacon.
225	80.	Titlepage, *The Whole Booke of Psalmes* (London, 1583).
226	81.	'Cupid' (or 'Archer') headpiece, *The Faerie Queene* (1609), by Edm. Spenser.
226	82.	'Cupid' (or 'Archer') headpiece, *The Consecration of Bishops* (1613) by Franciscus Mason.
226	83.	'Cupid' (or 'Archer') headpiece, *Novum Organum* (1620), by Francis Bacon.
226	84.	'Cupid' (or 'Archer') headpiece, Shakespeare 1st Folio (1623).
229	85.	'Pan' tailpiece, *Philostrate* (Paris, 1578) by Blaise de Vigenére.
229	86.	'Pan' tailpiece, *Histoire du Monde* (Lyon, 1583), by Plinius.
229	87.	'Pan' tailpiece, *The Faerie Queene* (1606), by Edm. Spenser.
230	88.	'Pan' tailpiece, Shakespeare 1st Folio (1623).
230	89.	'Pan' tailpiece, *Opera Omnia* (Frankfurt, 1665) by Francis Bacon.
233	90.	First page of 'The First Booke' of *The Faerie Queen* (London, 1617) by Edm. Spenser.
234	91.	Titlepage, Part II, *The Faerie Queene* (London, 1617) by Edm. Spenser.
235	92.	Last page, *The Faerie Queene* (London, 1609/11) by Edm. Spenser.
236	93.	'A Letter of the AV', *The Faerie Queene* (London, 1617) by Edm. Spenser.
247	97.	Titlepage, *Lucrece* (London, 1594) - anonymous but ascribed to Shakespeare.
250	98.	Titlepage, *General and Rare Memorials pertayning to the Perfect Arte of Navigation* (1577) by Dr. John Dee. *Reproduced by permission of the British Library (Ref. 48h18 TP 714394).*
256	99.	Page 69, *General and Rare Memorials pertayning to the Perfect Arte of Navigation* (1577) by Dr. John Dee.
257	100.	Page 53 from *General & Rare Memorials pertayning to the Perfect Arte of Navigation* (1577) by Dr. John Dee.

Page	Ref.	**Headpieces:**
15	i.	'Arcadian Bear' headpiece to Raleigh's *History of the World* (1665); also to be found in *Ælii Autonii Dictionarium* (Madrid, 1683).
21	ii.	'AA' headpiece to Baptista Porta's *De Furtivis Literarum* (1563), *The Essayes of a Prentise in the Divine Art of Poesie* (1584, 1585), De Loqué's *Single Combat* (1591), Shakespeare's *Taming of a Shrew* (1594), Hartwell's *Warres* (1595), Heywood's Works (1598), Hayward's *Of the Union* (1604), Paschal's *Coronae* (1671); design also to be found in *Striaticos* (1590), Shakespeare's *Contention of York and Lancaster,* Part I (1594), *Romeo and Juliet* (1599), *Henry V* (1598, 1600), *Sir John Falstaffe* (1602), *Richard III* (1602), *Regimen Sanitatis Salerni* (1597).
54	iii.	'AA' headpiece to Philip Sydney's English translation of Du Plessis Mornay's *De la Verité de la Religion Christienne* (1587); design also to be found in reduced form in Hardy's *Le Theatre* (1624).
57, 259	iv.	'AA' headpiece to *Arte of English Poesie* (1589), Mayer's *Praxis Theologica* (1629), Ben Jonson's Works, Vol.2 (1640).
71	v.	'AA' headpiece to Sir John Harrington's translation of *Orlando Furioso* (1591), Shakespeare's *Venus and Adonis* (1593), *Unnatural conspiracie of Scottish Papists* (1593), Spenser's *Faerie Queene* (1596), *Historie of Tamerlane* (1597), Barckley's *Felicitie of Man* (1598), *Nosce te ipsium* (1602).
87	vi.	'AA' headpiece to *The Florentine History* (1595); design also used in Barclay's *Argenis* (1636).
100	vii.	'Pan and Satyrs' headpiece to Spenser's *Faerie Queen* (1617).
110	viii.	'AA' headpiece to Shakespeare's *Richard II* (1597), *Richard III* (1597), *Henrie IV* (1600), *Hamlet* (1603), *Sonnets* (1609), and Matheieu's *Henri IV* (1612). Design also to found in double form in Seneca's Works (1614).
129	ix.	'IHVH' headpiece, Spenser 1st Folio (1611).
143, 271	x.	'AA' headpiece to *Daemonology* (1603), Spenser's *Faerie Queene* (1609). Design also found in enlarged form in Shakespeare 1st Folio (1623).

Page	Ref.	
159	xi.	'AA' headpiece to Spenser's *Shepheard's Calender* (1617), *The Rogue* (1622), Barclay's *Argenis* (1636), Bacon's *Remaines* (1648) and *The Mirrour of State* (1656); design also to be found in *Discourse upon the meanes of well Governing* (1602).
181	xii.	'AA' headpiece to Hardy's *Le Theatre,* Vol.4 (1626), Barclay's *Argenis* (1625-26), Aleman's *Le Gueux* (1632). Design also to be found in Sydney's *L'Arcadie* (1625).
207	xiii.	"Boar's Head' headpiece to Martyn's *Lives of the Kings* (1615), Seneca's Works (1620), Slatyer's *Great Britaine* (1621), Bacon's *Resuscitatio,* Part II (1671).
214	xiv.	'AA' headpiece to "Upon the Lines and Life...", Shakespeare 1st Folio (1623); *c.f. Daemonology* (1603), etc..
227	xv.	'Pan and Conies' headpiece to Bacon's *Henry VII* (1629), *New Atlantis* (1631) and Essay *Of Fame* in *Resuscitatio* (1657).
245	xvi.	'Pan and Cornucopiæ" headpiece to *Historia Regni Henrici Septimi* (1642) by Francis Bacon.
251, 296	xvii.	'IHVH' headpiece, Spenser 1st Folio (1611).

Page	Ref.	**Tailpieces:**
17, 109	xviii.	'Father Time' ('Pan') emblem, *L'Arcadie* (1625) by Philip Sydney.
53, 56, 66	xix.	'Caduceus and Cornucopia' emblem from titlepage, *Florilegium Epigrammatum Grecorum Eorumque Latino Versu a varijs redditorum* (1629), by Thomas Farnaby, which contains Francis Bacon's poetic parody on the Latin version of *The Seven Ages of Man.*
85, 299	xx.	"Tempore Patet Occulta Veritas" ("Time brings forth Hidden Truth") emblem from titlepage, *New Atlantis* (1627) by Francis Bacon.
98	xxi.	'Skull and Birds' tailpiece, Spenser 1st Folio (1611).
126	xxii.	'Skull and Fruit' tailpiece, Spenser 1st Folio (1611).
179	xxiii.	'Skull and Cornucopiæ' tailpiece, Spenser 1st Folio (1611).
205	xxiv.	'Skull and Baskets' tailpiece, Spenser 1st Folio (1611).
213	xxv.	'Resurrection' tailpiece, Spenser 1st Folio (1611).
224	xxvi.	'Death enthroned' tailpiece, Spenser 1st Folio (1611).
243	xxvii.	'Time enthroned' tailpiece, Spenser 1st Folio (1611).
249	xxviii.	'Cherub' tailpiece, Spenser 1st Folio (1611).
255	xxix.	'T' emblem, page 54, *General and Rare Memorials* (1577) by Dr. John Dee.
270	xxx.	'S' emblem, page 65, *General and Rare Memorials* (1577) by Dr. John Dee.
295	xxxi.	'L' emblem, page 100, *General and Rare Memorials* (1577) by Dr. John Dee.

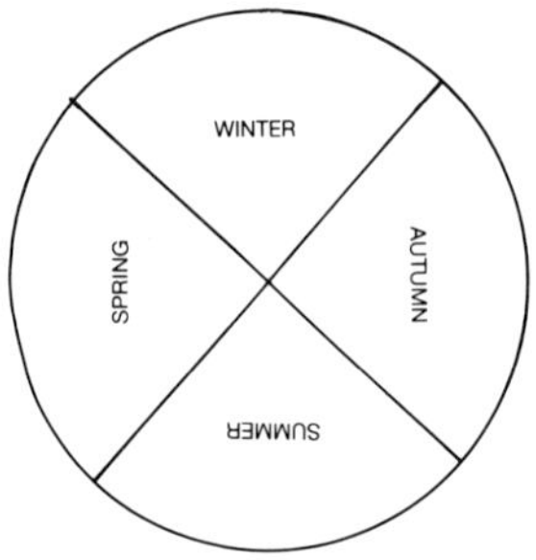

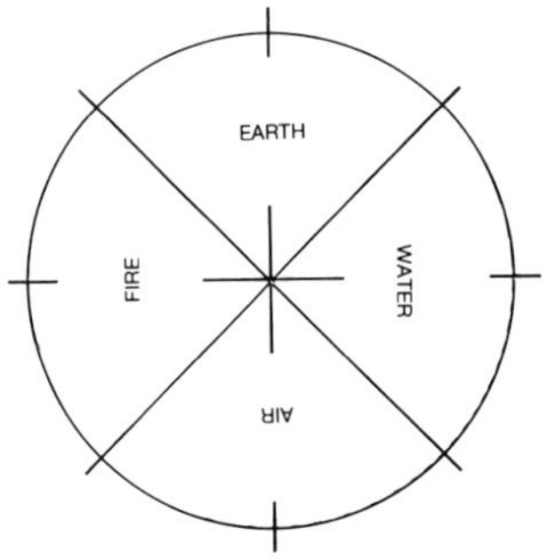

The F.B.R.T. Journals

The underlying purpose of the F.B.R.T. Journals is to give an insight into the complete initiatory path that all the great Religions or Wisdom Traditions have taught from time immemorial, as being the Way, the Truth and the Life. The scheme of the Journals is to cover the whole path of initiation in a series of sixteen volumes or Journals. Each Journal will mark a particular phase in the double-cycle of initiation (i.e. of the Lesser and Greater Mysteries). There are eight phases or initiatory points in each cycle, which the eight major Solar Festivals of each year also celebrate - four of them marking the end of a season and the beginning of the next, and four marking the mid- or high-points of the seasons. The degrees of initiation correspond to the seasons. (See FBRT Journal I/1: The Pattern of Initiation.)

The series of Journals includes studies of:

(A) the festivals or initiatory focal points that each Journal commemorates.

(B) the various principal Wisdom Traditions, which, although clothed in differing guises, all teach and practice the same fundamental knowledge of truth.

(C) the life and times of the great initiate who is known to the world as Francis Bacon, whose life demonstrated the initiatory path, and whose work specifically and purposefully laid the foundations for the eventual human achievement of a planetary civilisation living in peace, prosperity and enlightenment.

(D) Baconian teachings, and interpretations of the Ancient Wisdom teachings.

(E) The Great Instauration or Rosicrucian scheme for the complete renovation of all science and art the world over, and the subsequent instauration of a Golden Age of Illumined Peace.

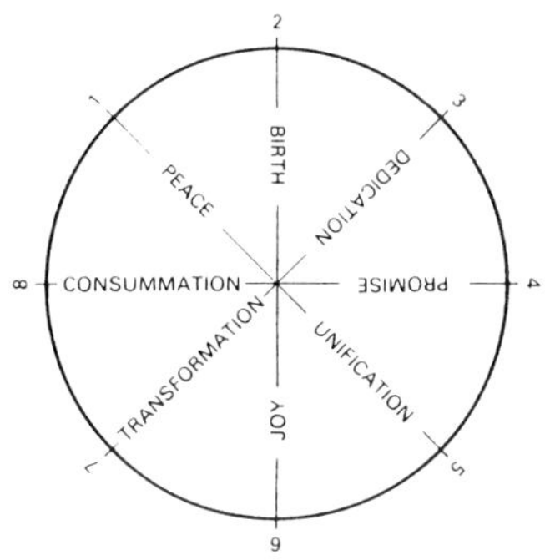

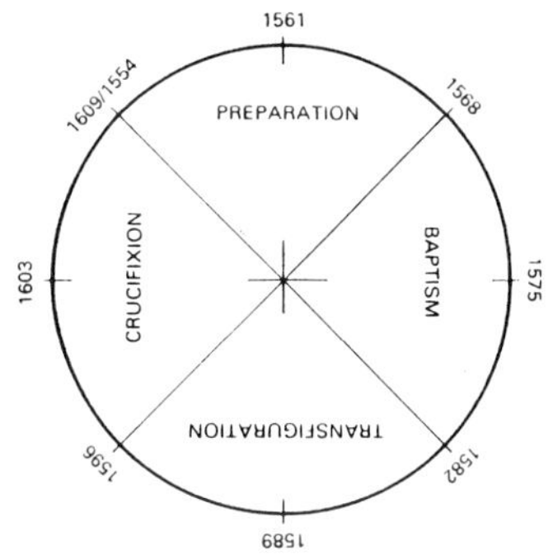

FBRT Journal I/1: ***The Pattern of Initiation,*** provides an overall view of the pattern of initiation in the evolution of human consciousness.

FBRT Journal I/2: ***The Virgin Ideal,*** is a study of the meaning and role of virginity in human life - a role which Queen Elizabeth I of England personified in allegorical terms.

FBRT Journal I/3: ***Dedication to the Light,*** introduces the Bardic Mysteries, with the allegorical story of the birth, adoption and initiation of Taliesin, the wonder-child, and parallels this with the story of Francis Bacon's secret royal birth, adoption and special upbringing.

FBRT Journal I/4, ***The Great Vision,*** deals in depth with the Judaic-Christian Mysteries, and continues the story of Francis Bacon's life, with an account of his great vision, his discovery of his royal birth, his subsequent banishment to France, his romance and his adventures with the French Court, and the revival of Arthurian and Grail traditions in Tudor England in the form of grand pageantry.

FBRT Journal I/5, ***Arcadia,*** in the first part of the book introduces the Ancient Egyptian Mysteries and Hermeticism, and the landscape temple in which those alchemical Mysteries took place. In the second part the story of Francis Bacon's life is continued, dealing with the secrets of Arcadia, the English Areopagus or brotherhood of poet-initiates who largely created the English Renaissance, the founding of the first group of new (or Baconian) Rosicrucians in Arcadian Britain, and the secret beginnings of modern Freemasonry.

(3) THE HERMETIC STAFF - illustration by Hans Holbein from *Das Buchdruckerzeichen des Johannes Froben* (1523).

PREFACE

> *"Speaking of human wisdom only, I do certainly for my own part incline to this opinion, - that beneath no small number of the fables of the ancient poets there lay from the very beginning a mystery and an allegory........ For so they must be regarded as neither being the inventions nor belonging to the age of the poets themselves, but as sacred relics and light airs breathing out of better times, that were caught from the traditions of more ancient nations and so received into the flutes and trumpets of the Greeks."*

> *"Parables have been used in two ways, and (which is strange) for contrary purposes. For they serve to disguise and veil the meaning, and they serve also to clear and throw light upon it........"*

This Journal, *Arcadia,* is the fifth in the series of FBRT Journals. The first part of the Journal provides an introduction to the Ancient Egyptian Mysteries, giving a rare insight into the greatness of that long-lasting Pharaonic civilisation, and of its land which was the living temple in which the alchemical mysteries of life and initiation took place according to recognised law. The second part of the Journal deals with the life of Francis Bacon between the years 1579-1585, when he was aged 18-24 years old, during which time he set in motion the first stages of the vast Rosicrucian scheme for the eventual enlightenment, prosperity and peace of the whole wide world.

This scheme, for which Francis Bacon devised a practical method to enable it to be carried out, was the dream of the 'Rosicrucian' sages and their disciples since the golden days of Atlantis. The ancient world civilisation of Atlantis gradually degenerated and perished in a series of cataclysms that destroyed large portions of their inhabited landmasses; but, throughout this period of disintegration of a once glorious civilisation, and afterwards, when the new races of humanity grew up in other portions of the globe, the sages held the dream and gradually prepared themselves and those who would listen to them for the restoral of a universal Golden Age of peace and concord.

One of the earlier civilisations to arise from the ashes of Atlantis, although not on such a large scale, and which seriously attempted to create a Golden Age for itself, was that of Ancient Egypt. Guided by great sages and an enormous body of proven traditional knowledge concerning the laws of nature and divinity, pharaonic Egypt achieved a special level of civilisation and illumination that is a shining example to us today. From the Egyptian sages the Greek poets and philosophers learnt the best of their wisdom; but, as Francis Bacon pointed out, the knowledges were veiled in allegorical fable, first by the Egyptians and then even more so by the Greeks and other nations. This was a method which worked, and still works to this day. Bacon's own method incorporates this system of veiling, which performs like a game of hide and seek - a glorified treasure hunt: for, as Solomon so wisely perceived, God hides the truth so that man might find it out.

Arcadia was one of the primary mystery names given to the state of the world in a Golden Age, and the whole Arcadian imagery and allegory is that of a Utopia or Paradise on earth that might be achieved. The world is on that very threshold of achievement now, according to the way the evolutionary cycles of divine law work. Jesus Christ and the Buddha prepared us for it in an outstanding way: Francis Bacon (alias the Master 'R') and the Rosicrucian Brotherhood have inaugurated it. All have used the Arcadian imagery, the meaning of which pierces deep, right to the heart of all mysteries.

Francis Bacon and a small group of friends began their work using Arcadian symbolism from the outset, and conjoined this symbolism with that of King Arthur and the Rose (or Red) Cross Knight. FBRT Journal I/4, *The Great Vision,* discussed some of the Rosicrucian and Arthurian significances in depth. This Journal points the way to perceiving the Arcadian and Hermetic truths.

Noone lives or works in isolation, not even the greatest of souls, and Francis Bacon was no exception. He was, beyond any shadow of doubt and according to eye-witness testimony, a marvel of a man - a supreme genius and benefactor of the world, acknowledged founder of modern science (and hence of modern civilisation), prophet, philosopher, poet, philanthropist, and promoter of civil liberties and justice. There was no avenue, it seems, into which his great heart and mind did not stretch. The boundaries of his mind, his consciousness, are so broad that it is hard for us to comprehend the scope of his genius and what it is capable of. Disbelief is a fairly normal reaction to the powers and works of a master soul. But Francis needed assistance, as does everyone, and not all the inspired or divine ideas came first through his mind. Many other good and talented souls played their part; some to prepare the way for Francis, some to work with him, and others to carry on and develop the work afterwards. Dr. John Dee was a notable figure who both taught and assisted Francis Bacon in the earlier stages of the Baconian-Rosicrucian work. Robert Fludd was another who exerted an important influence. But so did many others, their talents and work often unsung or unnoticed by posterity, or perhaps deliberately veiled.

I have tried in this Journal to present as broad a picture as I possibly can of this quite amazing period in the life of Francis Bacon and the related history of Elizabethan England, my aim being to focus upon that history which is more veiled or secret from public knowledge rather than to deal with matters which are published elsewhere in abundance. My hope is that, whilst revealing some things previously veiled, other veils to the truth will be pointed out, to help lead you, the reader, further along the path of the great Treasure Hunt of life.

Out of this crucible of English history has sprung modern science, modern society, an international language and the real possibility, if we get our love right, of a lasting and universal Golden Age - a "New Atlantis".

Love is the key. Love opens the eyes to see, to pierce the veils that conceal the truth. Love, in fact, is the truth; but we don't see or realise that until we truly learn to love. Four hundred years ago Francis Bacon spelt out for the first time, in scientific terms, the truth which the sages of all times have known and tried to teach the largely unsympathetic, uncomprehending world. Now, thanks to the inductive method which Francis taught mankind, we may be on the verge of comprehending the supreme truth properly: namely, that God is love; that love is emotion; that emotion is movement; that movement is energy; that energy exists as two types of energy polarized to each other - radiant energy and the vortex of energy - the one tending to infinite speed and the other to inertia; that these polarities of energy or movement constitute mind and matter; that mind and matter constantly interact with each other, producing the universe and all living forms, physical and metaphysical; that perfection is the eventual harmony or 'marriage' of mind and matter, in a balance of movement which reveals the original love out of which all manifest life originates.

The allegory of St. George and the Dragon hides (and reveals) this secret. The Holy Grail signifies the perfection of the life form. Arcadia is where the perfection takes place. Alchemy is the life process by which this perfection is gradually achieved.

Gentle reader, read on.

PART I

THE ANCIENT EGYPTIAN MYSTERIES

Introduction

The Hermetic Art of Alchemy

The Egyptian Mysteries of Initiation

(4) "SAPIENTIA PACEM: PAX OPULENTIAM" ("Wisdom brings Peace: Peace brings Prosperity") — emblem from titlepage to *Florilegium Epigrammatum Græcorum* (London, 1629).

Introduction to The Ancient Egyptian Mysteries

THE MAY FESTIVAL OF UNIFICATION AND THE 'AIR' INITIATION

May Day - the Festival of Unification, called by the Celts *Beltane* - celebrates the ending of Spring and the beginning of Summer. It is the time when the sprouting (*i.e.* springing) forth of new buds and plants from the ground or from the parent tree, together with the whole 'greening' or leafing process, turns into the general blossoming of flowers. Whereas the greening marks the Spring, flowering marks the Summer, and it begins in May with a bursting forth of colour. It is a time of happy celebration, for most people love flowers, colour and sunlight. Thus, in Britain, the celebrators at the May Day festivals used to bedeck themselves with blossom from the May tree, and dance around the gaily attired May Pole.

Because *Beltane* is a Quarter-Day festival (*i.e.* it is one of the four solar festivals which divide the year into its four seasons or quarters), it is not so easy to fix its exact date in the calendar. Whereas the Solstices and Equinoxes are precisely fixed by the position of the rising or setting sun on the horizon, the Quarter-Days are not so apparent. Similarly it is easier to establish the mid-point of a season, when its particular energies are at their highest, than to tell when one season changes into the next. The older societies, such as the Celts and Ancient British, solved this problem beautifully by choosing the time of the nearest full-moon to celebrate the Quarter-Day festival. Thus they married the solar cycle with the lunar cycle at these four points in the year - *Imbolc, Beltane, Lugnasadh and Samhain.*

Astrologically speaking, the full-moon which marks *Beltane* is therefore the one which is seen gleaming in the astrological sign of Scorpio, opposite to the astrological sign of Taurus in which the sun shines at that time of the year, as Spring passes into Summer. For this reason the May Day festival is known as a Taurean festival, and the bull, garlanded with flowers, is the primary symbol of this happy birth of Summer in many cultures. More than this, the bull is thereby associated with the meaning of the May Day festival. For the Ancient Egyptians, the bull was a major symbol of their whole culture, for many reasons.

In terms of human initiation, the Festival of Unification marks the completion of the first degree of initiation and the commencement of the second. Just as Winter is symbolised by the element of earth, and Spring by the element of water, so Summer is represented by the element of air, and the second degree of initiation is known as the Air Initiation. Winter, and the element *Earth,* signifies the dark, cold matrix out of which we emerge into the light of initiation. It is a time of germination and of preparation for initiation. Spring, and the element *Water,* represents for us the first degree of initiation, in which we learn to master our emotions and to love truly and strongly from the heart - water being a symbol of our

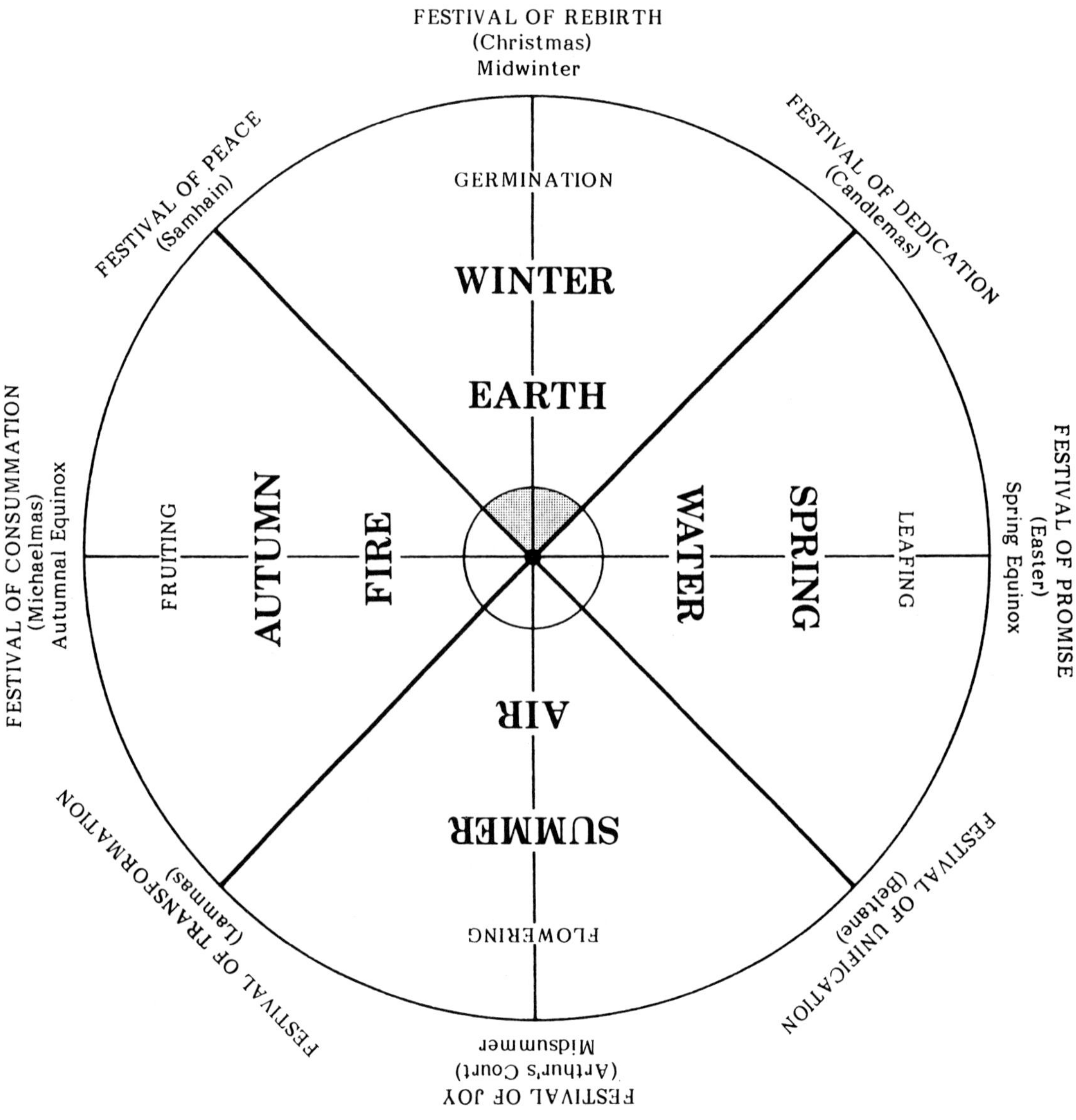

Diagram A: THE CYCLE OF SEASONS

emotional nature. Summer, and the element *Air,* portrays for us the second degree of initiation, in which we learn to perceive clearly and understand truly - air being a symbol of our intellectual nature, with colours, scents and flowers signifying beautiful thoughts.

Summer, therefore, marks the period of life in which light pours into the minds of men, bringing understanding or illumination. This is not full illumination, which only occurs after the ideas which understanding produces have been put into action (*i.e.* in the Fire Initiation), experienced to the full and a real practical knowledge obtained. Full illumination belongs to the fourth degree of initiation (*i.e.* the Earth Initiation). But a partial experience of illumination - clear vision, a perception of the glory of light, or life, and an intellectual understanding of truth - is the blessing of the second degree.

All thoughts are based upon desires or emotions, which, if they are pure, radiate into the mind as an enlightening energy. This radiation of energy from the heart vibrates with love, and we both hear it and see it. The hearing comes first, for as we listen to this vibration within us so our awareness of its existence increases. The more we listen, the more sensitised or sensitive to this life force we become. We hear it as the inner voice, the intuition, which is constantly speaking to our mind. If our emotion is true love, emanating from the heart, then what we hear will be the "still, small voice of God" - true intuition (*i.e.* the inner teacher). However, if our emotions are not of the purest, then the 'voice' that we hear will not necessarily be one of truth.

Hearing is the first stage of thinking. Almost instantly after the hearing comes the seeing. The inner voice stimulates the imaginative faculty of our intellect to see a picture or vision. This vision is a thought-form, which itself is a reflection in the mind of the emotional vibration or 'voice'. We see this energy as a form of light. It is this light which illuminates the mind. Whether we call it light or sound depends on whether we are seeing it or hearing it; it is still the same energy.

Once we have heard this energy, which is life itself, we become aware of it and sensitive to it. Being aware of it we can see it; and once we can see it, we can begin to understand it. In order to understand this light, we employ the other two aspects of the intellect - the reason and the memory - to help us create a clear picture or idea in our mind. The intellect is that part of the mind which receives the light and reflects it, giving it a form (*i.e.* as a thought form) which can be analysed, compared and comprehended. The three major faculties of the intellect are the imagination (*i.e.* the conceptual faculty, related to the right-hand hemisphere of the brain), the reason (*i.e.* the analytical faculty, related to the left-hand hemisphere of the brain), and the memory (related to the central brain ventricles and spine of the body). It is this process which is referred to as *illumination.* It bestows upon us knowledge of truth - first in the form of intellectual knowledge, in the second degree of initiation, and secondly full or experiential knowledge, in the fourth degree of initiation.

The first initiation is a process of learning how to love more and more strongly, in a balanced, peaceful and totally unselfish or unpossessive way. As we do this, our heart chakra opens out like a flower, pulsating with pure love. Together with this experience goes an increasing sensitivity and response to real love, as we learn to listen to it, thereby expanding our awareness through the functioning of our throat chakra. The culmination of the first initiation, which is marked by the Festival of Unification, occurs when we have sufficiently mastered our emotions and translated them into pure heart love, and can actually hear and clearly recognise the inner voice.

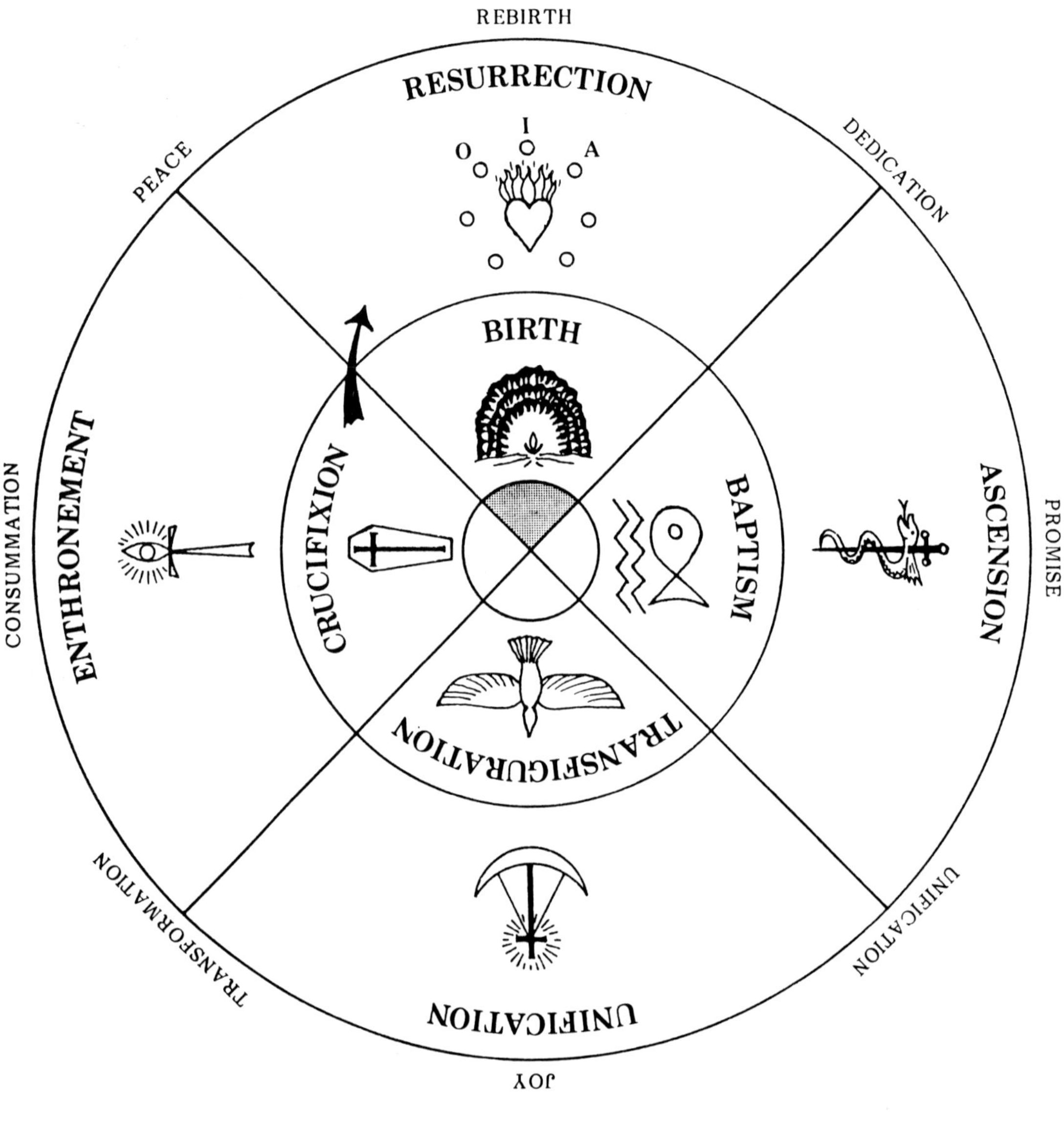

Diagram B: THE CYCLE OF INITIATION

In Christian teachings this event is marked by the story of Jesus' baptism in the river Jordan, followed immediately by Jesus seeing the heavens opened unto him and the Spirit of God descending like a dove and lighting upon him. At the same time he heard a voice from heaven saying, "This is my beloved son in whom I am well pleased." The dove is a symbol of the perfect thought-form or 'Holy Ghost' of light. Because the intellect is a reflector of light, like a mirror or an eye, what we perceive is actually a reflection or image of truth; hence the experience of seeing the heavens opened and the Spirit (or Light) of God descending like a dove is in reality how the intellect perceives the radiation of love issuing from the heart. Although the light originates from the heart, it appears (because of reflection in the dome of the mind) to come from above the head. The words "lighting upon him" is another way of saying that the light illuminated Jesus' mind. The words that were spoken are the classic words of confirmation that we all hear, in one way or another, when we realise that we have truly passed the first stage of initiation in whatever particular undertaking we are involved. The words not only refer to the person who is passing from the first to the second degree of initiation, giving confirmation and encouragement to proceed further, but more especially do they refer to the divine Son of Light - the radiance of love from the heart which appears a 'Sun' (or glorious thought-form) of light in the eye of the mind.

All this takes place at the so-called Festival of Unification, when our minds become united with the love-light of our hearts - a marriage of heart and mind (or of sun and moon) represented by the May Day marriages of young men and women in the temple circles of light. Then the second degree of initiation begins, in which we learn to understand what we feel and see. The next experience of this second degree is one of testing. Our intelligence immediately challenges the conceptual picture, testing the veracity of it and tempting the individual to alter his or her apprehension of truth to suit any lower desires that might still be hanging around. This testing faculty, which sometimes appears as scepticism, is a healthy attribute provided we keep it in its rightful place, as Jesus did, and not allow it to lead us back to selfish or impure thoughts that are based upon the desires of our lower nature. Always we have to make the effort of will to love purely and then to think clearly, taking care to avoid falling into the trap of glamorous or arrogant thoughts. It is our own thoughts, based upon selfish or degenerate desires, which constitute our 'tempter'. *Satan,* or Saturn, is the traditional name given to the testing faculty of our minds. It is no more nor less than our own intelligence, as it goes through the process of conceptualisation, analysis and recollection from the memory. This is the divinely given process whereby man is enabled to perceive right from wrong, good from evil, truth from falsehood. *Satan* ('the Adversary'), whose other name is *Lucifer* ('the Light bearer'), is pr[illegible]inently the name of Man in Western tradition.

As 'Man' himself means 'the Thinker', and is created in (or as
mankind is especially associated with this second degree of initi
good from evil is man's especial function and duty. The 'Eye' is
and of man's soul (which used to be described as man's mind
'Man' himself), and the 'All-Seeing Eye' signifies the enlighten
and comprehends all truth. The Ancient Egyptians represente
winged (or horned) Solar Orb of light, whilst the glorious thoug
truth which illuminated that 'Eye' they personified as *Rê* (or
directly associated with the intellectual functions, the human h
as the principal symbol of man, and the word *hu-man* desc
mind - *hu* (or *shu*) being the word describing the radiance c
heart into the mind as an arrow from its bow. These secre
found symbolised in the Great Sphinx at Giza, Egypt.

As mentioned previously, the Festival of Unification, which begins the process of understanding or enlightenment of the mind, is referred to as a Taurean festival. The sign of Taurus is taken as representing the initiating impulse that starts the process of illumination. Taurus, the Bull, is the great Bestower of Light, who fertilises the minds of mankind with seeds of light. In the East the festival of the Buddha, called the *Wessak* festival, is celebrated at the same full moon as the *Beltane* festival in the West - for they are one and the same. The festival is described as being the time when "light is gathered, concentrated and poured out to stimulate the mental state of humanity with wisdom, and incline men towards love and brotherhood." The festival is associated with the birth of the Buddha, for *Buddha* means 'the Enlightened One', which is a description of the enlightened mind or soul of man.

But astronomically the zodiacal sign of Taurus is not always the sign which marks the May Day festival, for the movement of the Earth (described unkindly as the 'Earth's wobble' about its theoretical or archetypal axis) creates what is known as the 'Precession of the Equinoxes'. This means that every 26,000 years the sun, when rising at the Spring Equinox, is observed to move (in its relationship to the Earth and the stars) through all twelve of the signs of the zodiac, remaining in each sign for about 2,160 years before passing on to the next. Right now, in the year 1987, the sun rises at the Spring Equinox on or near the cusp of the signs of Pisces and Aquarius, and the sunrise on May Day would now be seen to occur in the middle of the sign of Aries, not the sign of Taurus. But approximately 2,000 years ago the Spring Equinoctial sun arose on the cusp of Aries and Pisces (which cusp is known astrologically as 'the Point of Aries'), and the May Day sunrise was firmly in the centre of the sign of Taurus.

The chart of the heavens in relation to the annual cycle of seasons and festivals, as it was 2,000 years ago, is for some reason taken as the Archetypal condition for our world, and the changing appearances of this chart as the cycle of precession takes place are seen as being variations to this Archetype, colouring it rather than changing it. Hence the May Day festival is still referred to as a Taurean festival, although in reality it is Arian at present. At the time that the pharaonic civilisation of Ancient Egypt was supposed to have begun, *c.* 3,240 BC, the Spring Equinoctial sun arose in the middle of the zodiacal sign of Taurus, and the May Day festival sun was on the cusp of the signs of Cancer and Gemini.

This scheme, by which the Ages are usually calculated, is according to the 'equal sign' zodiac generally used by astrologers, in which the twelve signs of the zodiac are of equal duration, each spanning about 2,160 years. This zodiacal circle is known as the circle of the ecliptic, and our sun and the major planets of our solar system appear to travel around this circle, through the twelve zodiacal constellations, in the course of their movements. The pole of the ecliptic, for the northern hemisphere, is the occult north pole - the true or archetypal pole which would be the celestial north pole of our Earth all the time if our Earth did not 'wobble' about its occult axis. This occult pole lies in the heart of Draco, the Dragon. On this basis the Taurean Age began approximately 4,320 BC, the Arian Age about 2,160 BC, the Piscean Age about 0 BC/AD, and the Aquarian Age will begin about 2,160 AD, as calculated from the position of the sun at sunrise at the Spring Equinox.

…ilst this is interesting and significant, nevertheless to fully understand the Ancient …an civilisation we need to look at the astronomical zodiac of constellations as well, …ms to have been the main governing factor of what took place then and since. … Egyptians were renowned astronomers, and had built up a detailed and …f astronomy over a long period of time.

The astronomical zodiac, calculated from the actual positions of the twelve constellations which give their collective name to the zodiac (*i.e.* 'circle of creatures'), does not give signs of equal length or duration. Some zodiacal constellations take up more room than others on the zodiacal circle. Thus, for instance, according to the astronomical zodiac, the Age of Taurus began about 3,240 BC, the Age of Aries about 1,550 BC, the Age of Pisces about 0 BC/AD, and the Age of Aquarius will begin about 2,600 AD. It should be noticed, therefore, that the so-called 'Unification of the Two Lands' in Ancient Egypt, which marked the start of the first cycle of pharaonic dynasties, began with the commencement of the astronomical Age of Taurus, *c.* 3,240 BC, and that the 18th dynasty of pharaohs began at the start of the astronomical Age of Aries, *c.* 1550 BC. The symbol of the bull was of particular importance for Egypt throughout its history, but especially from the 1st to the 17th dynasties. When the 18th dynasty pharaohs came to the throne of Egypt the symbol of the ram was introduced - the head of the ram replacing the head of man on the sphinxes at Thebes, for instance.

The first and main cycle of Egyptian pharaonic civilisation, which ended with Pharaoh Akhenaton at the culmination of the 18th dynasty, spanned therefore more than just the astronomical Taurean Age. These astronomical ages were recognised by the Egyptians, and their civilisation adapted accordingly, but the cycle of their civilisation was not governed primarily by such ages. There was something even more significant in the heavens which governed that particular cycle of Egyptian civilisation - the association of two great constellations that have a special significance to each other, one near the north pole and the other in the zodiac. These two constellations, which determined the whole length and course of the first cycle of Egyptian pharaonic civilisation, were Ursa Major, the Great Bear, and Leo, the Lion. This particular Journal is not the place to go into detail about why this is so; but it is enough here to give the hint. Suffice it to say that Leo is also a symbol of the heart, and Ursa Major was considered by the Egyptians to be the magical 'thigh' (or right foreleg) of Taurus, the latter symbolising the illumined mind and referred to as the 'Eye' of Horus. With this 'thigh of the Eye of Horus' *Ptah* (the Creator god signifying the Voice of God) 'opened' the mouths of divine entities and the souls of the dead.

EGYPTIAN PRE-HISTORY

The known historical period of Ancient Egypt appears to have begun some 5000 years ago with what is called the Unification of the Two Kingdoms of North and South Egypt. This unification was achieved under the leadership of a pharoah known as Menes. Menes is further identified as Pharoah Aha. From this unification the whole historical period of Ancient Egypt is dated, and its marvellous civilization unfolded over the course of 2000 years - or 3000 years if one includes the foreign rulership by Nubians, Persians, and finally the Greek Ptolemies and Romans. In terms of known history, it is the longest lasting and probably greatest civilisation of man on record, although it is possible that there have been other comparable or greater civilisations that we at present have no knowledge of. By comparison, the Roman civilisation that so much has been written and talked about appears paltry.

The Ancient Egyptian civilization appeared on the scene *c.* 3000 BC (although some commentators push the date back to *c.* 4000 BC). Quite remarkably it appeared fully fledged, in the sense that it seemingly possessed all the knowledges and organisational structures which were to guide and govern its civilisation step by step over the centuries. This fact suggests quite clearly that either there had been a long evolutionary period of growth preceding the historical pharaonic civilisation, in which the various tribes living in the Nile valley developed increasingly more civilised ways of life and profound knowledge, until it reached a peak that gave rise to and was marked by the Unification of Egypt, or that the knowledges and plan of civilisation were imported into the Nile valley from another part of the world where they had been developing for some considerable time, and imposed upon the aboriginal tribes of the Nile valley. So far archaeological findings have not been able to help much in determining the predynastic history of Ancient Egypt, although they do indicate that a high degree of culture and craftsmanship did exist before the days of Menes.

Sacred tradition does, however, give an account of what took place, and other discoveries and records augment the historical (as well as symbolical) picture. Briefly, it would appear that the civilisation which we know as Ancient Egypt was the result of a pre-planned and carefully controlled experiment in human civilisation, the organisation and knowledge for which had been carefully prepared over many thousands of years. The original knowledge was given to the first sages of Egypt by the 'Gods', and those sages are themselves said to have come as colonists from Atlantis during the last few thousands of years of that 'Titanic' period of man's development. In other words, there was a sacred tradition of wisdom knowledge handed on from the past *via* a line of sages or high initiates, that was said to have come from the great Atlantean sage, Enoch (as he was known to the Hebrews), or Thoth (as he was known to the Egyptians), who "walked with God" and was raised up into the highest heaven to become the Messiah or Teacher of all mankind.[1] He, in his turn, had received a sacred tradition of wisdom knowledge *via* a line of sages stemming from Adam, the first incarnate man, who had been instructed by God after his fall from Paradise. Enoch was known mystically as "the seventh from Adam."[2]

At the same time, quite naturally, the sages of each generation were making their own contact with the realm of Divinity and, together with the results of their own experience, were continually adding to the sum total of this traditional wisdom knowledge. But the essence of the teaching has always been the same in age after age, even though the outer interpretations and manifestations of it change and differ from one age to another. It is this essence that I would like to deal with, as far as I am able to.

According to Egyptian sacred lore, there were initially **Ten Divine Dynasties** - 'the Reign of the Gods'. These 'Gods' (*Neters*) should be understood as being divine creative principles and archetypes, equivalent to the *Sephiroth* of the Hebraic Cabbala, or the supreme Spirits (*Ælohim*) and Archangels of the Judaic-Christian tradition. It should also be understood that Moses, the great law-giver of the Hebrews, was learned in all the knowledges of the Egyptians,[3] having been the high priest of On (Heliopolis) and second only to the pharaoh in initiateship and rulership of all Egypt, and that it is essentially the same wisdom knowledge that the Egyptians had which was passed on to the Hebrews *via* Moses. **The Ten Divine Dynasties** form a spiritual 'Tree of Life'[4] (*i.e.* a divine Idea) that brought life and civilisation into being on earth.

Issuing from the Unknown and Hidden Truth (***TAT,*** or **THOTH**), the Ten Divine Dynasties are enumerated as follows:-

(1) **PTAH** - the Voice and Creative Spirit of the Cosmos,[5] which created:-
(2) **RA** - the Archetypal Idea and Cosmic Light of God, whose aspects are:
(3) **SHU** and **TEFNUT** - the Radiant Energy and Fiery Dew (or Form) of Light;
(4) **GEB** and **NUT** - the Earth and the Heaven;
(5) **OSIRIS** and **ISIS** - who brought fertility, civilisation and culture;

(6) **SETH** and **NEPHTHYS** - who brought insurrection and division;
(7) **HORUS** (the Elder) - who subjugated Seth and reunited all things.

These were followed by:-

(8) **THOTH** - who first established a common tongue, named the objects, invented letters, instituted the cult of the Gods, and gave man the first principles of astronomy and music;
(9) **MAAT** - who brought perfect order, justice, harmony, illumined consciousness and peace;
(10) **HORUS** (the Second) - the King to reign over all.

These Ten Divine Dynasties conceal and reveal what is known as **the Tree of Life** and its archetypal principles of life (the *Neters,* 'Gods'), that are created (or established) in the heavenly Mind by the Voice of Command (***Ptah***). The geometric form of the 'Tree' constitutes the fundamental archetypal Vision of Life, known as ***Râ***, the spiritual 'Sun' or thought-form of Light. This 'Tree' or thought-form of Light is the Archetype of all temples.

Within the overall Archetype of Light are various integral principles or ideas concerning the various aspects of divine Life. These principles begin with the idea of the wholeness or macrocosm of Light (***Râ***) which is both invigorating (***Shu***, the radiant life force) and enabling (***Tefnut***, the reflective life form). This basic Trinity has a pattern or order, which is a divine law, that gives rise to spatial form and purpose. This seven-fold pattern is equated as follows:- the 'above' (***Tefnut***, 'heaven') and the 'below' (***Geb***, 'earth'), the 'front' (***Osiris***, the giver of life, which is light) and the 'back' (***Isis***, the receiver and reflector of light), the 'right' (***Seth***, heterogeneous emotions and will) and the 'left' (***Nephthys***, heterogeneous reason and thoughts). In the centre, uniting them all, is the heart or 'throne of glory' (***Heru-ur***, 'Horus the elder', also known as ***Herakhty***, 'Horus of the Horizons', the solar principle of harmony and beauty).

From all this is born (or generated), through the fusion of spirit with matter and the alchemical process of initiation that this fusing generates, the full manifestation of God as the fully developed and perfect human soul. The three stages of initiation required to reach this perfection, and which form an integral whole, are pictured as being the expression of (1) ***Tat*** (Thoth, perfect desire or love), (2) ***Maât*** (perfect understanding or cosmic consciousness), and (3) ***Heru-pa-Khret*** ('Horus the younger', or Harpocrates,[7] the Son of ***Râ***, the true kingship obtained through perfect sacrifice or service).

The Ten Divine Dynasties of Egypt constitute the archetypal Wisdom or Light that brought Egypt into manifestation - the knowledge and correct interpretation of which gives rise to all other knowledges. It is these gods (or archetypal principles) that created and governed the evolutionary history of Egyptian civilisation - as indeed they create and govern all civilisation everywhere, although they may be referred to by other names elsewhere and in other epochs of history.

The Ten Divine Dynasties were followed by the **Thirty Dynasties of the Sages** (or Demi-Gods). It is with these dynasties that we reach a 'historical' period of life in Egypt, as the gods (or archetypal principles) begin to manifest and mould matter in order to express their spiritual laws and forms of being. The first aspect of matter that is impressed and formed by the spiritual principles is the mind or intellect - the mind of nature and of man. Hence these Thirty Dynasties are called the Dynasties of the Sages.

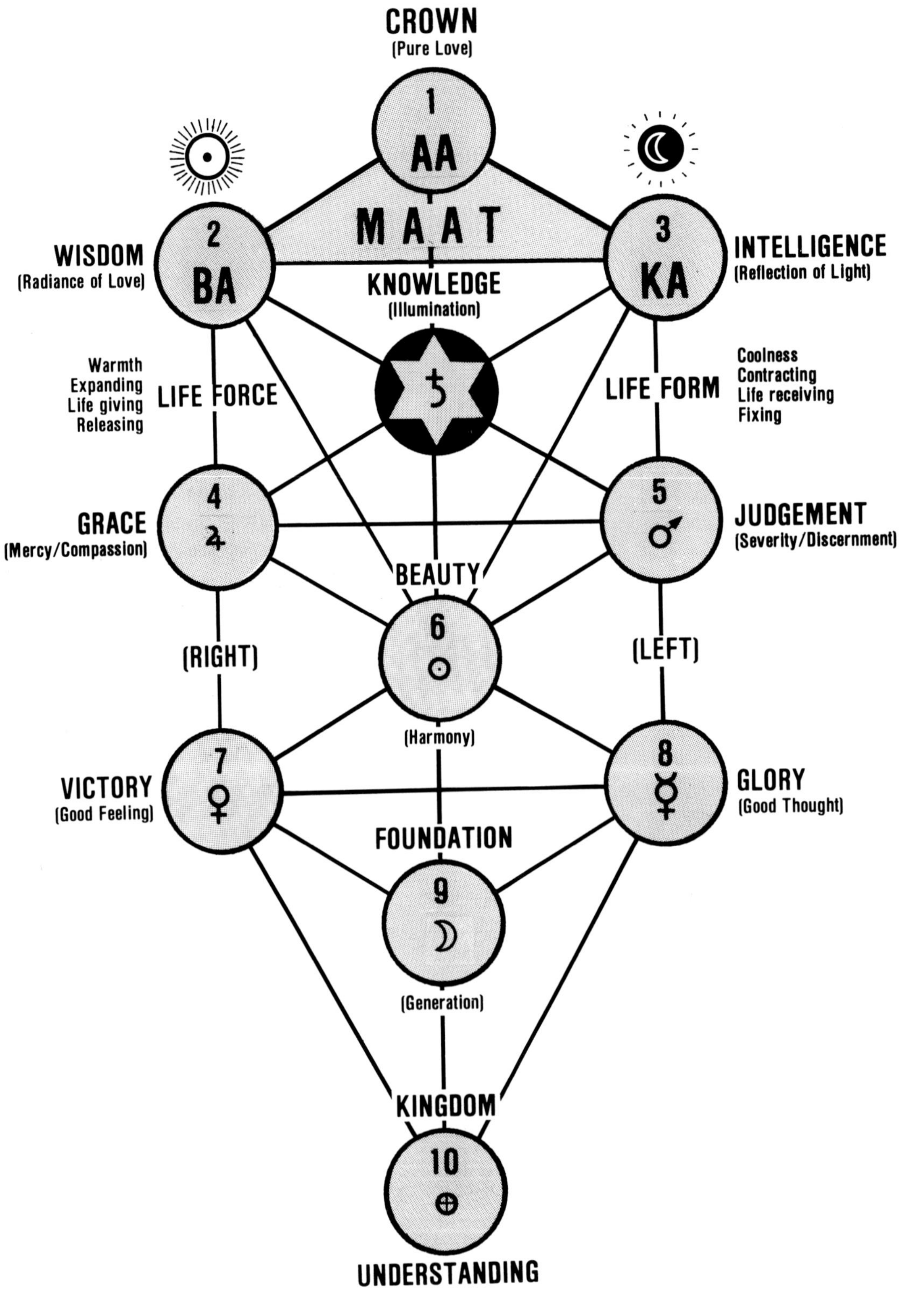

Diagram C: THE TREE OF LIFE

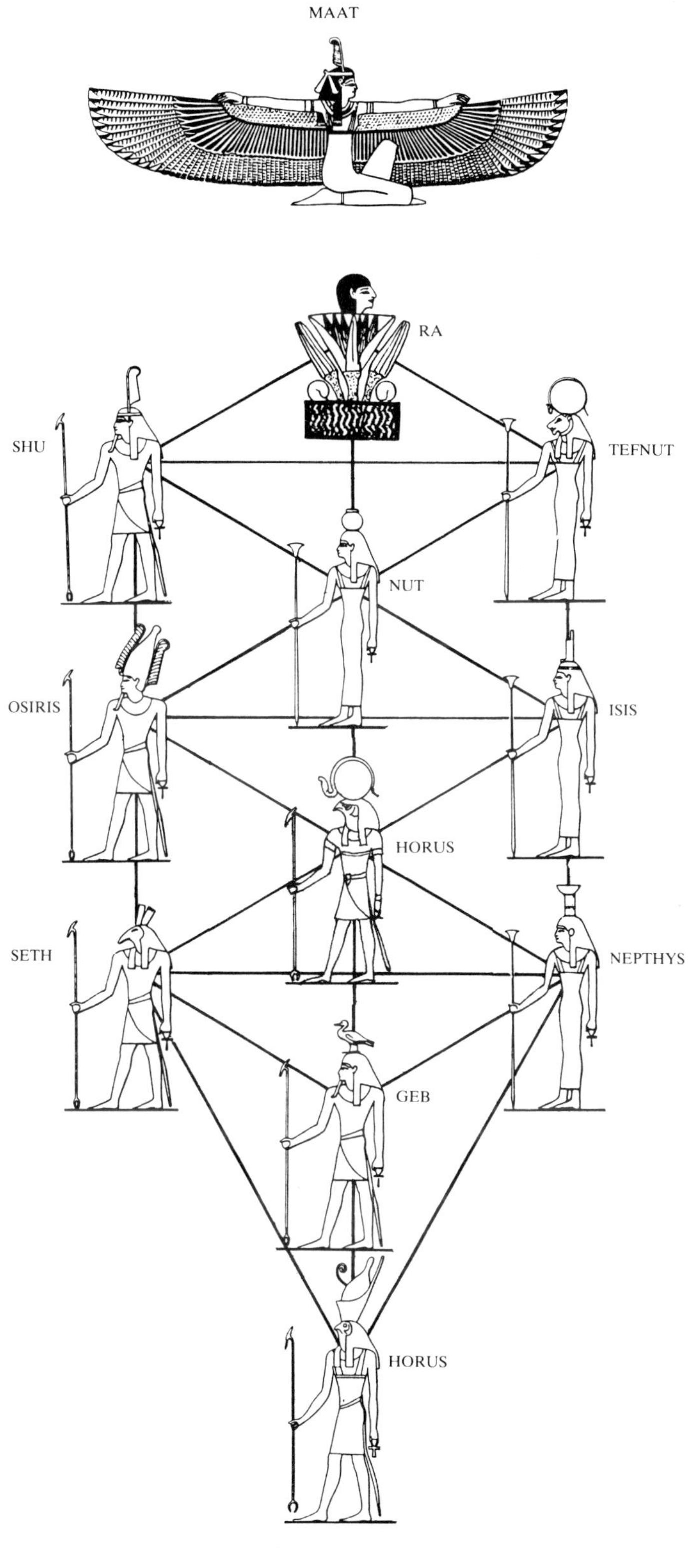

Diagram D: EGYPTIAN TREE OF LIFE

The first of these dynasties is said to have begun about 42,000 years ago. Their total duration is recorded as being twenty-five Sothic cycles, each Sothic cycle consisting of 1,461 years. This gives a total of 36,525 years, although a little more -36,620 years - is stated. **The Thirty Dynasties of Sages** were grouped into three principal groups, the first two groups existing for approximately 23,200 years and the last group for 13,420 years:-

(1) **The Venerables of Memphis;**
(2) **The Venerables of the North;**
(3) **The *Shemsu-Hor* ('Companions of Horus').**

Then, and only then, does the known historical period of Ancient Egypt actually begin, and it begins with the so-called 'Unification of Egypt'.

THE MASTER PLAN AND HIGH COUNCIL OF ANCIENT EGYPT

What should be realised from the start is that the Ancient Egyptian civilisation was no accident, no chance happening, no gradual evolution from an initial primitive state to a complex final state. On the contrary, the pharaonic civilisation, with its sovereignty, religion and basic philosophical concepts, was fully constituted from the beginning. The art of writing was fully developed and practised, an exact calendar was in use, most of the technical methods that Egypt was ever to invent or use were already known, and the country's political organisation was firmly laid out. Most important of all, a master plan for the whole development and progress of Egyptian civilisation, that was to cover nearly two thousand years, was in existence from the start.

The Ancient Egyptian civilisation was carefully planned and guided from start to finish, as a statement and expression of the Ageless Wisdom. There was always, from generation to generation, an inner council of Egyptian sages - the 'prophets' or highest initiates of the land - who knew the plan and were guardians of it, and responsible for seeing that the plan was carried out as well as possible. Collectively they were known as *Tat* (Thoth),[8] the 'Word', for they were the living representatives of the Divine Word of Truth which the Egyptians called ***TAT***. This high council was composed of the pharaoh plus the dignitaries[9] and prophets of the four principal teaching and initiatory centres of Egypt. These great centres of light and learning were also the principal seats of government. They were located at *Un* (Hermopolis), *Mem-Nefer* (Memphis), *An* (On, or Heliopolis) and *Waset* (Thebes).

Of this high council, the pharaoh was the chief and 'essence'. He was elected by the council from out of the ranks of the 'blood royal', but only after he had himself achieved the degree of prophet. In addition the heir apparent had to have a good working knowledge of the eight further grades of dignity that a prophet can attain to, and have 'seen' and understood the archetypal plan in respect of the whole Egyptian civilisation, including especially his own part in it if he became pharoah. He, on his own, was also referred to as Thoth, as the whole council was seen to be summed up in him, and he was the ruler of that council; for he, above all others, was intended to be the living and visible representative on earth of the spiritual Truth.[10] He was the 'Son of the Sun of God' - the 'Great House' or 'Royal House' (*i.e.* Temple) in which the Light of God resides and through which that Light is made manifest - which is what *Per-âa* (Pharaoh) means, and also what *Is-Râ-El* means. The Greeks later translated the name Thoth as Hermes, and the Great Lord or 'Word of Truth' was referred to by them as Hermes Trismegistus ('the Thrice-greatest

Word'). The Romans used the name Mercury. Horus (Egyptian *Heru*), meaning 'the Visible Appearance of God', was another name or title of the pharaoh.[11]

This all-male council was counter-balanced by its feminine equivalent, which played a quite different but complementary role in the leadership of Egypt. Acting out the symbolic roles of spiritual principles, the male initiates of Egypt were primarily concerned with what is known as the initiating, teaching, spiritualising 'Pillar' of Life Force (*Shu*), whilst the female initiates were primarily concerned with the complementary nourishing, enabling, form-making 'Pillar' of Life Form (*Tefnut*). Male and female initiates were expected to marry in order to fulfil all spiritual and natural functions and responsibilities, and thereby to manifest Truth. Usually only those who had married and raised children were accounted fit and sufficiently experienced to be able to teach the Mysteries to others.[12]

This did not mean that men only learnt and practised one aspect of life, and women another. On the contrary, during actual temple training the sexes were normally kept apart and celibate, the purpose being to achieve purity, concentration and mastery of creative powers which could then be used either in temple work or in community and family life. During this period each sex had to develop within themselves and become proficient in both force (male, *fire*) and form (female, *water*) aspects of life, and be able to 'pass between the pillars' in their progression through the degrees of initiation. But the Egyptian philosophy recognised the normally clear distinction between sexes in terms of their particular purposes and functions, as manifested in the physical form and psyche of each sex. They respected this fact, took heed of it as being the expression on earth of a divine principle or law, and sought to enable that divine law to manifest in the best possible way. Hence these major roles, as played by the sexes, were allowed for and enhanced in every way possible and known to them. Such training culminated in a marriage of some kind. In the pharaoh and his queen the two roles were (supposedly) blended and united in the highest possible way, spiritually and naturally, in a full symbolic expression of what is known as the divine or mystical marriage.

The conscious and deliberate observance of divine law, as understood by the Egyptians, is the key to their whole civilisation - its organisation, teaching and history. They worked to a plan, divinely given, which provided them with an archetype of national as well as individual life. They saw all living things as being formed out of matter by the Spirit or Thought of God, which Thought provided the Archetype for all living form and its evolutionary life. Perfect form they understood and referred to as the '**temple**' of God or, more accurately, as ***Is-Râ-El***, 'the temple of the Light of God' - *Râ* ('Light') being a term for the Spirit or Thought of *El* ('God').

Furthermore, the Egyptian sages clearly understood that such a temple exists in terms of **time** and **space**. That is to say, a life form that is built out of matter has a spatial organisation and existence, and at the same time it has a period of life from its birth to its death (or dissolution). Both the spatial organisation and the life period of any living form was seen as being governed by the same divine law, and as being two complementary and inseparable manifestations of that archetypal Thought of God. The temporal aspect constitutes the inner 'movement', which can be described as a process of life, whilst the spatial aspect is the outer form in which the process takes place. The former gives meaning and purpose to the life form, whilst the latter enables that purpose to be fulfilled. **Time** is related to the pillar of force, and **space** to the pillar of form.

Not only did the Egyptians see this truth in terms of individual man, but they also perceived a whole nation to be but a larger expression of the same Archetype. The spatial manifestation of a nation is its **country**, and the temporal manifestation of a nation is its historical **cycle of life**. The **civilisation** of a nation, which results from the unified expression of these two, is the nation's process and manifestation of life from birth, through childhood, to maturity and death - a process in which sciences and arts[13] of various kinds are developed and flourish. It follows a definite law that is the same law for everything throughout the universe.

This law also determines what is known as the **path of initiation**, which in fact is but the process of life consciously and deliberately lived by man in as perfect a way as possible. After an initial stage of preparation there are a further seven recognisable stages which are known as degrees of initiation. The first three degrees lead the human soul to the point of conscious death, in a state of purity and total surrender (or obedience) to divine law. Such an initiatory death transmutes the natural soul into an immortal and illumined state of perfection (Horus), that rises above (or transcends) all boundaries of the natural world and is a 'light' to that world. The final four degrees are to do with the sublime stages of life after such an initiatory death.

The three basic stages of the life process, which are the same for all living creatures, but in particular man, are:-

(1) **desire,** followed by
(2) **thought,** followed by
(3) **action.**

The results of the action give rise to further desires or feelings, which in turn create further thoughts in the mind, and these thoughts in turn generate further actions. And so the 'wheel' turns - the wheel that is variously called the Wheel of Life, the Wheel of Fortune, the Wheel of Karma, the Wheel of Fate, the Medicine Wheel, *etc*.. The completion of an action is the completion of a cycle, and that cycle is then dissolved - it dies, and another cycle is born from its ashes as the wheel turns inexorably on. The human soul is caught up in these continuing cycles - and, indeed, is composed of these cycles. The soul is caught up in and made up of the conditions of each cycle, which may be far from perfect if the soul has bad desires, thoughts and activities.

Initiation is the term given to the perfecting of each cycle, until all desire becomes pure **love**, all thought becomes clear **understanding**, and all actions become an act of **service** for the benefit of the whole of life. The Wheel of Fortune then becomes the Royal Path or Way of Initiation, called *TARO* by the Egyptians, in which the soul is able to transmute its conditions into something higher and more beautiful and perfect. When perfection is reached, the Wheel is transcended, for the spatial and temporal conditions of perfection are eternity and infinity. The soul is free, because it is no longer limited, yet it is still subject to and expressing the same divine law.

This process or 'wheel' of life takes place within an outer form which is spatial in nature. The two are inseparable from each other in all ways. Just as the process of life is clearly determined by the immutable law, so the spatial layout of the soul is also clearly defined. For instance, the seat of desire, if it be true love, is in the **heart** (although if it is imperfect desire then its seat is in the solar plexus), the seat of thought is in the **head**, and the seat of action is in the **abdomen**. These three main areas of the outer form are in a distinct and unvarying relationship to each other; and neither would we expect them to be otherwise, because (instinctively and from experience) we know that it must be so.

Thus it is that all the organs associated with **loving** (which is analogous to life itself, and the breath of life) are to be found in the central 'heart' area of the body, and protected in animals and human beings by a bony rib cage; all the organs associated with **thought** lie in the head or 'upper' part of the body, and (in animals and human beings) are protected by a bony skull; and all the organs associated with **bodily activity**, that provide the energy and driving power for all forms of generation and procreation, lie in the abdomen or 'lower' part of the body, and are without a protective bony construction. Not only do these three main areas exist spatially as upper, middle and lower regions of the body, linked together by a spinal column or central axis, but it should be readily perceived that the energies involved in the process of life move through this space in a predetermined and unchanging way - *i.e.* from heart (or solar plexus) to head to abdomen, and then returning again to the heart to complete the one cycle and begin the next.

This whole truth is summed up in an ancient symbol for Man (and for any temple of life), ☿. This symbol is commonly known as the **Sign of Mercury** (or Sign of Hermes, Sign of Thoth). The symbol portrays the archetypal design of the temple or living soul. In the centre is the solar emblem- the central dot within a circle - signifying the heart of love which emanates its love (or heart-desire) as radiant, life-giving light. This is known as the '**Supreme**' realm. Above the sun-sign is the lunar emblem - the crescent set as a cup or receptacle - signifying the head and mind that receives the love-light, reflects it as an image or vision, and then reflects or meditates upon this vision in order to comprehend it. This is known as the '**Superior**' realm. Below the sun-sign lies the earthly emblem of the four elements - the four-armed cross, signifying the elemental realm of matter in which and by means of which the thought concerning the love-light is put into action and final expression. This is known as the '**Inferior**' realm.

These three stages of life and their areas of operation express not just the Thought of God, but the whole Being of God (for the divine Thought is the perfect and complete thought concerning divine Being). Thus the sages knew and taught that all existence began when the unknown, unmanifest Divinity desired to live, to become, to be manifest. This Desire was an emanation of the Heart of God in perfect love, which is life. The Desire was a movement, which instantaneously set up a polarity of 'Heaven' and 'Earth', or Mind and Matter, movement and inertia. Thus the three basic spheres of divine Existence were established: **Heart, Mind** and **Matter** - Supreme, Superior and Inferior.

Then the divine Mind thought about the Desire which emanated into and filled the Mind. It heard the vibration of this Desire as a Voice (or Sound) and saw it as Light. The archetypal Thought, which was perceived as Sound and Light, is known as the Holy Spirit.[14] This Spirit then incarnates into matter in order to express itself in matter as the natural living soul. This expression takes place in terms of time and space. When the natural soul is brought to perfection, wherein it manifests the spiritual Archetype perfectly and completely, it is transmuted (or 'raised') into a state of eternity and infinity (*i.e.* 'eternal life') as the celestial or spiritual soul, wherein it has become the perfect and direct expression of the very Heart of God. In this final state, Mind and Matter (*i.e.* Heaven and Earth) have been reunited or 'married', but in such a way as to make manifest and known the whole Being and Life of God.

Mind is known as the **Heavenly Father**, 'the Most High', and Matter is known as the **Earthly Mother**, 'the Deep'. The harmony or relationship between the two polarities (which is basically the manifest expression of love between the Divine Parents) is called the **Son**. The Son has a heavenly (or spiritual) aspect as the holy spirit, an earthly aspect as the natural soul (or body), and a totally synthesised aspect as the celestial soul (the true soul). Each of these three principal aspects of spirit, soul and body is also three-aspected, *ad infinitum*. It is the Son that is represented by the three-fold Sign of Thoth. It is for these reasons that Thoth, or Hermes, is called 'the Thrice Greatest'.

The divine Son is a spirit, a body and a soul - three-in-one. The three natures of spirit, body and soul are inseparable. The spirit progressively incarnates into and builds the body which, through the alchemical process of life, generates (or becomes) the true soul that manifests the Truth fully. Naked Truth is always to be found at the Centre - the point of balance, harmony and synthesis.

The same truth, which expresses itself in this three-fold pattern, is found at every level of life but expressed according to the nature of that level. Hence the body, the soul and the spirit each have a three-fold nature or composition. Thus it is that the natural soul or body of man has a generative nature, an emotional nature and an intellectual nature. These three natures are ascribed to the roles of **king, priest** and **philosopher** respectively, and relate to the divine qualities of Power, Love and Wisdom. In the celestial soul these three are expressed perfectly. The Pharoah was the visible representation and (it was hoped) manifestation of these three, as Thoth. Therefore, as Francis Bacon said in hopeful reference to King James I of England:

> "....your Majesty standeth invested of that triplicity, which in great veneration was ascribed to the ancient Hermes; the power and fortune of a king, the knowledge and illumination of a priest, and the learning and universality of a philosopher."[15]

THE TEMPLE OF EGYPT

The Venerable Sages realised that, just as an individual human being can consciously and deliberately choose to enter the path of initiation that leads to mastery of life, rather than remain in the semi-conscious, unredeemed state of mortal fate, so also can a whole nation make this choice and walk the Royal Way. Pharoanic Egypt, with its civilisation and country, was planned by them to become a veritable Temple of the Light of God.[16]

> "Do you not know, O Asclepius, that Egypt is the copy of Heaven, or rather, the place where here below are mediated and projected all operations which govern and actuate the heavenly forces? Even more than that, if the whole truth is to be told, our land is the temple of the entire World."[17]

Just as Nature, or a gardener, has to turn the earth into a rich dark compost in order that any seed planted therein will grow into a strong and healthy plant, so it is that the natural soul of man needs to prepare the soil of his own nature until it too becomes a rich compost in which the seed of his spiritual self may become well bedded, and from which it may grow strongly and healthily through the stages of initiation to reach its ultimate perfection. This rich compost is known as the *Prima Materia* or 'First Matter' - the starting point of all initiation. The Egyptian word for it was ***Kemit*** ('the black earth'), and this is one of the names that they gave to their country. The Greeks knew this name as *Khemia*, from which we derive our word 'chemistry' and, more particularly, 'alchemistry' - the chemistry of the 'all' or whole being. The Hermetic science is founded upon this *Kemit*. **Alchemy** is the process of initiation that leads the natural soul from the state of *Kemit* to a condition in which it is transmuted into the 'golden' state of the spiritual soul.

Another name that the Egyptians had for their country was *Ta-nutri* ('the land of the *Neters*"), for a dedicated and holy land is the kingdom and temple of the gods. Similarly the name *Ta-meri* ('the beloved land') was used, for the temple is the beloved bride of the Spirit as well as being beloved by its inhabitants.

Besides *Khemia*, the Greeks also knew Egypt as *Aigyptos*, which was (according to Diodorus of Sicily) originally the Greek name for the river Nile and derived from their word for a vulture, *aegypta*. The importance of this we shall see later. *Aigyptos* is also thought to be derived, by transcription, from the Egyptian *Haikuptah* (*i.e. Ha-Ka-Ptah*, 'The Temple of the *Ka* of Ptah'), the name of the great temple of Ptah at Memphis. From *Aigyptos* we derive our modern English word, 'Egypt'. The Hebrews called Egypt *Mizraïm*, from which is derived the Arabic name for the country, *Misr*.[18]

But Egypt was also known to the Egyptians as '**The Two Lands**', and in this description we have a very important allusion to the Royal Way which builds the Temple and fills it with life and light.

THE COUNTRY OF EGYPT

Egypt, as a country that can be inhabited by man and nurture a flourishing civilisation, owes its very existence in natural terms to the river Nile. This great river, which issues from Lake Victoria (and many other tributaries), flows northwards for 4000 miles, cutting a long and narrow valley through desert mountains and hills until it comes to the end of the desert uplands. Emerging from the constraints of the upland territory, the river fans out into several branches and crosses the low lying territory of northern Egypt to reach the Mediteranean Sea. In pharaonic times there were seven branches of the river, but nowadays only two branches of any importance remain. This branching of the river forms a great delta of extremely fertile alluvial land, which projects itself out into the Mediteranean in a great arch.

Once a year the Nile used to flood, with two of its major tributaries (the Blue Nile and the river Atbara) carrying the run-off water from the winter rains on the Ethiopian plateau and discharging them into the Nile in a great torrent of water and silt. Soon after the summer solstice the Nile used to start rising, gradually flooding its banks and the lower ground of the Nile valley - the water level reaching its greatest height between the end of September and beginning of October, after which it gradually subsided again. First came the 'green Nile', green coloured and foul, but this was followed by the 'red Nile' which carried a valuable reddish-brown silt. As the waters subsided this rich thick silt was deposited on the banks of the river and the surrounding valley floor, and over the ages built up the great delta of fertile land in the north of Egypt which became renowned as the granary of the Middle East.

The annual inundation of the Nile was the primary source of fertility to the land, building up a bed of fertile soil and making it possible to cultivate the Nile valley and delta, and to sustain a remarkable civilisation over a long period of time. Three seasons were recognised in the civil year - '**Inundation**', '**Seed**' and '**Harvest**'. The inundation ended the yearly cycle, literally dissolving and washing away (or purifying) the harmful salts and refuse of the previous year's labour and harvest. At the same time it transformed the fields once again into the 'dark earth', *kemit*, from which the new growth could appear; and so the inundation was not only the death of the old but also the fertile womb and starting point of the new, in which the seeds of 'light' were sown. During the inundation the Nile was called *Hapi*. **Hapi** is symbolised either by an androgynous figure, having the body and genital organs of a man but the breasts and womb of a pregnant woman, or by two figures, one having a papyrus plant upon his head (emblem of Southern Egypt) and the other having a lotus plant on his head (emblem of Northern Egypt). He carries the *ankh* ('cross of life') and a grail vessel filled with the 'gifts of the Nile'.[19]

The inundation of the Nile was particularly associated with the heliacal rising of the so-called dog-star *Sothis* (Sirius), which occurred annually at the summer solstice. This was the Egyptian New Year. Between 3000-2000 BC this midsummer heliacal rising took place during the period when the midsummer sun rose in the zodiacal sign of *Leo*, the Lion.[20]

Sirius ('Prince', 'Guardian') is known as the 'Dog Star' and marks the heart of the 'Great Dog' constellation (*Canis Major*). This 'dog' signifies the Egyptian god, Anubis, the 'pathfinder' or 'light of reason', which is a name for the secondary aspect of Thoth as messenger, teacher and conductor of souls on the path of initiation (*i.e.* the mercurial mind or intellect), who accompanies Osiris upon his conquest of the world. Another name that the Egyptians had for the constellation was *Apes* ('the head'), symbolised by a hawk, a sign of Horus.[21] The Egyptians associated Sirius, the heart of the Great Dog or Hawk, with Osiris, Isis, Horus and the whole pantheon of gods, and the star was referred to as 'the Archer', signifying the divine heart and its love-light that creates all things.[22] In many parts of the world, besides Egypt, Sirius is known as the great 'Leader' and 'Teacher', whose light fertilises the world and brings forth all things. The Romans called it 'the Star of Mercury', whilst in the East it is known as 'the Star of Buddha', the great Instructor of mankind and Chief of all other Buddhas.

Lying just outside the circle of the zodiacal constellations, and associated with *Leo*, is *Crater*, the Cup or Grail. *Leo* is symbolic of the heart and of the desires of the heart, and it is in the heart that all things begin and end. Around 3000 BC, when the Egyptian civilisation began with the Uniting of the North and South Lands, and the celestial north pole was marked by the star Thuban, the midsummer sun and heliacal rising of Sirius was directly linked in a celestial line with *Leo* and *Crater*. Thus the beginning of the Egyptian new year, and the beginning of the pharaonic civilisation, was marked by the conjoining of Sol and Sirius plus the Lion and the Grail on the eastern horizon at midsummer. The perceptive should be able to see the enormous significance in all this. Its mystery is concealed in the symbolism of the great Sphinx at Giza and the falcon-headed Horus.

Not only was the Nile the single source of natural fertility for the country, but it was also the main channel of movement and communication. It was quite literally the spine of Egypt. This, together with the annual inundation of the Nile, was the great central truth apparent to all Egypt's inhabitants. Life was centred around and depended upon this vivid reality.

The Two Lands or Kingdoms of Egypt, as far as the general public were concerned, were:-

(1) the Nile valley itself and the surrounding desert uplands, that extend from the first cataract of the Nile (24° latitude) northwards to the mouth of the river valley (30° latitude). This kingdom is known as 'Upper Egypt' or 'Southern Egypt'.

(2) The low-lying Nile delta, extending from the mouth of the Nile valley to the coast of the Mediterranean Sea. This kingdom is known as 'Lower Egypt' or 'Northern Egypt'.

The Unification of the Two Lands, under Menes (Pharoah *Aha*), consisted of setting out the boundaries of Egypt in the north and south, founding a unified civilisation within those boundaries that was ruled over by the one Pharaoh and his high council, and establishing the primary centres of government and teaching according to the divine plan. Chief among these acts was the building of *Mem-nefer* (Memphis)[23] on an island in the middle of the Nile, at the point where the Nile begins to emerge from its valley and the delta begins. The island was actually created by the Egyptians, by diverting part of the river, and great dykes were raised around the perimeter of the island, fortifying the city against the Nile's floods and possible human aggression. These walls must have gleamed white in the sun, for the oldest name that we have on record for Memphis and its surrounding district is *Ineb-hedj* ('The White Wall'). This city served as the capital or principal seat of government throughout the Early Dynastic Period and Old Kingdom (1st-8th Dynasties, *c*. 3000-2134 BC). Perhaps the most apt name that we know of for Memphis was a term used to describe the city, that appeared in the Middle Kingdom (11th-14th Dynasties, *c*. 2040-1640 BC) - *Ankh-tawy* ('That which binds the Two Lands') - for this special city lay on and marked the boundary between the Two Lands.

But Memphis was not the only principal centre of government and teaching of Ancient Egypt. As mentioned before, there were four such centres altogether, each playing their 'capital' part in the history of the nation as and when appropriate. In addition, but subsidiary to these principal four, there were three other 'great' centres and a host of smaller centres of importance. An understanding of this is a key to the history and behavior of the Egyptian nation and their Pharoahs.

THE SEVEN GREAT CENTRES OF EGYPT

The Land of Egypt was seen and understood as being a **living temple** that was being built out of Nature by the Spirit of God - a work in which man played an important and key role, not only being an integral part of the temple but also able to be a conscious assistant to the divine Spirit in the alchemical process. Like all temples, it is patterned upon an eternal archetype which is true for all life forms, and which is exquisitely expressed in the human form of man himself - the crowning achievement of Nature. To understand the human form is a key to understanding all life forms, including the great temples of the landscape in which man lives and moves and has his existence.

Egypt is a landscape temple - one of the many landscape temples of the planet Earth, which itself is but one of the 'landscape' temples of the Solar System, *etc*.. Its **spine** is marked by the river Nile. Its **head** is to the north, and its **body** is to the south. **The seven great centres** that the Egyptians established along this Nile marked the seven major focal points of energy (or *chakras*)[24] in the body, which manifest physically as certain endocrine glands, ganglia and plexuses, and groups of organs that serve particular functions that can be identified in seven main groups. Each of these great centres had teachings concerning life and the particular function which that centre played in Egypt, the principles of which were identified and referred to as the *Neters* (or 'Gods'); and those that were initiated into and worked in those centres were serving the purpose of that centre in respect of its role in the whole Kingdom of Egypt.

These seven main centres or chakras can be called (in ascending order) Root, Sacral, Solar Plexus, Heart, Throat, Brow and Crown. They were marked in Egypt by:-

(1) **Root:** Island of Philae and the town of Elephantine.
(2) **Sacral:** Thebes.
(3) **Solar Plexus:** Abydos.
(4) **Heart:** Hermopolis (and *Akhetaton*).
(5) **Throat:** Memphis.
(6) **Brow:** Heliopolis.
(7) **Crown:** Behdet (and Heliopolis).

The key to the purpose and working of these centres in Egypt lies in knowing the function and purpose of the seven main chakras. Briefly these can be summarised as follows:-

(1) **Root:-** located at the foot of the spine, it is the material pole or 'earth' (the polarity to the crown), wherein matter is first stirred into motion and from where it is built up into the life form. The root can be likened to the quarry that provides the stones for building the temple. It is directly associated with the Earth Mother, Matter, known as *Anna* ('the Black Virgin'). It is the seat of the will-to-exist in material form. In the East it is called *Muladhara* ('Root Support'), as it supports all other chakras. It governs the adrenal glands, the sacral plexus, and the functioning of the kidneys and spinal column.

(2) **Sacral:-** located at the level of (or slightly above) the sacral bones, it is the generative or procreative centre and centre of conception in the material world. It is the power centre of the body, providing the power or driving force that enables thoughts to be put into action. It is the seat of all sexual urges, as well as the urge to build kingdoms, empires, *etc.*. In the East it is called *Svadhishthana* ('Her Special Abode'). It governs the gonad glands, the prostatic plexus, and the entire activity of the reproductive system.

(3) **Solar Plexus:-** located above the navel at the level of the solar plexus, it is the centre of personal desires and feelings. It is the seat of the lower self or ego - the personal will and emotional energy. It is a centre of inner digestion and assimilation, of thoughts and feelings, as well as ruling those functions in physical terms. It is also a centre of synthesis and transmutation, in which all the energies below the diaphragm can be gathered up and centred in the lower self, and then that lower self transmuted into (or 'offered unto') the higher self whose seat is in the heart centre. The Egyptians knew the liver as the specific physical seat of the lower self. In the East the chakra is called *Manipura* ('City of the Shining Jewel'). It governs the pancreas gland, solar plexus and navel, and all the organs of digestion, assimilation and excretion, including the stomach, liver and gall bladder.

(4) **Heart:-** located in the centre of the chest, this is the great 'royal' centre - the centre of Being and seat of the divine Self, the pure Will-to-Be. It is the centre of inspiration - of perfect love and life - the divine desire or emotion which is the life force that emanates forth to create all things. Here all things begin and end. It is the Alpha and the Omega. In the East it is called *Anahata* ('Not Struck'), for this is "the place where the Sound is heard that is not made by any two things striking together". It is the centre of Truth - the divine Self that is called the 'Word' - the Will

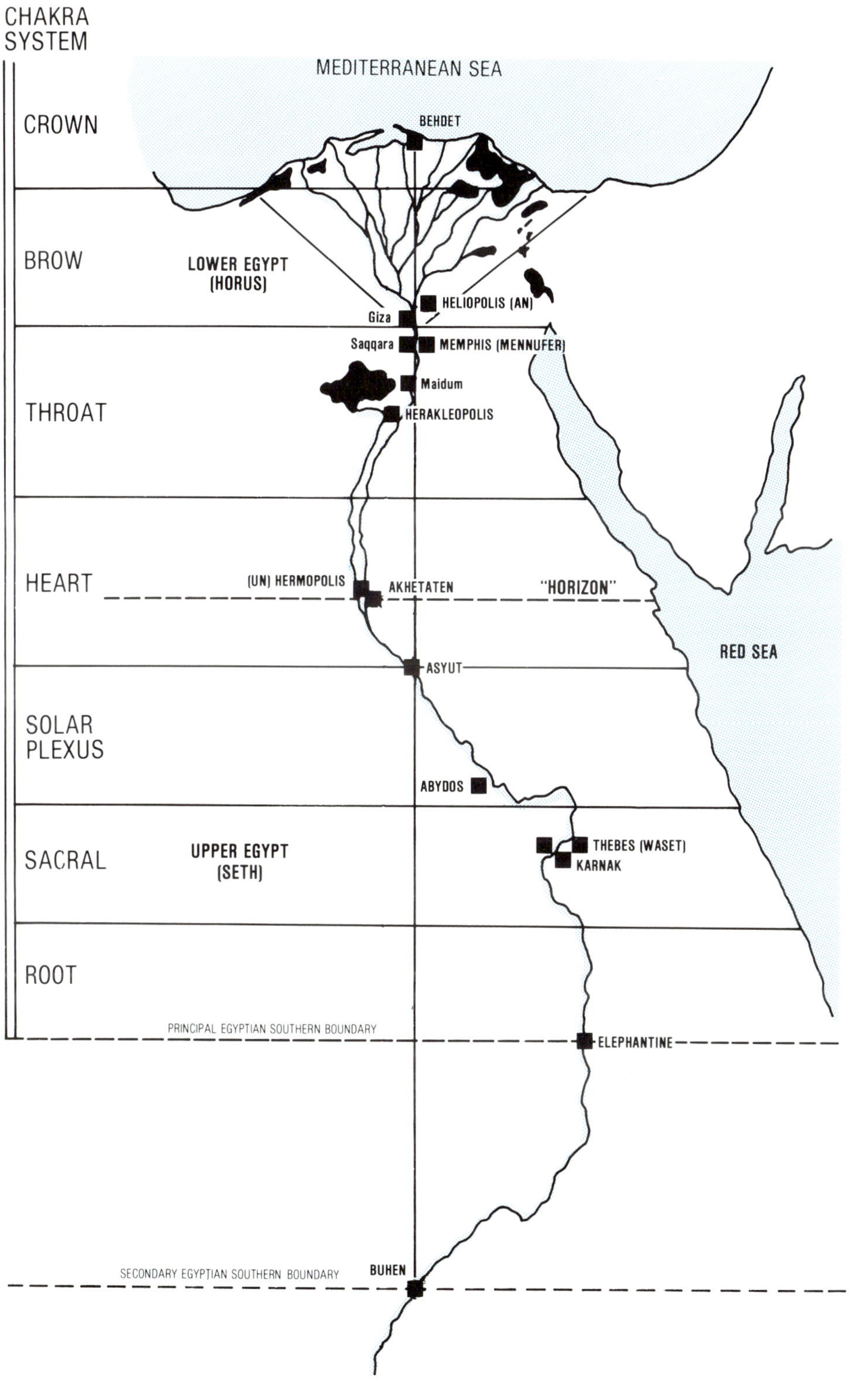

Diagram E: MAP OF EGYPTIAN LANDSCAPE TEMPLE

(or Desire) of God. This centre governs the thymus gland and cardiac plexus (the 'plexus of life'), the heart, blood and whole system of blood circulation, and the vagus nerve. The Egyptians specifically placed the seat of the divine Self in the heart.

(5) **Throat:-** located in the throat, this is the centre of hearing and awareness (listening), as well as the centre of speech and spiritual creativity. It is the centre of intuition, where the Voice of God is sounded and can be heard. In the East it is called *Vishuddha* ('Purified'), for to hear the Voice of God purifies the soul and leaves it only with a longing to hear God, talk to God and know God. It governs the thyroid gland, pharyngeal plexus, the vocal and bronchial apparatus, alimentary canal, ears, lungs and respiratory system. It is closely associated with the heart centre, and conducts the 'breath of life' that bears the Word of Love as a Vibration or Voice to the mind and whole body, 'speaking' to it. All desires speak to the mind *via* the throat centre. The Egyptians particularly associated the focus of the throat chakra with the tongue, mouth and back of the throat.

(6) **Brow:-** located in the centre of the head, at the level of the brow (between the two physical eyes), this is the centre of perception, reason and revelation. It is also the centre of decision and direction - the seat of judgement and overall government. It is the 'eye' of the mind, where the Voice of God is seen as a thought-form, so that it can be comprehended and a decision made as to how to put the thought into practice. It is the centre of the intellect, which has three major aspects - analysis, conceptualisation and memory. The whole is summed up as imagination - the faculty of perception. However, what is seen here is qualified, restricted by the individual form that our reason allows the image to take, which is determined to a large extent by our individual state of experience, understanding, and ability to understand. In the East it is called *Ajna* ('Command'), for it is the over-seeing centre of government. The chakra governs the pituitary gland (and hence all the endocrine glands, stimulating inactive or inefficient glands when necessary), the pineal gland, the cavernous plexus of brain, the eyes and nervous system.

(7) **Crown:-** located at the top or crown of the head, this chakra is the spiritual pole or 'heaven' (the polarity to the root). It is closely related to the brow centre, and completes the vision of truth. It is the centre of unqualified, universal consciousness, of unconditioned rapture, where the thought-form is unlimited and the visual experience is one of boundless light. It is the centre of pure knowledge - a knowledge or awareness of God that is beyond (*i.e.* 'above') understanding, wherein the perfect Will of God may be known and enjoyed. It is directly associated with *An* ('the Most High'), the Heavenly Father or Mind of God. It is the seat of the will-to-exist in spiritual form. In the East the chakra is called *Sahasrara* ('Thousand Petalled'). It co-governs the pineal gland, and controls the higher brain and its ventricles (in the top or crown of the head).

The seven great centres in Ancient Egypt were established in their appropriate places along the river Nile (the spine of the country), their positions being located by the sages

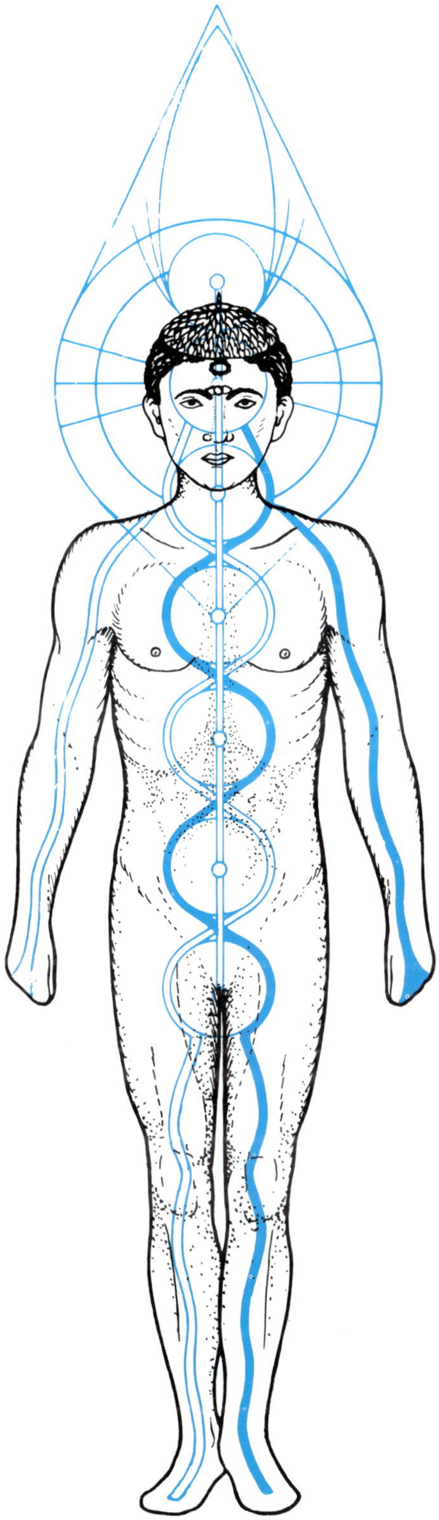

Diagram F: THE CHAKRAS IN MAN

through intuition, reason and revelation. The other centres of civilisation in Egypt, such as the 42 nomes (*i.e.* administrative districts - 22 in Upper Egypt and 20 in Lower Egypt) and their capitals, were subsidiary to the seven great centres, each playing their appropriate part. But of the seven great centres, four were the 'principal' centres - **Hermopolis, Memphis, Heliopolis** and **Thebes**. These four were the major teaching and initiatory centres, principal seats of the high priesthood and kingship, and each had their own cosmology.

The four cosmologies together constituted the wisdom knowledge and teachings of Egypt. Each cosmology was unique, with its own pantheon of gods and accounts of creation, for each described the cosmos and its laws in terms of the purpose and function of the chakra that the cosmological centre expressed. Thus **Hermopolis**, the heart centre of Egypt, was concerned with the pure Truth or Life of God - the Divine Love or Word of God, represented by ***Tat***;[25] **Memphis**, the throat centre of Egypt, was concerned with the Sound or Voice of God, represented by ***Ptah***,[26] and with purification and teaching; **Heliopolis**, the brow centre of Egypt, was concerned with the Vision or Archetypal Thought of God, represented by Râ,[27] and with the overall direction of the country and its civilisation; and **Thebes**, the sacral centre of Egypt, was concerned with the generative Power of God, represented by ***A-Min***,[28] and with the whole procreative function of Egypt and its drive to build a perfect kingdom or empire. Cabbalistically these four principle functions relate to what are known as the '**Four Worlds**' of **Emanation, Creation, Formation** and **Action**. Thus the four cosmologies are in fact a statement of the 'Tree of Life' in each of the Four Worlds.

INCARNATION AND RESURRECTION

From heart to head to abdomen; from supreme to superior to inferior; from absolute being in the centre and source of life, to creation as an idea in the heavenly mind, to generation as a natural living form in matter: this is the process of life. The end result of generation is the natural living soul, but this soul has to evolve towards a conscious and active perfection. To begin with, it is unconscious and inept, but gradually, through the alchemical process of initiation (whose stages are symbolised by earth, water, air, fire and æther), the natural soul is evolved to its eventual glory, in a return to its heart or source. This process involves birth, death and rebirth: or successive and progressive change, transforming from one state into another. This is mortality, and it leads to immortality.

> "It is in this manner that these stories are to be understood; they do, after all, conform to all revelatory texts. 'Evil' will stand for all that reduces the spiritual to the corporeal, and this corporeal must die to allow the return to its source of the spiritual. But the *fall* of the divine into perishable Matter has tarnished it, and these blemishes must be eliminated for the return to be complete: this requires the Passion. Indeed, the divine Word of the beginning is All - what we call 'spiritual' as well as *what caused it to fall into the mortal*. This is the cause of the work and struggle indispensable to the attainment of final deliverance."[29]

The fall of the spiritual and immortal into the corporeal and mortal, is the fall (or incarnation) of divine thought from the superior realm (the Mind) into the inferior realm (the Body) of the Universe. Matter materialises spirit, imprisoning it and binding up with its inertia the free movement of the spiritual thought. But at the same time the spirit begins to spiritualise matter, gradually making it move and order itself according to the spiritual idea: in other words, giving matter life, intelligence and form. The living form is the soul,

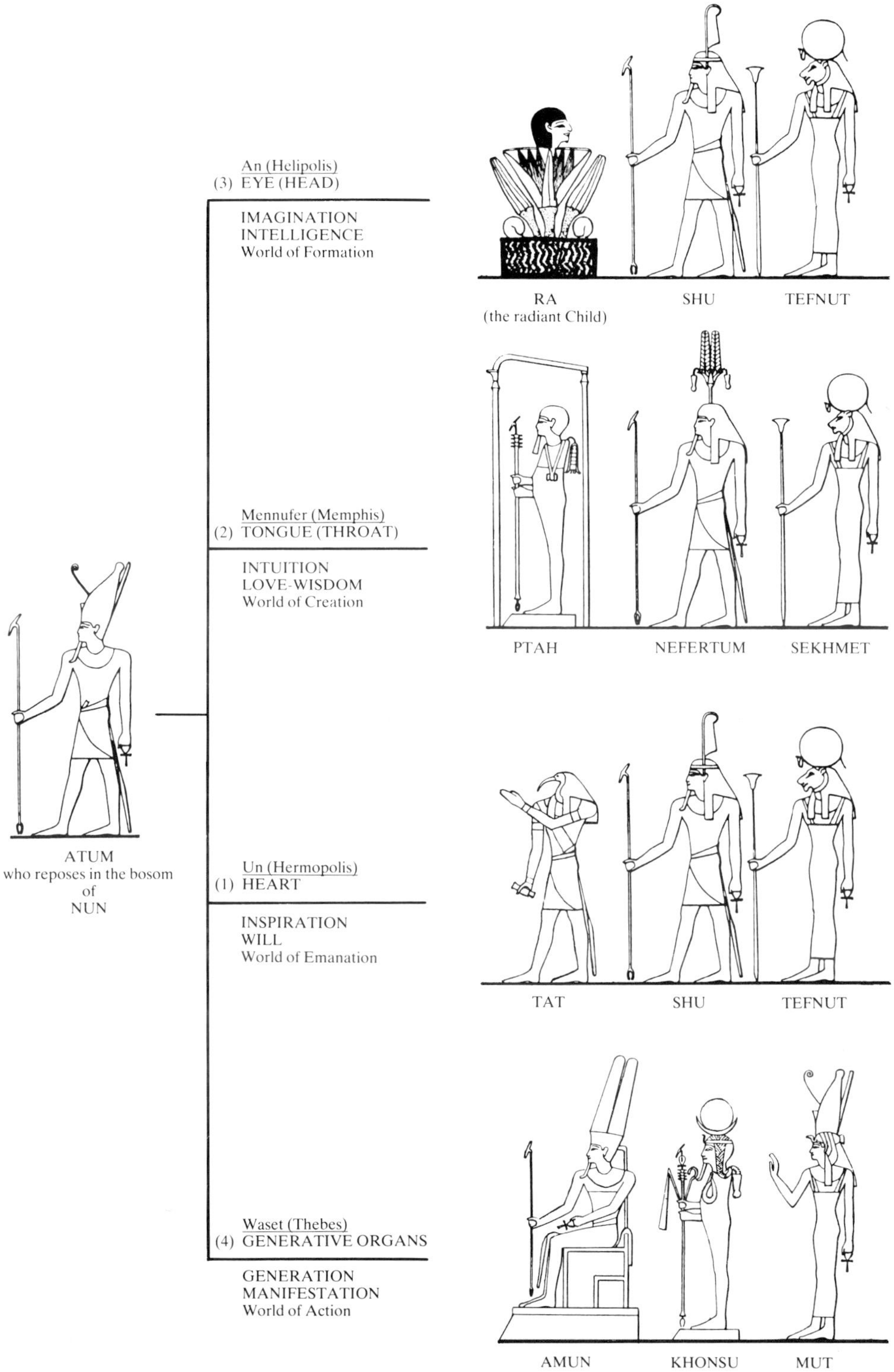

Diagram G: EGYPTIAN COSMOLOGY

and its growth from an almost inert, unintelligent, simple materialistic form to a quick, highly intelligent and complex celestial form is accomplished through the spiritualisation of matter. The process is gradual - that is to say, it proceeds in steps or stages which are defined by spiritual law; and this process is known as Alchemy (*i.e.* the chemistry of the All), or as Initiation.

The alchemical-initiatory process is defined by the very same law that brings all things into being in the first place. That is to say, *there is only One Law*. Thus every living soul has to proceed as God proceeds, with **desire**, followed by **thought**, culminating with **action** (*i.e.* the thought put into action in matter): the only difference being that the living soul begins as a simple 'prototype' expression of God's desire and thought, and has to gradually evolve towards a fuller and more complex expression that will eventually fulfil the whole divine desire and thought perfectly.

As the real purpose of the creation and evolution of the soul is to express God in all God's manifold ways, and to thereby know God intimately (or 'face to face') - so that God might know God - it therefore means that the gradual evolution of conscious and active intelligence is a fundamental part of the evolution of the soul. Furthermore, growth in complexity and growth of intelligence go hand in hand. Out of all the types of life form that exist, man is the designated leader in this question of intelligence, and thus is designed to be the 'crown' of creation, the one who may come to know God fully, face to face.

Because everything follows the One Law, we all desire, think and act. We all use first our centres of desire (in solar plexus or heart), then our heads, then our abdomens, and return again to the centre of our desires, ready for the next cycle of desire-thought-action.

Man, in order to really know God, or Good, has also to know Evil. This has entailed the human soul in the discovery and practice of things which are not good, which are imperfect and even distorted expressions of truth. Man cannot break the fundamental law, but he can express it badly and corruptly. He can desire, think and act badly, or even evilly. Bad or selfish desires are a trespass against true love; corrupt or distorted thoughts are a temptation to do the wrong things; whilst wrong doing is the real evil, the final result of trespass and temptation. These are the three sins: sin in word (*i.e.* desire, which is motive), sin in thought, and sin in deed. The last is the worst; but behind every deed is a thought, and behind every thought is a desire. Thus we should look to our desires first, to make them pure; then to our thoughts, to make them pure; then to our actions, to make them pure also. To raise ourselves from impurity to purity of soul is the first task of the human soul, and the preliminary stage before real initiation can begin.

To reach a purity of heart, mind and body (or desire, thought and action), the human soul undergoes endless cycles of desires, thoughts and actions, making mistakes, doing badly, learning from the mistakes, doing better, and gradually, step by step, transforming his or her desires, thoughts and actions into something good: in fact, transforming them into **love, understanding** and **service**. When this level of purity is reached - even if it is only for one small desire, thought and action - initiation begins. Naturally the extent of initiation depends upon how much of the soul reaches the level of pure unselfish love, understanding and service. But always that which reaches the required level of purity builds itself, as it were, around the essential spark of incarnate light that every soul has in its heart. This purity forms the heart of the soul, symbolised by the Egyptians as a white lotus bud that gradually grows and opens to the sunlight. This heart has its own microcosm of supreme, superior and inferior realms. Each Pharaoh symbolised, for the people, a human soul who had reached the stage of purity (or 'whiteness') completely - a soul who had become not only 'pure in heart' but also 'all heart'. The *hedjet* (white) crown that they wore on their heads signified this.

> "Blessed are the pure in heart, for they shall see God."[30]

> "Let your heart therefore be perfect with the Lord our God, to walk in his statutes, and to keep his commandments, as at this day."[31]

Once the 'whiteness' is achieved, the human soul proceeds through initiation. His loving nourishes and fortifies his soul, causing it to grow more and more glorious and creative in its purity. His understanding is sublimated or elevated into states of ecstacy and profound illumination. The light of his illumination fills his body and stirs it into such service that he acts as a leaven both in his own life and within society. This leads to the creation of an initiate soul, who is a shining example and powerful helper to others. In the fulfilment of his service, his sacrificial act takes him through a transmutation wherein his shining being is exalted, elevated to an immortal, celestial form of beauty and activity (*i.e.* the alchemical 'gold'), in which he knows truth face to face and enjoys the joy that such knowledge brings. Filled with this joy the adept soul is transformed to a life-bestowing master soul, who then intensifies and multiplies his own love energies or life forces. This 'multiplication' is known as the 'reddening' - the pouring out of the heart love, 'reddening' the golden-white heart. The Pharoahs of Egypt symbolised, for their people, this final perfection of the 'Philosopher's Stone' in its fullness. (the 'Red King') as well as the whiteness of purity (the 'White Queen') from which their 'redness' was born. The 'Red King' is Horus, the king to rule all and to heal all. This 'reddening' was signified by the *deshret* (red) crown of the pharoah.

Armed, as it were, with a strong bow and a quiver-full of arrows (each a ray of love-light), the master soul projects his arrows of love-light into the hearts and minds of many souls, and into the heart and mind of the world, to help heal all ills and bring all to perfection. He acts as God acts, and is a 'god' to others.

It is the 'blackness' (*Kemit*) that starts off the process of initiation, for out of the death or dissolution of man's imperfections is born the 'whiteness' of purity. Egyptian civilisation was a deliberate alchemical experiment. It was recognised that the lives of a large number of people would be lived in the pre-initiatory states of imperfection, but that they could be brought to the portals of initiation through the help and example of others who were on the path of initiation. The civilisation was begun by initiates and, throughout its history, there were always many people who entered the initiatory path, who were carefully helped and encouraged, and given every facility in the many temples of Egypt to pass through initiation. First they were helped to purify their nature, in the preliminaries to initiation; then, when sufficiently pure and strong in heart, they were carefully conducted through the grades known as the Lesser Mysteries. Out of these disciples and initiates there were always a few who were able to go beyond the Lesser Mysteries and enter the Greater Mysteries of Light - the 'reddening'. Those who entered the Greater Mysteries (in the four great temples) represented the 'exalted' or 'chosen ones' of *Is-Râ-El* - the 'golden suns' who were raised up and, 'in the twinkling of an eye', transmuted into the 'fiery light' of celestial being.

To step from the preparatory degrees into the Lesser Mysteries, and then to step from the Lesser into the Greater Mysteries, involves a complete change or transmutation of the soul from one state of being into something higher and more spiritual. In the Lesser Mysteries the soul is concerned with developing its feminine 'form' aspect, acting as a pure receptacle which the spirit increasingly incarnates into and illuminates. In the Greater Mysteries the soul has reached such full illumination that it now becomes a source and bestower (or force) of illumination - a god-like creator and illuminator. The change from the one state into the next is in fact death - a death or dissolution of the first state followed by a rebirth in the new state. The first death and transmutation brings to birth the "White Stone' (*i.e.* pure soul). The second death and transmutation brings to birth the 'Red Stone' (*i.e.* celestial soul). The 'White Stone' is born from the 'Blackness', whilst the 'Red Stone' is born from the 'Whiteness'.

In archetypal terms the 'Blackness' is that of *Nut*[32] (also known as Anna), the Black Virgin. The 'Whiteness' is that of *Ast* (Isis, also known as Mary), the White Virgin. The 'Redness' is that of *Heru* (Horus, also known as Jesus), the Red King. This is the spiritual pattern or archetype that the soul follows in its expression of divinity.

Death and rebirth takes place in the heart, to which the natural soul returns for its final judgement - the Last Judgement. From the time of its birth the natural soul is step by step evolving its way towards its place of origin - the centre of Truth. It cannot escape this fate or destiny. At the time designated for its death the soul will enter the Hall of *Maât*, in the heart centre, and there be judged and weighed in the scales. It is the heart of a soul that is weighed in the balance, for the heart is always that part of us which is the most perfect part, as well as being the most central part of us in which all the elements of our being are to be found in microcosm. The rest of our psyche and its corporeal forms are wrapped around this heart.

Whatever the natural soul has managed to bring to perfection during the course of its life will be built up around the original spark of divinity that is the soul's heart centre, thereby increasing (or causing to grow) the pure 'lotus' of the heart. If it is truly pure then it will be light enough (*i.e.* spiritual enough) to balance exactly the feather of Truth. Whatever is impure of it will weigh too heavy, and the impure part will be cut off from that which is pure and cast away into the darkness of oblivion (*Nun*). Only that which is truly pure is ever left, and that is known when the scales at last balance. The true pharaoh or master soul is one who has become 'all heart', and thus there is nothing impure in his nature to cut off and cast away. For this reason Horus, the ideal king or master soul, is known as 'the Great Heart'. When we refer to 'the Great Heart of Jesus' we are referring to this total perfection of a whole personality - complete mastery in its fullest sense. The Egyptian pharaohs tried to demonstrate this truth as fully as each was able. Jesus of Nazareth actually did it, totally.

In the Lesser Mysteries the pure soul is an ever-developing vessel for the golden light of Wisdom, which fortifies, then illuminates and finally impregnates the soul, becoming fully embodied within it. The soul becomes as a shining golden body or sun, called the *Aton* ('solar body'), which enters the heart for its Judgement. This shining soul passes through death untouched by death's destroying hands, for it is pure and holy. It is raised victorious over death and exalted (or resurrected) to become the *Sahu* ('Spiritual Body', or 'Celestial Soul') of the Greater Mysteries. To reach this perfection the soul must "live in truth".

The Egyptians used to teach these truths in their temples and by means of the symbology of the temple. For instance, each perfect soul was likened to a true and living ashlar that has been carved out of the dark quarry of matter, cut, shaped, ground and polished according to the archetypal design for it, submitted to the overseer for checking, passed as being good and correct - a true and living Aton - and then raised up and built into the shining temple of everlasting life - the temple that is the celestial soul. When exalted the

Aton becomes an 'eye' that sees all and knows all, for the perfection of the human soul brings with it full illumination or seership. This 'true eye' or 'eye of Horus' is the celestial temple, which every earthly temple represented - in particular the pyramids (hence the names given to them, such as 'shining', 'beautiful', *etc.*).[33] The celestial temple or 'eye', which is the *Sahu*, is referred to as the 'face' or 'visible appearance' of God. It is a joyous experience and expression, for the love of God is expressed as joy in its perfect manifestation - that joy being inspiring and creative light that has the divine power to help and heal all others. Thus it is said that, at the end of time, the sons of God sing for joy.

Besides the temples, the boats on the Nile were another vivid reality of life in Egypt, and so the *Aton* was also represented as a solar boat in which the spirit is carried as the helmsman, navigator and fisherman. Raised to heaven it not only becomes the celestial boat of *Râ*, but also a navigator and fisherman of others, brandishing his spear of light untold. Yet another appropriate symbol of the *Aton* was that of the golden chariot driven by the spiritual hunter or god. Resurrected to its celestial form it becomes the 'chariot of fire', which is itself a god or hunter in its own right, bending his bow of burning gold and projecting his arrows of heart desire.[34]

> "Praise be unto thee, O Osiris, lord of eternity, Un-nefer, Heru-Khuti, whose forms are manifold, and whose attributes are great, who art Ptah-Seker-Tem in Anu (Heliopolis), the lord of the hidden place, and the creator of Het-ka-Ptah (Memphis) and of the gods [therein], the guide of the Underworld, whom [the gods] glorify when thou settest in Nut. Isis embraceth thee in peace, and she driveth away the fiends from the mouth of thy paths. Thou turnest thy face upon Amentet, and thou makest the earth to shine as with refined copper. The dead rise up to see thee; they breathe the air and they look upon thy face when the Aton riseth on its horizon; their hearts are at peace, inasmuch as they behold thee, O thou who art eternity and everlastingness."[35]

> "Homage to thee, O thou glorious Being, thou who art dowered [with all sovereignty]. O Temu-Harmachis[36], when thou risest in the horizon of heaven, a cry of joy cometh forth to thee from the mouth of all peoples. O thou beautiful Being, thou dost renew thyself in thy season in the form of the Aton within thy mother Hathor[37]; therefore in every place every heart swelleth with joy at thy rising for ever. The regions of the North and South come to thee with homage, and send forth acclamations at thy rising in the horizon of heaven; thou illuminest the two lands with rays of turquoise light. Hail Râ, thou who art Râ-Harmachis[38], thou divine man-child, heir of eternity, self-begotten and self-born, king of the earth, prince of the underworld, governor of the regions of Aukert![39] Thou didst come forth from the water, thou hast sprung from the god Nun, who cherisheth thee and ordereth thy members. Hail, god of life, thou lord of love, all men live when thou shinest; thou art crowned king of the gods. The goddess Nut doeth homage unto thee, and the goddess Maât embraceth thee at all times. Those who are in thy following sing unto thee with joy and bow down their foreheads to the earth when they meet thee, thou lord of heaven, thou lord of earth, thou king of Right and Truth, thou lord of eternity, thou prince of everlastingness, thou sovereign of all the gods, thou god of life, thou creator of eternity, thou maker of heaven, wherein thou art firmly established. The company of the gods rejoice at thy rising, the earth is glad when it beholdeth thy rays; the peoples that have long been dead come forth with cries of joy to see thy beauties every day. Thou goest forth each day over heaven and earth, and art made strong each day by thy mother Nut. Thou

> passeth through the heights of heaven, thy heart swelleth with joy; the abyss of the sky is content thereat. The Serpent-fiend hath fallen, his arms are hewn off, and the knife hath cut asunder his joints. Râ liveth in Maât the beautiful. The Sektet boat draweth on and cometh into port; the South and the North, the West and the East, turn to praise thee, O thou primeval substance of the earth who didst come into being of thine own accord. Isis and Nephthys salute thee, they sing unto thee songs of joy at thy rising in the boat, they protect thee with their hands. The souls of the East follow thee, the souls of the West praise thee. Thou art the ruler of all the gods, and thou hast joy of heart within thy shrine; for the Serpent-fiend Nak hath been condemned to the fire, and thy heart shall be joyful for ever.[40]

That which has been raised perfect and immortal becomes a god in its own right, for the celestial soul is a soul transmuted to become radiantly creative and spiritually active. Such a soul is the fulfilment of the divine Idea for man, wherein man becomes the perfect and complete embodiment of divinity - 'like unto God' - the true *hu*-man (*i.e. Shu*-man, or Christ-man).

> "But some man will say, 'How are the dead raised up? and with what body do they come?'
> Thou fool, that which thou sowest is not quickened, except it die:
> And that which thou sowest, thou sowest not that body that shall be, but bare grain, it may chance of wheat, or of some other grain:
> But God giveth it a body as it pleased him, and to every seed his own body.
> All flesh is not the same flesh: but there is one kind of flesh of men, another flesh of beasts, another of fishes, and another of birds.
> There are also celestial bodies, and bodies terrestrial: but the glory of the celestial is one, and the glory of the terrestrial is another.
> There is one glory of the sun, and another glory of the moon, and another glory of the stars: for one star differeth from another star in glory.
> So also is the resurrection of the dead.
> It is sown in corruption; it is raised in incorruption:
> It is sown in dishonour; it is raised in glory:
> It is sown in weakness; it is raised in power:
> It is sown a natural body; it is raised a spiritual body.
> There is a natural body and there is a spiritual body.
> And so it is written, 'The first man Adam was made a living soul;' the last Adam was made a quickening spirit.
> The first man is of the earth, earthy: the second man is the lord from heaven.
> As is the earthy, such are they also which are earthy: and as is the heavenly, such are they also that are heavenly.
> And as we have born the image of the earthy, we shall also bear the image of the heavenly.
>
> "Now this I say, brethren, that flesh and blood cannot inherit the kingdom of God; neither doth corruption inherit incorruption.
> Behold, I shew you a mystery; We shall not all sleep, but we shall all be changed, in a moment, in the twinkling of an eye, at the last trump.
> For the trumpet shall sound and the dead shall be raised incorruptible, and we shall all be changed.
> For this corruptible must put on incorruption, and this mortal must put on immortality.

> So when this corruptible shall have put on incorruption, and this mortal shall have put on immortality, then shall be brought to pass the saying that is written, 'Death is swallowed up in victory.'
> 'O death, where is thy sting? O grave, where is thy victory?'
> The sting of death is sin; and the strength of sin is the law.
> But thanks be to God, which giveth us the victory through our Lord Jesus Christ."[41]

THE A-B-C OF LIFE

The fundamental sound and letter that signifies the primordial **Beginning** is '**A**', pronounced '**aaah**'. This is the sound of the heart expressing itself in pure love. It is the vocal representation and echo of the emanation of God's heart in love, when God desired to be. It is the natural sound that we all make when we are struck with love for something or someone - particularly for a baby who draws out the tender love in us, or for a sight of rare beauty that inspires us. It is the sound of love, and the sound of wonder. Love is the beginning of all life - and is life. Wonder is the beginning of all knowledge.[42] '**A**' is the beginning of all sounds, and thus of all else.

'**A**' is above all the unique sound and letter, the first of the Alphabet. As a sign or symbol, its form is derived from the equilateral triangle, mystic symbol of the Holy Trinity - the Divine All. Everything is contained in and unfolds from '**A**', and hence the letter '**A**' is used to represent the whole Alphabet.

For instance, besides indicating the beginning of all things as an expression of pure heart-love, '**A**' is also the sound of initial vision - a natural sound (in the form of a gasp, 'Aah!') which we make when we see something beautiful that fills us with wonder. Then, following on from this, '**A**' is the sound of perception (in the form of an understanding 'Ah!' or 'Ah-ha!'), which we make when we eventually grasp the meaning of the vision or idea that we have been thinking about. Finally '**A**' is the sound of revelation or knowledge (in the form of a definitive "Aaah!'), which we make when we have completed doing something good and have paused for a moment to take in what we have done, enjoying the fruits of our labours. It is the sound of both seeing and knowing a truth, which is a revelation of God, a revelation of the Beginning.

All in all, '**A**' summarises the whole divine process, from beginning to end; for the eternal **Beginning** is also the unfolding **Middle** and the joyful **End**. The Greek word *Arche*, which means 'Ever-Beginning', also means 'End' and all that is between the two, and it is signified by a circle. The Ancient Egyptians knew this and used this knowledge, as did other nations.

'**A**' is the Beginning and the End, the Alpha and the Omega. This is precisely what the sound and letter stood for in Ancient Egypt. It was often written as ***AA*** (*i.e.* the 'double-A'), representing the Beginning and the End - Genesis and Revelation.[43] As the Word of Truth, the ***A*** was built into the word ***TAT*** ('Truth', the 'Word' of God) by enclosing this sound of divine love within the Tau Cross sign for divine wisdom. The cross symbolises wisdom, which is light, by virtue of it representing the rays of light that radiate from the heart and establish the cosmos. In the form of the '**T**' (or Tau Cross) it portrays the three basic principles or aspects of light (the Holy Trinity) that come from the One Originating Principle. The One Originating Principle is divine love, the principle of illumination. The use of the double-T (***TT***) is to signify (a) that love is enfolded or hidden within that wisdom or light, and (b) that light has two complementary aspects or polarities - radiant and reflective, giving and receiving, masculine and feminine, or force and form. The harmonious relationship between these two polarities, which is their unifying principle, is the perfect expression of love (***A***) as balanced light, and for this the Sun-symbol 'O' is sometimes used, hence the word ***TOT*** which is another form of ***TAT***. In ***TAT*** (or ***TOT***) the idea of the Holy Trinity is expressed.[44]

The form of the letter '**A**' is chosen to represent the heart sound, as it symbolises geometrically (a) the Triangle of the Holy Trinity - the Three-in-One, (b) the Hierarchy of Creation, which is the order and pattern of light in the mind of God, and (c) that Hierarchy as it manifests in matter and builds up the Pyramid - the holy hill or mountain, symbol of the perfect soul that manifests light.

The two polar aspects of living light, signified by the double-T, express themselves in the mind of God, and then in matter, as the ***BA*** and ***KA***. These are the masculine-feminine creative principles that emanate from the ***A*** and are the dual expressions of that ***A*** for the purpose of creation. Hence the Holy Trinity, in Egyptian terms, was described as being ***AA-BA-KA*** (cabbalistically equivalent to our **A-B-C**). Rearranged in its more symbolic order, corresponding to ***TAT***, it would be **BA-AA-KA** (or ***B-A-C*** in the English alphabet).[45] These three are known as the **Three Rays** or **Three Pillars**, and comprise the manifestation of Life in the three realms of Spirit, Soul and Body (*i.e.* Divinity, Humanity and Nature).

BA is the animating spirit - the life force - which bestows the breath of life. It is creative, life-giving and illuminating. Its characteristic is non-fixity, which needs support (or fixing) in order to manifest and do its work. It is pure, formless energy- a force or vibration - requiring the *KA* to give it form and stability. In terms of consciousness it is the pure formless light of intuitive (or bliss) consciousness - the essence of all forms of light. As such ***BA*** is known as the '**Son**' of the Divine Father-Mother (Truth). Written as ***AB***, this same life force and light-essence is called the '**Father**', for it is the Father of everything that proceeds from it. ***Ab*** is also the Egyptian name for the vitality and love-light of the human heart. The ***BA*** exists in the three 'worlds' of Spirit, Soul and Body as divine *BA*, human *BA* and natural *BA*. It corresponds to the right-hand 'pillar' of the Temple or Tree of Life and, on the Egyptian version of the Tree of Life, the divine archetypes for its manifestation in the three worlds of Spirit, Soul and Body are *Shu, Osiris* and *Seth* respectively.

KA is the formative, ensouling aspect of life - the life form - which attracts, receives, fixes and stabilises the *BA*. It is formative, life-receiving and capable of being illumined. It gives form and definition to the formless essence that is the *BA*. It provides the environment in which and the means by which the *BA* can work. It is the principle of fixity, and conditions the universality and impersonality of the *BA*. The *KA* 'names' or identifies the *BA,* dividing the universal and undifferentiated *BA* into individual and differentiated forms that define and reveal the various hidden aspects of *BA*. The word ***Kabbala*** (Cabbala - 'the Received Light or Wisdom') is derived from *KA*. Thanks to the *KA,* the *BA* is able to become an individually recognisable entity. The soul is generated by the form-making powers of *KA* which assimilate the life-giving powers of *BA*. In terms of consciousness *KA* is the thought-form (or forms) of light, reflected and thus conceived in the mind - the intellectual consciousness. It is the principle of all form, and the sensitivity of form, which is animated and acted upon by the *BA* in all three worlds. The ***KA*** is known as the '**Daughter**' of the Divine Father-Mother (Truth). ***Ka*** is also used to describe the form of the heart, sometimes referred to as the 'dweller in the body'. Written as ***AK***, it is called the '**Mother**', for it is the Mother of everything that proceeds from it. It exists in the three worlds of Spirit, Soul and Body as divine *KA,* human *KA* and natural *KA,* which are respectively the spiritual (or divine) body, the celestial body and the natural body. It corresponds to the left-hand 'pillar' of the Temple or Tree of Life and, on the Egyptian Tree of Life, the divine archetypes for its manifestation in the three worlds of Spirit, Soul and Body are *Tefnut, Isis* and *Nephthys* respectively.

AA also exists in the three worlds of Spirit, Soul and Body as divine *AA,* human *AA* and natural *AA,* which are respectively the spiritual self, the celestial self and the natural self. It corresponds to the central 'pillar' of the Temple or Tree of Life and, on the Egyptian Tree of Life, the divine archetypes for its manifestation in the three worlds of Spirit, Soul and Body are *Ra, Horus* and *Geb* respectively.

MAAT, the goddess who is represented by a white ostrich feather or a jar, represents the fully illumined consciousness of the human soul - the state of divine knowledge and exquisite joy that a soul is able to experience whilst living and practising truth to the highest extent possible. ***Maât*** is **Truth revealed**. It is the illumined state of Horus the king, who "lives in *Maât* " - the perfect manifestation of the 10th principle or *Neter* on the Tree of Life - the fulfilment of God's will.

The Hermetic Art of Alchemy

THE SCARAB

The scarab beetle vividly illustrated and symbolised for the Egyptians the various phases of the process of life. In this process there are three main phases, known alchemically as (1) the Lesser Work, (2) the Greater Work, and (3) the After Work. In terms of the Mysteries of initiation, these are referred to as (1) Preparation or Purification, (2) the Lesser Mysteries, and (3) the Greater Mysteries. In colour symbology these three phases are identified as (1) the Black Phase, (2) the White Phase, and (3) the Red Phase.

The scarab was the particular symbol of the god ***Khepri*** ('He who came forth from the earth', or 'He who came into being of himself'), the Self-created. *Khepri* stood for the idea of the whole process of life, which is alchemy. The scarab illustrated this as follows:-

(1) **Black Phase:** The scarab lays its egg, which is black, in a pellet of dung which it buries underground in absolute darkness. (This phase lasts for one month, until the egg reaches maturity.)

(2) **White Phase:** From this egg a white larva is born. During the second phase (lasting a further month) this larva feeds off the egg substance. This phase culminates with the larva entering its nymphal state, during which the 'mummified' larva works its self-transformation.

(3) **Red Phase:** From this mummy emerges the new scarab. Its head and thorax are red, its abdomen and wingsheaths are white, whilst the remaining parts of its body are black.

THE THREE WORKS OF ALCHEMY

The Lesser Work is concerned with purifying the natural soul or life form, and returning it to the state of *Prima Materia* ('First Matter' - *i.e. Kemit*) - the state of primordial darkness, pure and perfect - a rich humus containing the seeds of all new growth, and from which the **White Stone** of the Philosophers is born. The White Stone is the pure (*i.e.* virgin) soul, who is the philosopher (*i.e.* the bride and lover of wisdom, the spirit or light of God). She is also known as the **White Queen**.

The Greater Work is concerned with the nurturing, growth, strengthening, refining and leavening of the White Stone, incarnating more and more life and light (*i.e.* spirit) into her. The spirit impregnates its bride with its vitality and light, transfiguring her and stirring her to the highest expressions of devotion, understanding and service. In her is conceived (*i.e.* embodied) the love, wisdom and power of God, which eventually transmutes her highest and most spiritualised nature into alchemical **Gold** - the radiant child that is born from her - which becomes the **Red Stone** by virtue of its own activity. The Red Stone is the celestial soul, which is also known as the **Red King**.

The After Work is the multiplication of the Red Stone, increasing its size, strength and abilities, and the projection of the powers of the Red Stone onto nature in order to help the work of the Spirit. The celestial soul becomes as a god or spirit to others, helping to purify, inspire, enlighten and lead others to perfection. The **Red Stone** or **King** has the power to transmute all other substances to gold, and he wields this power.

THE THIRTEEN STEPS OF ALCHEMY

Each of the three main phases or works of alchemy has four steps or degrees of alchemical transformation within it, making twelve steps or degrees altogether. The four steps are patterned on and expressions of *Thoth, Ptah, Râ* and *Amen* - *i.e.* Emanation, Creation, Formation and Manifestation. They relate to the four seasons of the year (winter, spring, summer and autumn), and to the degrees of initiation that are associated with the alchemical elements (earth, water, air and fire). The work begins with the calcination of the original body or life form.

THE LESSER WORK

(1) EARTH — **Calcination**:- burning to ashes; purifying by consuming the grosser part in fire and leaving only the calx (*Sulphur*).

(2) WATER — **Solution**:- the calx dissolved into liquid form by the addition of *Aqua Ardens* (*Mercury*).

(3) AIR — **Separation**:- distillation; dividing the subtle from the gross; analysis; extraction.

(4) FIRE — **Conjunction and Putrefaction**:- 'Chymicall Wedding'; energetic union between subtle and gross, *Mercury* and *Sulphur*, *etc.*, leading ultimately to death - disintegration of the form into a formless unity; decomposition; rotting into a compost of 'blackness' (= the First Matter).

THE GREATER WORK

(5) EARTH — **Coagulation**:- crystalisation. Death leads to a triumphant return. Out of the dark womb of the First Matter is born a white star (= the White Stone).

(6) WATER — **Cibation**:- nourishing; feeding the matter (with love); fortifying; growing more and more glorious; developing love.

(7) AIR — **Sublimation**:- vapourising; transforming into something higher, more sublime; refinement; elevation of the mind to ecstatic states; illumination of the mind; transfiguration; developing understanding.

(8) FIRE **Fermentation**:- leavening; excitation; stirring up; thoughts put into action; developing service. This results in the creation of the *Sol Flavus* ('Golden Sun'), the celestial body. The 'yeast' dies when the 'bread' (*i.e. Sol Flavus*) has leavened.

THE AFTER WORK

(9)EARTH **Exaltation**:- lifting up of the *Sol Flavus;* resurrection; elation; knowledge; exultation; joy. The Golden Sun is tinctured by joy to become the Red Stone. "In a moment comes the Tyrian colour."

(10) WATER **Multiplication**:- propogation; generation; procreation; magnification. The Red Stone becomes willingly creative, actively loving and sounding its note of love. Its love activity increases its fire, feeding the Red Stone and multiplying it.

(11) AIR **Projection**:- "the proving of the perfect, strong Red Stone"; casting forth the Red Stone upon other substances in order to transmute those substances; the projection of light or wisdom into the world, healing all diseases, expelling all poisons, transmuting all to gold.

(12) FIRE **Completion**:- the Glorification and Sabbath of the Great Work.

The Egyptian Mysteries of Initiation

INTRODUCTION

The Egyptian Mysteries were a dramatic and symbolic presentation of the alchemy of life, enacted in the seperate temples of Egypt and, eventually, in the whole landscape temple of Egypt. Through drama, in which the person undergoing initiation was given the role of the main character, followed by careful instruction and training, each soul was given a series of progressive experiences and teachings that were physically, psychologically and spiritually profound. By playing upon all the senses, expanding and educating the mind, and awakening and strengthening the heart and will, each candidate was taken carefully through the degrees of initiation, first in the Lesser Mysteries and then, if suitable, the Greater Mysteries.

The story of the gods is the archetype for all manifestation of life in matter and the development of nature and the human soul. The story of Osiris, Isis, Horus, Seth and Nephthys is, in particular, concerned with the human soul and its initiatory path. Several variants of the basic story were developed, depending upon the time and place and purpose for the teaching being given, but the essence of the story remained (and still remains) constant. The Mysteries were developed as expressions of the archetypal story, their essential form and teaching always remaining constant, but with variations according to time and place and individual purpose.

An account of one such practice of the Egyptian Mysteries has survived in the form of the ***Crata Repoa***. Manley P. Hall, in his erudite presentation of the *Crata Repoa* in his book *Freemasonry of the Ancient Egyptians,* has this to say:-

> "The *Crata Repoa* is a work made up of fragments from many ancient authors. The unknown compiler had as his purpose a restoration of the ancient Mysteries based upon the hints and allusions contained in the classical writings. Most of the great philosophers of antiquity were initiates of State Mysteries. In their writings, these men frequently allude to some fragment of the initiatory rites. These hints have been carefully organised, placed in sequential order, and connected with an editorial tissue. The result is a restoration founded upon the actual words of the initiates themselves."[1]

The *Crata Repoa* illustrates the usual **seven major degrees of initiation** that follow on after the initial period of preparation. These seven degrees of initiation are the last seven stages in the Alchemical Work of twelve stages. The first four stages are preparatory, whilst the fifth is the threshold in which the 'White Stone' is born from the darkness of *Prima*

Materia. The 'threshold' plus the first three degrees of initiation form the Lesser Mysteries of Virginity. The final four degrees comprise the Greater Mysteries of Light.

Each set of four stages or degrees is based on the pattern of (1) birth (or rebirth), (2) desire or learning to love, (3) thought or learning to understand, and (4) action or learning to serve.

The following is a condensed rendering of the *Crata Repoa,* and I would refer the reader to Manley Hall's book for a fuller account. Purposely an attempt is not made here to interpret the symbology in any great detail (although notes are provided), but to allow the symbols and story to speak for themselves, which they should do - especially when compared with the initiatory stories and Mysteries of other traditions, and with the Hermetic science of Alchemy.[2]

PREPARATION FOR INITIATION

(1) EARTH — Having asked, been recommended and then accepted for temple training, the aspirant was circumcised and placed on a special diet.

(2) WATER — The aspirant was compelled to spend several months in an underground chamber, at the end of which he was strictly examined to ascertain the limit of his intelligence.

(3) AIR — If found suitable, the aspirant was led into a columned hall or gallery, every column engraved with certain maxims of Hermes which he had to learn by heart.

(4) FIRE — When he had accomplished this, the aspirant was approached by the *Thesmophores* (Introductor), who took him to a grotto. Having had his eyes bandaged and his hands bound, the aspirant was left in darkness for some time.

THE LESSER MYSTERIES

(5) EARTH — **Birth.**
In the grotto, still in darkness, the aspirant was prepared by the Thesmophores as a candidate for initiation.

(6) WATER — **1st Degree - *Pastophoris* (Apprentice, or Neophyte).**
The candidate was led by the hand from the grotto by the *Thesmophores,* eyes still bandaged, to the Gate of Men. He was announced, questioned (the 'password' asked and given), the gate was opened and he was admitted to the temple.
Inside the temple, the candidate was questioned anew by the *Hierophant.* He then had to travel around the Birantha, being subjected to artificial lightning, thunder, hail, rain and tempest, in order to test his courage and fearlessness.
If he was not too discouraged, the *Menes* (Reader of Laws) read to the candidate the Constitution of the Society. The candidate promised to conform himself thereto.

The candidate, with uncovered head, was then conducted to the *Hierophant,* before whom he knelt. A sword was placed at his throat whilst he took the oath of fidelity and discretion as a newly entered *Pastophoris.*

Having taken the oath, light was restored to him. His blindfold was removed and he was placed between two square columns. Directly before him, between the columns, was a ladder with seven steps and an allegorical figure portraying a set of eight doors. The *Hierophant* addressed the *Pastophoris,* exhorting him to guard himself from prejudices and passions that might draw him from the road to happiness, and to continually fix his thoughts and eyes upon the One Divine Being in order the better to govern his heart and senses.

The *Pastophoris* was then made to ascend the ladder, being taught that it was a symbol of *Metempsychosis* (*i.e. alchemical transformation and* transmutation).

The *Pastophoris* was given the word of recognition (***Amoun*** - "Be discreet"), taught the hand grip, and dressed with a hood, apron and collar with tassels falling on his breast.

The instruction of this grade was natural science, including anatomy and healing, symbology and the common hieroglyphic writing.

The duty of the *Pastophoris* was to guard the Gate of Men.

(7) AIR

2nd Degree - *Neocoris* (Craftsman, or Disciple).

After a year of apprenticeship, if suitable, the *Pastophoris* was severely tested to see if he had mastered the first degree. He was placed in a chamber (the *Endymion*) and given a delicious feast served by beautiful women (temple virgins). Invited to make love to them, he had to prove the command that he had over his desires.

If proved worthy, the *Thesmophores* came to him and propounded a variety of questions. If the *Pastophoris* answered satisfactorily, he was introduced into the Assembly. Having been sprinkled with water by the *Stolista* (Baptiser), he affirmed that he had conducted himself with wisdom and chastity, and was accepted as a *Neocoris.*

At this point the *Thesmophores* ran at the *Neocoris* and flung a living serpent over his body. The chamber then appeared to the *Neocoris* to be filled with such reptiles. This was to test his courage and to show the power of mind over matter. He was then led towards two high columns, between which was the figure of a griffin pushing a four-spoked wheel, signifying the sun and the four seasons[3] or elements.

He was given the insignia of a caduceus, taught the word of the degree (***EVE***),[4] and instructed as to the meaning of the fall of the human race.

The sign of the degree consisted of crossing the arms upon the breast.

The learning which followed included architecture, geometry, mathematics, and geographical and other measures of various kinds.

The duty of the *Neocoris* was to wash the columns of the temple.

(8) FIRE

3rd Degree-*Melanophoris* (Master Mason, or Initiate).

When the *Neocoris* merited the next degree he was conducted by the *Thesmophores* into an underground vestibule, entered *via* the Gate of Death. This place was filled with mummies, coffins, and with those who laboured with the dead. In the midst of the crypt was the coffin of Osiris, bearing traces of blood. Osiris was supposedly recently assassinated. After earnestly denying that he had any part in the murder, the new *Melanophoris* was led into a hall where the other *Melanophoris* were assembled, all dressed in black, together with the *Pharaoh* who wore his robes of state.

The *Pharaoh* offered a golden crown to the new *Melanophoris,* asking him if he considered himself to have sufficient courage to sustain the trials ahead. Knowing that he should reject the crown (symbol of material power and wealth) if he was to attain the spiritual crown of Truth, he threw it down and trampled it underfoot. Crying 'Outrage! Vengeance!", the king seized the sacrificial axe and struck[5] the *Melanophoris* upon the head, causing the initiate to swoon (or feign death). The *Melanophoris* was then wrapped in mummy bandages and conveyed on a bier to the Sanctuary of Spirits, where he was laid in an open coffin and surrounded with fire, thunder and lightning.

Following this experience, the *Melanophoris* was led by Charon to the Hall of Judgement, where the judges condemned him to wander in the subterranean galleries.[6] He was divested of his mortuary apparel and was given instruction concerning death and resurrection.

He was taught the sign of recognition - a peculiar embrace expressing the power of death - and given the word of the degree (***Monach Caron Mini*** - "I count the days of anger").

The learning which followed concerned drawing and painting, the Hierogrammatical alphabet, the history and geometry of Egypt, and the basics of astronomy. He was also trained in rhetoric. His work was involved with serving the dying and dead, and their relatives, decorating the mummies and tombs, and giving funeral orations.

THE GREATER MYSTERIES

(9) EARTH

4th Degree - *Chistophores* (Arch-Mason, or Adept).

After 18 months[7], if found suitable, the *Melanophoris* was visited by the *Thesmophores* who invited him to the next degree. The *Thesmophores* armed the initiate with sword and shield, and led him through dark labyrinthine galleries, in which they were attacked by hideously masked men carrying flaming torches and serpents. The *Thesmophores* had to withstand the danger, but was overcome. With eyes bandaged and a cord about his neck, he was led into the Hall of *Maât* where he was to receive the next grade.

He was there 'raised extended',[8] his blindfold was removed, and he was introduced to the whole assembly of adepts and masters, all robed in splendour and positioned in order in a magnificent and dazzling hall, presided over by the *Pharaoh* and Officers of the Society.

After the Orator had given a discourse of praise for what the new *Chistophores* had achieved, and encouragement for what lay ahead, he was presented with a cup containing a potion of bitter herbs, for him to drink. Then he was invested with a buckler, buskins, mantle, scimitar, helmet, spear, bow and quiver full of arrows.[9]

Thus armed, he had to enter a cavern and strike off the head of a very beautiful-looking 'woman'[10] and present the head to the king.

The *Chistophores* now changed his old garments for new clothing,[11] and his name was inscribed in a book along with the other adepts of Egypt. From thenceforth he was allowed to communicate freely with the *Pharaoh,* and was given nourishment (materially and spiritually) from the Court.

He was presented with the Code of Laws and the emblem of an owl,[12] together with a palm[13] and an olive branch.[14] He was given the word of the degree (***IOA***).[15]

He then learnt the Ammonitish (*i.e.* secret) language and the human sciences (*i.e.* the sciences concerning the human soul), preparing himself for the final three degrees in which he would actively wield the 'Word' with increasing power.

(10) WATER

5th Degree - *Balahate* (Philosopher, or Master).

The *Chistophoris* had the right to demand entrance to this degree. The *Demiurgos* could not refuse the request.

The adept was led to the Assembly, where he was received as a *Balahate.* From thence he was conducted to another hall, arranged as a theatre. For once he was the sole spectator, whilst the others played the parts in the drama.

Horus entered the hall with several masters, bearing torches, as if seeking something. They came to the entrance of a cavern, from which belched flames. In the cavern was Typhon. Horus drew his sword and fearlessly approached the hundred-headed dragon. In the ensuing battle, the dragon was overthrown and decapitated by Horus, and the corpse thrown back into the fiery cavern. In silence Horus exhibited thc dragon-heads.[16]

The new *Balahate* then had the meaning of the drama explained to him. The word of the degree was ***CHYMIA,***[17] and the learning which followed was in Alchemistry.

(11) AIR

6th Degree - *Maawr* (Seer, or Priest):
— Astronomer before the Gates of the Gods.''

This degree began by placing the candidate in irons and leading him down four steps to the Gate of Death. The crypt was now filled with water,[18] across which he was rowed in the Barque of Caron to be received by the Assembly.

There he took his new oath, and the history of the origin of the gods was explained to him, including how they emanate from the One Sole God.

After the reception the new *Maawr* was led up from the crypt and into a great hall *via* the Gate of the Gods. Here he saw the pantheon of gods represented and identified in order, with their meanings explained. The more esoteric nature of the history of Egypt, the world, the universe, and of the Society and its Order, was then taught to him by the *Demiurgos*. He was also given a list of all the Chief Inspectors of the Society, past and present, and a table of the members of the Society throughout the world. Finally he was taught the Priestly dance.[19]

The word of the degree was ***IBIS***.[20]

The work of the degree was to become proficient in astronomy.[21]

(12) FIRE **7th Degree - *Saphenath Pancah* (Prophet):**
"The Man who knows the Mysteries".

The reach this degree, the *Maawr* had to be invited by the *Pharaoh* and *Demiurgos,* with the consent of the Inner Fraternity.

Following the invitation, a great reception was arranged for the new prophet, followed by a public procession known as the *Pamylach.*[22]

The procession over, an assembly was held by the initiates during the night, in special houses called *Maneras.*[23] Here he was given a beverage called *Oimellas,*[24] the ambrosial drink of the gods, and he was informed that he had arrived at the end of all the proofs.

He was then invested with an *Ankh*[25] and a very beautiful white striped robe called *Etangi,*[26] His head was shaved, forming a square coiffure.[27]

The principal sign of this final degree was made by carrying the hands crossed in the wide sleeves of the robe.

The word of the degree was ***ADON***.[28]

In this degree was given the most complete and detailed explanation of the Mysteries, and the prophet had permission (and the ability) to read all the mysterious books (of Hermes), including the scriptures of other nations, to which he was given the 'royal key'.

The greatest prerogative of the 7th Degree was that the prophet could vote in the election of the *Pharaoh,* and it was from the ranks of the prophets that the officers of the Society were drawn.

THE OFFICES

The offices, and those who fulfilled them as officers or principals of the Society, were patterned upon the principles of the Tree of Life - the Tree of Life signifying the Divine Order or Idea of God in which all the gods are to be found as archetypes arranged in their cosmic order. The grade of **prophet** was the official starting point, corresponding to the ninth principle of the Tree of Life, called 'Foundation'. The other eight principles were represented by officers ascending in order up to the highest temple office of *Demiurgos,* who corresponded to the principle called 'The Crown'. The supreme office of *Pharaoh* was above all the others and summed them all up in a unity as being their source and creative principle. The offices were as follows:-

0. ***Pharaoh*** (King)	THE CREATOR or VOICE
1. ***Demiurgos*** (Initiator)	CROWN
2. ***Hierophant*** (Teacher)	WISDOM
3. ***Stolista*** (Purifier/Guardian)	INTELLIGENCE
4. ***Hierogrammatista*** (Secretary)	MERCY
5. ***Thesmophores*** (Guide/Tester)	JUDGEMENT
6. ***Zacoris*** (Treasurer)	BEAUTY
7. ***Komastis*** (Overseer of Banquets)	VICTORY
8. ***Odos*** (Orator/Chanter)	GLORY
9. ***Saphenath Pancar*** (Prophet)	FOUNDATION
10. [The Temple]	[KINGDOM]

Level	Initiation	Stage
		Peace
7 FIRE "ADON"	SAPHENATH PANCAR (Prophet/Illuminated One) **ENTHRONEMENT** "The man who knows the Mysteries". Full illumination and unification with Divinity.	**Consummation**
		Transformation
6 AIR "IBIS"	MAAWR (Seer/Priest) **UNIFICATION** "Astronomer before the Gates of the Gods". Discovery and knowledge of the Divine Principles and their spiritual Plan.	**Joy**
		Unification
5 WATER "CHYMIA"	BALAHATE (Philosopher/Master) **ASCENSION** Defeat and mastery of the Typhonic energies of life. Mastery of the processes of life (Alchemistry).	**Promise**
		Dedication
4 EARTH "IOA"	CHISTOPHORES (Arch Mason/Adept) **RESURRECTION** Emergence into a higher life and consciousness. Defeat and mastery of all self-illusion and vanity (Gorgon). Investment with the powers and attributes of an adept.	**Rebirth**
		Peace
3 FIRE "M.C.M."	MELANOPHORIS (Master/Initiate) **CRUCIFIXION** 'Development and mastery of the will power and ability to act in service in any conditions of life. Experience of humility, death and the Underworld.	**Consummation**
		Transformation
2 AIR "EVE"	NEOCORIS (Craftsman/Disciple) **TRANSFIGURATION** Purification of the mind. Development of understanding and mastery of thoughts.	**Joy**
		Unification
1 WATER "AMEN"	PASTOPHORIS (Apprentice/Neophyte) **BAPTISM** Purification of the emotions. Development of love and mastery of the emotions.	**Promise**
		Dedication
0 EARTH	ASPIRANT (Candidate for Admission) **PREPARATION** Purification of the body and development of control over physical appetites. Preliminary testing in emotional response, intelligence, will power and obedience.	**Birth**
		Peace

Diagram H: LADDER OF INITIATION (Egyptian)

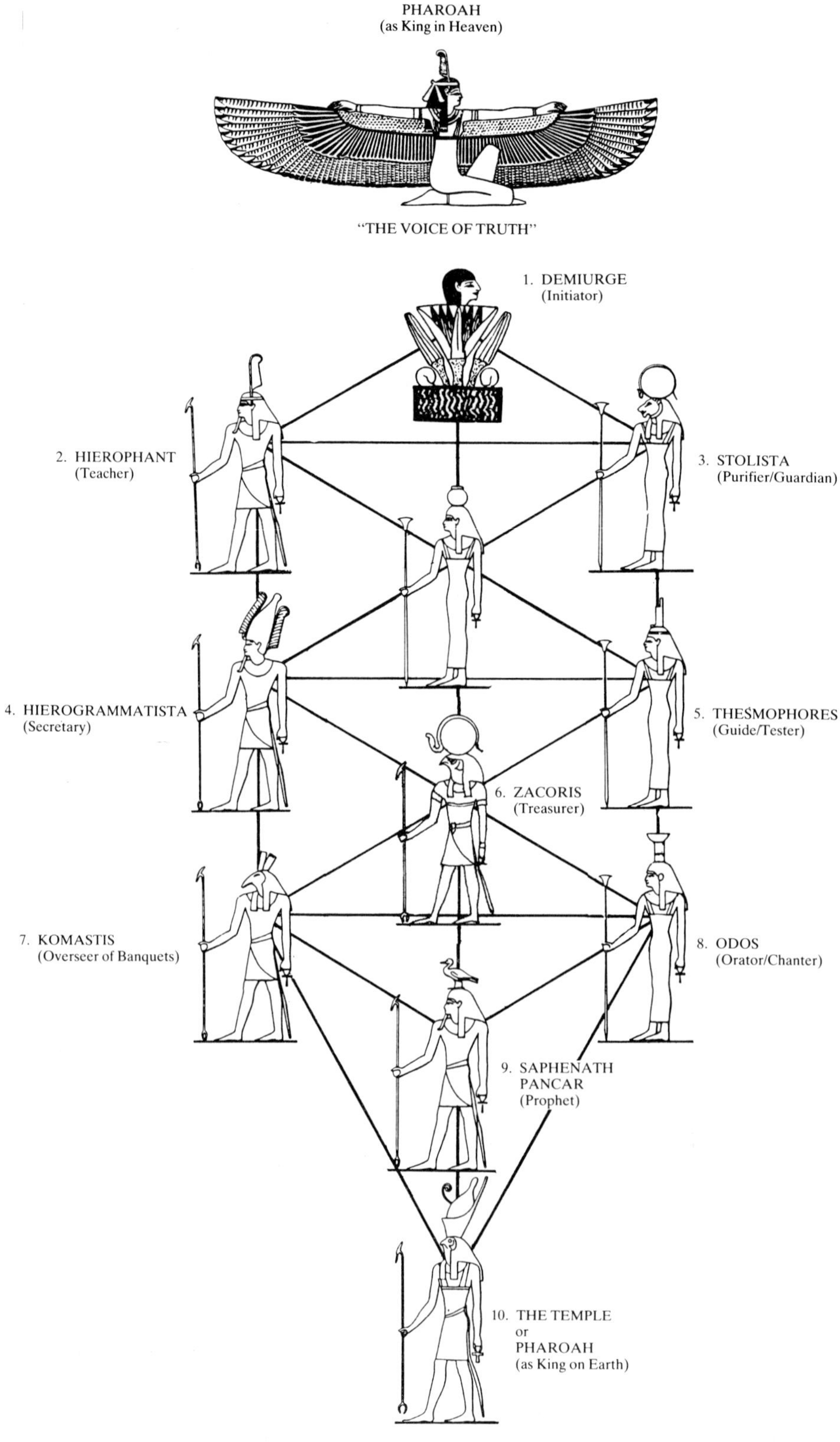

Diagram I: THE TREE OF LIFE (with officers)

The officers of the Society were those who governed the four principal temples of Egypt - at Hermopolis, Memphis, Heliopolis and Thebes. These four temples were associated with the cabbalistic worlds of Emanation, Creation, Formation and Action respectively. The pharaoh was supreme principal and ruler of all four great temples, and there were four sets of officers beneath him, each set of dignitaries representing a Tree of Life in one of the four cabbalistic spheres. However, since the pharaoh particularly represented *Ptah,* the Voice of God, whose wisdom illuminates the mind, so the official hierarchy of Heliopolis was the chief 'Tree of Life' of all the temples, representing the hierarchy of gods that form the overseeing divine idea, *Râ,* which reflects and gives form to the pharaonic light of Ptah. The demiurge of Heliopolis was thus known as the chief inspector of the Society and supreme judge of the land, second only to the pharaoh in dignity.[29] This demiurge was also known as the *Ur-maawr,* the 'great seer' of Egypt.

The *Demiurgos, Hierophant* and *Stolista* formed the first three principals of the temple hierarchy - a trinity that ruled the others. They were the direct expression of the *Pharaoh,* representing the three main aspects of sovereignty: will, love-wisdom and active intelligence. They were the high or 'arch' priests.

The ***Demiurgos*** wore a robe of sky blue spangled with embroidered stars, girdled with a cord of gold. From his neck he wore a sapphire jewel surrounded with brilliants, suspended from a gold chain. He dwelt always in the temple and was only seen by his disciples. The *Demiurgos* was the chief prophet and ruler of the temple, and, in the case of Heliopolis, the supreme judge of the land and chief inspector of the Temple. His office is symbolised by the *Primum Mobile* ('First Swirlings'),[30] which is a name for the initiating or originating power that gets things moving.

The ***Hierophant*** had similar robes, but wore a cross upon his breast. The *Hierophant* was the principal teacher of the temple and interpreter of the Law - the cross representing the key to the Mysteries as well as signifying the Law itself. His office is symbolised by the Zodiac, which is essentially a complete expression of divine will as perfectly ordered and harmonious vibrations of energy.

The ***Stolista*** wore a robe of white stripes and a special foot covering[31] which emphasised his role. The *Stolista* was the purifier or baptiser (with water), and the chief guardian or protector of the temple. His office is symbolised either by Saturn, whose orbit defines the limit of and gives protection to the other sacred planets, or by the stars disposed as constellations, which express and reveal the underlying zodiac of vibrational forces.

The ***Hierogrammatista***[32] wore a plume as his headdress, signifying the truth or illumination (*Maât*) that is bestowed *via* him upon others. He carried a vase of ink (*canonicon*) in his hand, signifying the illumination of heaven that can be written by him upon the earth, using the plume as a quill pen. The *Hierostolista* was the dispenser of the light of the three principals, and the instructor of all initiates. His office is symbolised by Jupiter, the Giver or Disposer of Light.

The ***Thesmophores***[33] had the task of introducing all aspirants to each of the degrees, and testing them. He was both a counsellor and also an examiner or tester. He personified the perception and judgement of the mind, and its questing nature, that leads the soul onwards and tests the truth of the matter. The office of *Thesmophores* is symbolised by Mars, the planet signifying the martial qualities of organisational ability, *etc.,* and the quest for justice, perfection and truth.

The ***Zacoris*** was the treasurer of the temple, looking after, organising and disposing wisely of not only its material wealth but also its spiritual and human wealth. The treasure of the temple is essentially its beauty and harmony, which is central to and unifies all the other aspects. The office of *Zacoris* is symbolised by the Sun, the unified expression of radiant life and beautiful form.

The ***Komastis*** was the organiser and overseer of the banquets, not only of physical food but also of spiritual food - of light-filled beauty and harmony. The office of *Komastis* is symbolised by Venus, the planet of love, wisdom, the sharing and enjoyment of life, and good feeling generally.

The ***Odos*** was the orator and chanter, recording and communicating the thoughts and experiences of the fraternity. He was the principal messenger and musician of the hierarchy, and his office is symbolised by Mercury.

The ***Saphenath Pancah*** was the foundation of the hierarchy of officials - the temple prophets being the ones who put the decrees of the hierarchy into operation and who saw that the work is done. Their role involved administration, organisation and guardianship of the temple work. Their office is symbolised by the Moon, the planet whose form reflects the light of the Sun and whose orbit affects and gives rhythm to the procreative cycles of animals and man.

VERITAS TVA ET VSQVE AD NVBES
SVBLIME DEDIT OS HOMINI

PART II
THE ARCADIAN ACADEMY

(The Life and Times of Francis Bacon, 1579-1585)

(5) "ANCHORA SPEI" ("Anchor of Hope") — emblem from titlepage to *The Art of English Poesie* (1589).

(6) FRANCIS BACON, aged 18 years (b.1561-d.1626) - miniature painted (1578/9) by Nicholas Hilliard, in France. Soon after this picture was painted, Francis returned to England.

Inscription: *"Si tabula daretur digna animum mallem."*

Translated: "It would be preferable if a picture deserving of his mind (soul) could be brought about."

Return to England

At the end of March/beginning of April, 1579, Francis Bacon arrived in London, home from the continent where he had lived in exile during the previous four years. He came armed with a special dispatch to Queen Elizabeth from Sir Amyas Paulett, the English Ambassador to the French Court, in which Sir Amyas praised Francis to the Queen, with the commendation that the eighteen-year-old youth was "of great hope, endued with many good and singular parts, and if God gave him life, would prove a very able and sufficient subject to do her Highness good and acceptable service."

Francis' delightful and instructive sojourn in France, where he had courted both love and learning, and faced excitements and dangers that well suited his ebullient and daring nature, had come to an abrupt end. On February 17th or 18th he had dreamt that Gorhambury, the country home of his wise and caring foster-father, Sir Nicholas Bacon, "was plastered all over with black mortar" instead of the white plaster with which its walls were faced.[1] Two weeks or so later he received news from England that Sir Nicholas had died on February 20th, just two or three days after he had received his prophetic dream. With the news came a royal summons home and, after tidying up his affairs in France and packing, by March 20th he had set out from Paris for his native country.

Francis did not arrive home in time for the funeral of the man who had been a good father to him. This had already taken place on March 9th, in a manner befitting the Queen's celebrated Lord Keeper.[2] He was just in time to see his foster-brother, Anthony, to whom he was much attached both in brotherly love and common purpose, before Anthony sailed at the end of April for the Low Countries, on his third secret mission as 'Edward Burnham'.[3] Probably Francis' first call, after seeing Her Majesty (his natural but secret mother), would have been to Gorhambury, to see Lady Anne, his fond and devoted foster-mother. At that stage Lady Anne would have only just moved her much reduced household and belongings from York House (the Lord Keeper's London home) to the much smaller but more attractive country house of Gorhambury, near St. Albans in Hertfordshire, that was now to serve as her dower house (although held in trust for Anthony, her son). The situation could not have been a happy one, as Lady Anne was hard hit by the death of her husband, and it is recorded that she fell into a "sadness"[4] as a result, from which she never recovered. Indeed, as the remaining years of her life went by (she died in 1610), she is said to have gone from "sadness" to a "fast" to a "watch" to a "weakness", which led tragically to a "madness" and finally death. Nevertheless, charged under the terms of her husband's will, and through the strength of her own love for her two sons (natural and adopted), she continued to be concerned about and to see to the welfare of Anthony and Francis for many years, as far as she was able to.

Anthony, working as a secret agent or 'intelligencer' for the Queen, under the instructions of Her Majesty's principal Secretary of State, Sir Francis Walsingham, was probably not away for more than a month or two on this particular mission. The mission involved

Anthony in penetrating the camp of Alexander Farnese, Prince of Parma, the successor to Don Juan, to discover and report back on the new situation in the Low Countries. On his return, which was probably in June 1579, he "drew up a relation approv'd of by the Queen and the Secretary."[5]

Anthony was not home for long, as in September that year he set out with a few attendants on a twelve-year continental tour - to all intents and purposes made at Anthony's own request, "to travel into France, to see the country and to learn the language, and to enable himself by learning good things there, to serve his country better."[6] However, this was only a partly true explanation. Anthony undoubtedly desired to travel more liberally, as many gentlemen of his time did, now that he was master of several manors and commanded a good income. But the suggestion to go was really Lord Burghley's, who was not only Elizabeth's Lord Treasurer but also Master of the Court of Wards, the real aim being to enable Anthony to gather political and other intelligence, and act as a key English agent or intelligencer abroad. At the same time Anthony would be able to help further the great scheme of Francis' that he was privy to, by building up an international fraternity or fellowship of like- minded people who would participate in and help develop the Great Instauration - the Baconian programme for achieving a Golden Age of virtue, brotherhood and illumination.

Lord Burghley, a friend and brother-in-law of the late Lord Keeper, was Sir Nicholas' nominee for overseeing the execution of his will, and the appointed guardian of Anthony and Francis until they each came of age. Anthony reached the age of twenty-one sometime that year (1579) and thus became his own master before he left for his prolonged continental travels; but Francis was then only eighteen and still had to wait until 1582 before he reached his majority. Under the terms of Sir Nicholas Bacon's will, Francis depended almost entirely on his 'uncle' and guardian, Lord Burghley, for any financial allowance derived from his inheritance; and even then he had only been left one fifth of whatever sum of money Sir Nicholas had laid by when he died, plus whatever Lady Anne chose to release to Francis from her dower, out of which Burghley probably took his commission as Master of the Court of Wards.[7] For many years Francis was to live in financial straits, trying to carry out his vast educational programme to regenerate all arts and sciences, and to build an English language and culture that would one day be the glory of the age, but without a lucrative position and, seemingly, without proper financial support from his real mother, the Queen.

> "Though she [the Queen] cheered him much with the bounty of her countenance, yet she never cheered him with the bounty of her hand."[8]

(Not until 1595 did the Queen provide Francis with anything substantial for his living, and she and Burghley constantly refused to give Francis a paid position in the State, despite his immense genius and talent. But in 1595 she indirectly allowed Francis the lease of Twickenham Park, and made him one of her counsel learned in the law.)

Francis' main financial support during the next sixteen years seems to have come from his foster-brother Anthony, from his foster-mother Lady Anne, from Anthony's half-brother Edward, and from various friends who supported his work in one way or another. Some financial help may have come from his real father, the Earl of Leicester, who was a great patron of scholars and who, like many other wealthy noblemen, would pay for the privilege of being named and adulated on the 'dedication' pages of books, and his name and memory thereby 'immortalised' as a patron of learning. Certainly it seems that Francis may have gone to live at first with his father, at Leicester House in the Strand, London,

(7) **QUEEN ELIZABETH I** (b.1533-cr.1553-d.1603) - illustration painted by Nicholas Hilliard, from charter authorising Sir Walter Mildmay to found Emmanuel College, Cambridge (dated 11th January 1534).

(8) ROBERT DUDLEY, Earl of Leicester (b.1532-d.1588) - oil painting, date and artist unknown.

Robert was created a Knight of the Garter and appointed the Queen's Master of the Horse and Privy Councillor in 1553, He secretly married the Queen in 1560, by whom he had two sons, Francis and Robert. He was created Earl of Leicester in 1564. The principal country residence of Leicester was Kenilworth Castle in Warwickshire.

and possibly also with his cousins, the Sydneys, at Penshurst Place in Kent, until he took up residence in Grays Inn, in May 1580.

The evidence for suggesting that Francis might have lived in the household of his natural but secret father during the latter half of 1579 is based on the knowledge that Francis was not only Leicester's eldest son by the Queen, but also the famed but secret poet, 'Immerito', who later used the name of Edmund Spenser to mask his poetic works. Leicester's household had been established at Leicester House (formerly Paget House) since 1569. Sir Philip Sydney, eldest son of Sir Henry Sidney and his wife, Mary Dudley, the scholarly sister of Leicester, and therefore the nephew of Leicester and cousin to Francis Bacon, spent much of his time at Leicester House where his uncle established a meeting place or centre for poets, scholars, writers and artists of various kinds. Principally these talented people who were patronised by Leicester were related to him by family ties, or were friends of his relatives; but, in association with the Walsinghams, several university scholars and poets were among the numbers, plus a professional acting company, 'Leicester's Men', that included James Burbage.[9] One of the scholars seems to have been, at one time, Edmund Spenser from Pembroke College, Cambridge, who had been introduced to Leicester by Francis and Anthony's old tutor and friend, Gabriel Harvey. When Francis returned home and came to live at Leicester House, Spenser was (seemingly) already living there as a secretary and messenger of the Earl. Later (in 1590),[10] possibly in return for advancement, lands or a pension, Spenser allowed his name to be used by Francis as a mask to the true authorship of Francis' famous works of 'poesie'.[11]

Robert Dudley, childhood friend, lover and eventually secret husband of Queen Elizabeth,[12] was created Earl of Leicester in 1564. Although he never achieved his dream of becoming publicly acknowledged as Royal Consort, nevertheless he became the most powerful man of the realm, with enormous riches and privileges heaped upon him by his royal wife and Sovereign of the realm. Although the Queen undoubtedly loved him and bore him two children (Francis, born January 1560/1 and adopted by the Bacons, and Robert, born November 1566 and adopted by Sir Walter Devereux and his wife 'Lettice'), the marriage was never publicly announced or celebrated. It must have been an extraordinarily difficult situation to have lived with; and, after Leicester's failure at the Kenilworth Festivities in 1575 to get the Queen to acknowledge him publicly as her husband, it is perhaps quite understandable that Leicester sought out other possible liaisons. As he had a natural affection for and interest in his children (yet was forced to remain aloof from them), besides being attracted to good-looking and intelligent women, it is perhaps not surprising that he eventually married Lettice, the Countess of Essex, after the death of her husband. Lettice was the daughter of Sir Francis Knollys and Catharine Carey (cousin to Queen Elizabeth I), and the foster-mother of Robert, Leicester's second son by the Queen. This marriage, which also took place in the greatest secrecy so that the Queen might not know, was performed on September 21st, 1578, at Wanstead House. At the time of the marriage Lettice was pregnant with a child by Leicester, which was probably why the marriage was forced upon him. It meant that Leicester had, in effect, two wives, both secret, and one with the power of life or death over him, thereby greatly multiplying his problems rather than reducing them. In addition he had to bear the emotional pain of having children that he wanted so much but whom he could not publicly acknowledge as his own.

Francis arrived home to this exceedingly difficult situation. Leicester had installed his new wife with as much secrecy as possible in his large London mansion, Leicester House; and to this house Francis seems to have gone, to be taken under his father's wings - at least for a while. This grand house leading off the Strand and facing the Thames waterfront, right

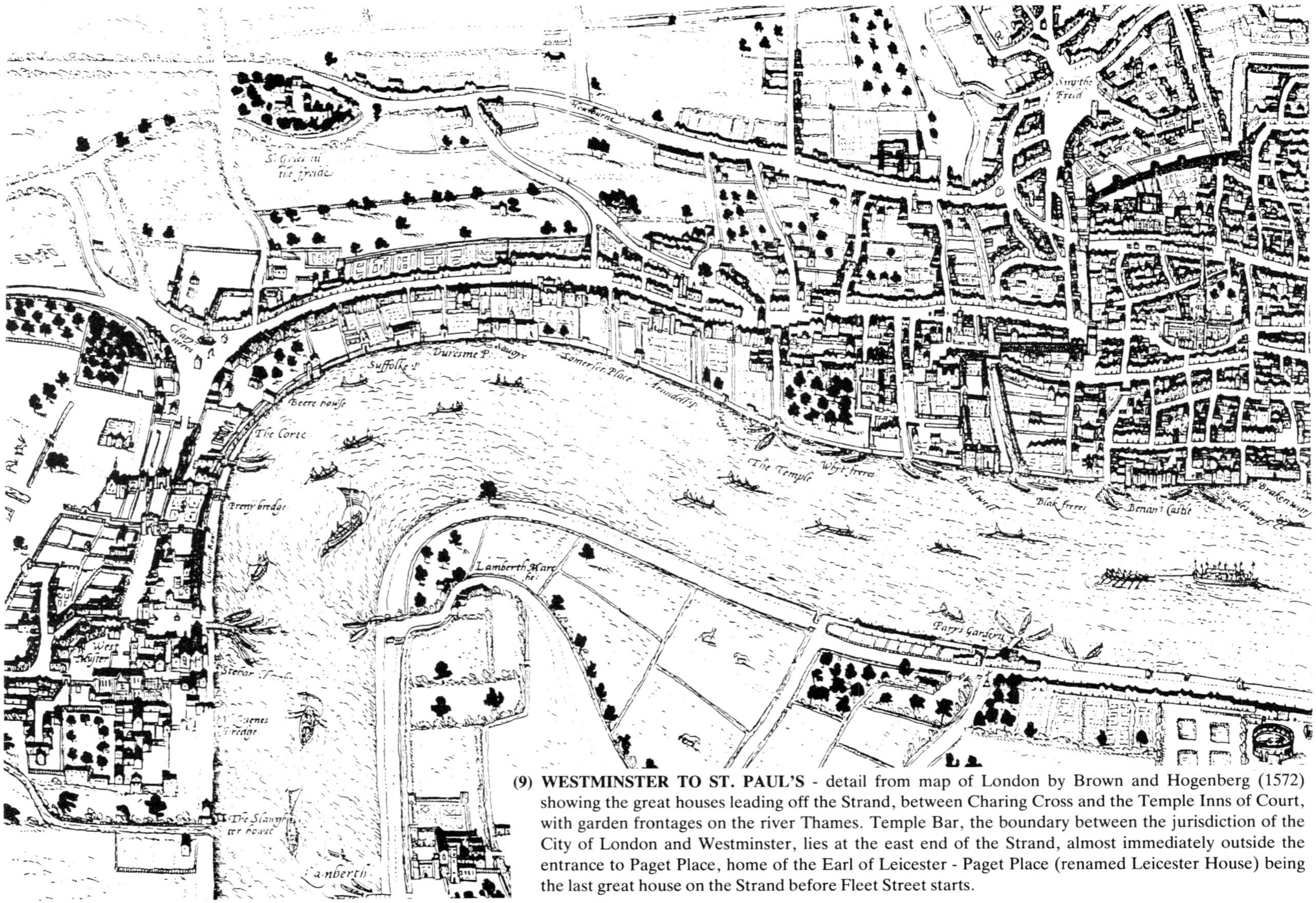

(9) WESTMINSTER TO ST. PAUL'S - detail from map of London by Brown and Hogenberg (1572) showing the great houses leading off the Strand, between Charing Cross and the Temple Inns of Court, with garden frontages on the river Thames. Temple Bar, the boundary between the jurisdiction of the City of London and Westminster, lies at the east end of the Strand, almost immediately outside the entrance to Paget Place, home of the Earl of Leicester - Paget Place (renamed Leicester House) being the last great house on the Strand before Fleet Street starts.

next door to the Temple Inns of Court and their beautiful Elizabethan hall, was once the great house which the first Lord Paget had bought from the Bishop of Exeter and subsequently enlarged. In 1569 Leicester acquired the lease and transformed it into "the stately place" that Francis, *alias* Edmund Spenser, recorded. Besides having its own great hall and two courtyards surrounded with buildings, in its gardens Leicester had built a delightful little banqueting house, two stories high, on the bank of the river. Outside its entrance off the Strand stood the Temple Bar, the gate or bar to the City of London at which the Sovereign always had to wait for permission to enter the City in acknowledgement to the ancient rights and freedoms of the people of London. By the time Francis came back to England and entered the house of his natural father, the mansion had become a well-known centre for poets, artists, scholars and *intelligensia* of many kinds, whom Leicester patronised.

Robert Dudley, Earl of Leicester, was a well educated and intelligent man, besides possessing great vivacity and charm. He was a very attractive man to the ladies, with a strong personality, commanding of others, and with a love of wealth, honour and fame. But he also loved learning and the arts, and spent more of his time and effort in encouraging and supporting the Renaissance movement in England than is generally realised. He was the fifth and youngest son of John Dudley, Duke of Northumberland and Lord Protector during the reign of Edward VI. His other brothers were, in order of birth, John, Ambrose, Henry and Guildford. He also had three sisters, Jane, Mary and Catharine. Their mother was Jane Guildford. After Northumberland had tried to place Lady Jane Grey on the throne (immediately following King Edward's death), having expediently married her to his youngest son, Guildford - an abortive attempt which resulted in Mary Tudor incarcerating Northumberland and all his sons in the Tower of London - he was executed for his blatant and treasonable attempt to seize power. Soon afterwards his fourth son, Guildford, was executed together with his unfortunate wife, Lady Jane. The other four Dudley brothers were eventually pardoned and released from the Tower. John Lord Lisle, the eldest son, is said to have died within a few days of his release, but the other three lived on.

Ambrose, Northumberland's second son, who had been twice married to heiresses by his father, inherited his father's estates that had been restored to him, thereby becoming the Earl of Warwick. With this title and lands, he also (presumably) inherited his father's valuable library, which was one of the largest in the kingdom at that time. For, despite being a ruthless and scheming man who sought regal power both for himself and his family, Northumberland was also an intelligent man who rated education highly. Not for nothing was his family crest the Rampant Bear and Ragged Staff, nor were their claims unjustified that the Dudley family were an important branch of the Tudors.[13]

The Tudors formed the ancient royal line of British kings, with a genealogy carefully preserved and recorded by their bards. They trace their line of descent back to the famous King Arthur, with direct links (*via* the daughter of Joseph of Arimathea) with the Judaic royal line of David, and beyond even the later Celts to Hugh the Mighty (also known as Brut, or Brit), who led the original Trojan settlers to 'the White Land' (*i.e.* Albion) in the second millenium BC, and gave these islands their name of 'Britain' (meaning 'The Chosen Land'). This family line provided the 'Crowned Princes' (*Cunos*) and 'High Kings' (*Arviragus*) of Ancient Britain, and carried with them the ancient initiatic knowledge. The High Kings, all of whom were elected from the "blood royal" (*i.e.* the 'Tudor' blood line) and chosen for their strength, wisdom and initiatic powers, bore the hereditary title of 'Merlin' (*Myrrdin*), as they combined in their office the roles of both high priest and king.[14]

(10) AMBROSE DUDLEY, Earl of Warwick (b.1528-d.1590) - oil painting, date and artist unknown. The principal home and seat of Ambrose, Earl of Warwick was Warwick Castle.

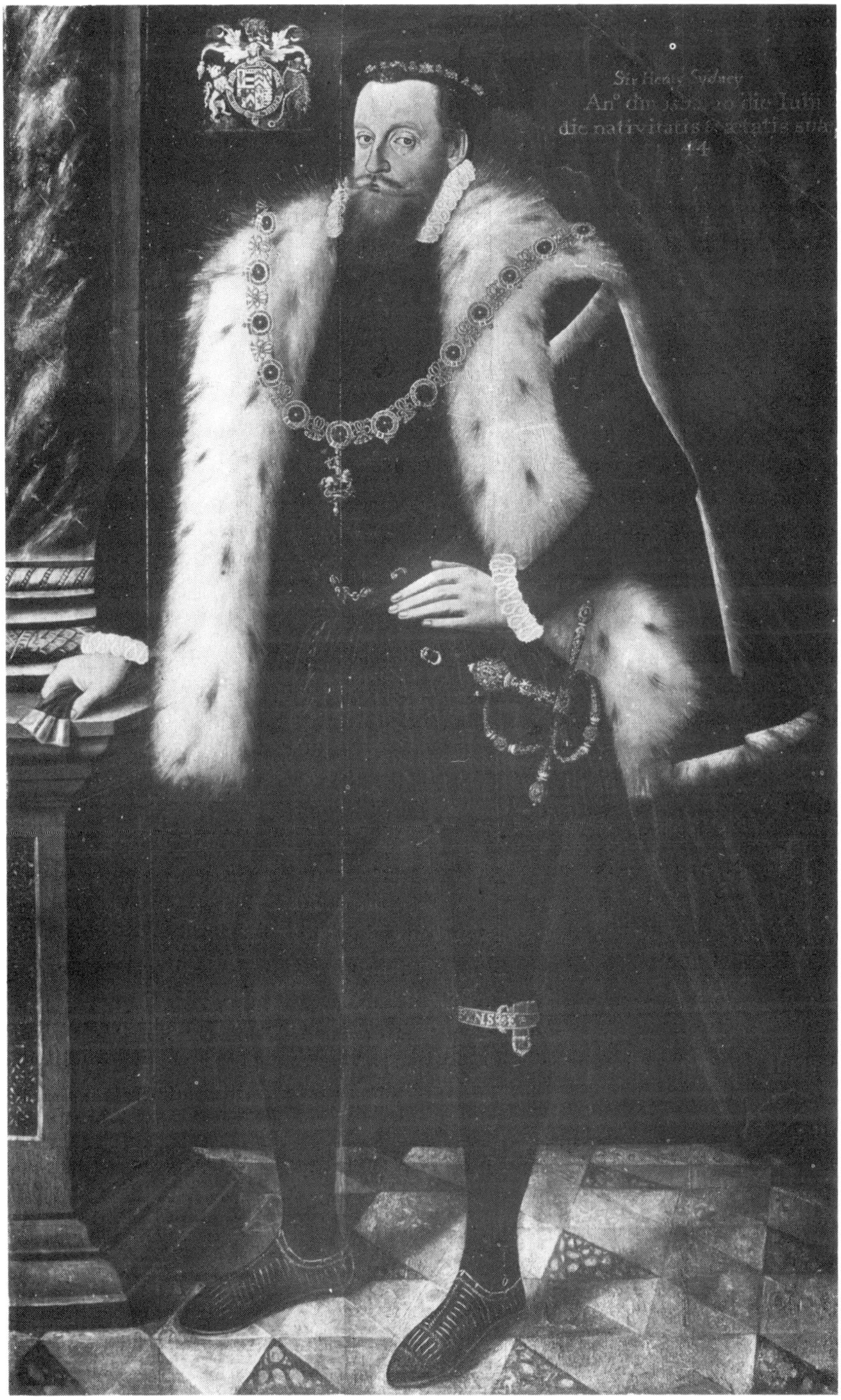

(11) SIR HENRY SIDNEY, aged 44 years (b.1529-d.1586) - oil painting (1573), artist unknown.

Sir Henry Sidney married Mary (*née* Dudley, sister to Ambrose and Robert Dudley) in 1550. They had four children - Philip, Mary, Robert and Thomas. Sir Henry was appointed Lord President of the Welsh Marches (from 1559) and Lord Deputy of Ireland (from 1565-1578). The Sidney's family home was Penshurst, Kent; but, as Lord President of the Welsh Marches, Sir Henry lived mainly at Ludlow Castle after his final return from Ireland until his death (1578-1586).

Associated in esoteric knowledge with this role of high kingship is the symbol of the North Pole Star - the 'crown' of the world - and the circumpolar constellations. Of these, the Lesser and the Greater Bear are the most important: the former providing the present North Pole Star. The 'Bear' is a veil to the real name, which means 'Dove', or 'Sheepfold', or 'Chariot'. As the Chariot, it is Arthur's vehicle of manifestation, as also the fiery chariot in which Elijah ascended to heaven. Hence Arthur is known as 'the Bear'. As the Sheepfold, it is the heavenly place of peace and enlightenment that Jesus spoke of, to which the 'sheep' (*i.e.* initiates) of mankind are led, and from which the 'shepherds' (*i.e.* the adepts and masters) come from. As the Dove, it is known as the vehicle of the Holy Spirit, by means of which that Spirit descends to the earth and governs its kingdom. It is signified by the royal crown of a High King (*i.e.* a crown with arches) - a halo of light which descends upon a truly worthy and anointed (*i.e.* illumined) soul, bestowing full knowledge of Truth.[15] The Great Bear is known as the mother of the Little Bear, which is her son. The Rampant Bear and Ragged Staff emblem of the Dudleys depicts just this - the Bear constellation tied to the Pole of the earth's axis, and dancing around it.

With these things in mind, one should reconsider exactly what was going on behind the scenes when (a) the Dudley family, led by the Duke of Northumberland, tried to seize power by marrying Ambrose Dudley to Lady Jane Grey and attempting to set her on the throne of England; (b) Elizabeth Tudor was placed on the throne of England - an occurrence that was carefully planned and executed, and which enabled the reformation of the English Church to continue, so as to return it to the early model of the original Celtic or British Church; and (c) the marriage of Robert Dudley took place with Elizabeth I, thereby forging a strong union in the Tudor blood line through what was undoubtedly a major dynastic marriage. Consider also the implications when Elizabeth felt unable to (or refused to) proclaim her husband, or publicly recognise her sons of that marriage.

The Dudleys were a power to be reckoned with in Tudor England, for reasons that are not all generally known or realised; not just the men, but the women also - particularly Mary Dudley, who married Sir Henry Sidney. There were others also, such as Sir Nicholas Bacon, whose authority came almost solely from his immense wisdom, hard work and integrity; although the Bacon family has some traditional weight to its name, being derived (it is thought) from the Bascoigne family, who were the Lords of Molay in France. Jacques de Molay made this family famous, by being the last publicly known Grand Master of the Knights Templars. But 'Bacon' might also be a synonym for 'Tite', which is a form of 'Tudor'; in which case the Bacon family might well be another branch of the spread-eagled Tudor bloodline.[16]

Sir William Cecil, created Baron Burghley in 1571, was another exceedingly influential man who, like the Tudors, was also of Welsh descent. Together with Sir Nicholas Bacon, Cecil was largely responsible for seeing that Elizabeth Tudor was placed safely on the throne of England. Both these men, together with those most closely connected with them, were intimately involved with the Reformation and, particularly in the case of Bacon, with the Neoplatonic esoteric currents of the Renaissance.[17] Their roles in these respects have already been discussed in *Dedication to the Light*[18] and *The Great Vision.*[19]

As Elizabeth's reign progressed, two powerful factions gradually emerged into the open, whose politics increasingly diverged after the death of Sir Nicholas Bacon. Eventually these factions became polarised into what one might call the daring Dudley camp and the cautious Cecil camp: the former being a powerful driving force which included interest and involvement in Renaissance esotericism and its chivalric (or Arthurian) expression, especially suited to the noble classes; whilst the latter was more an expression of the

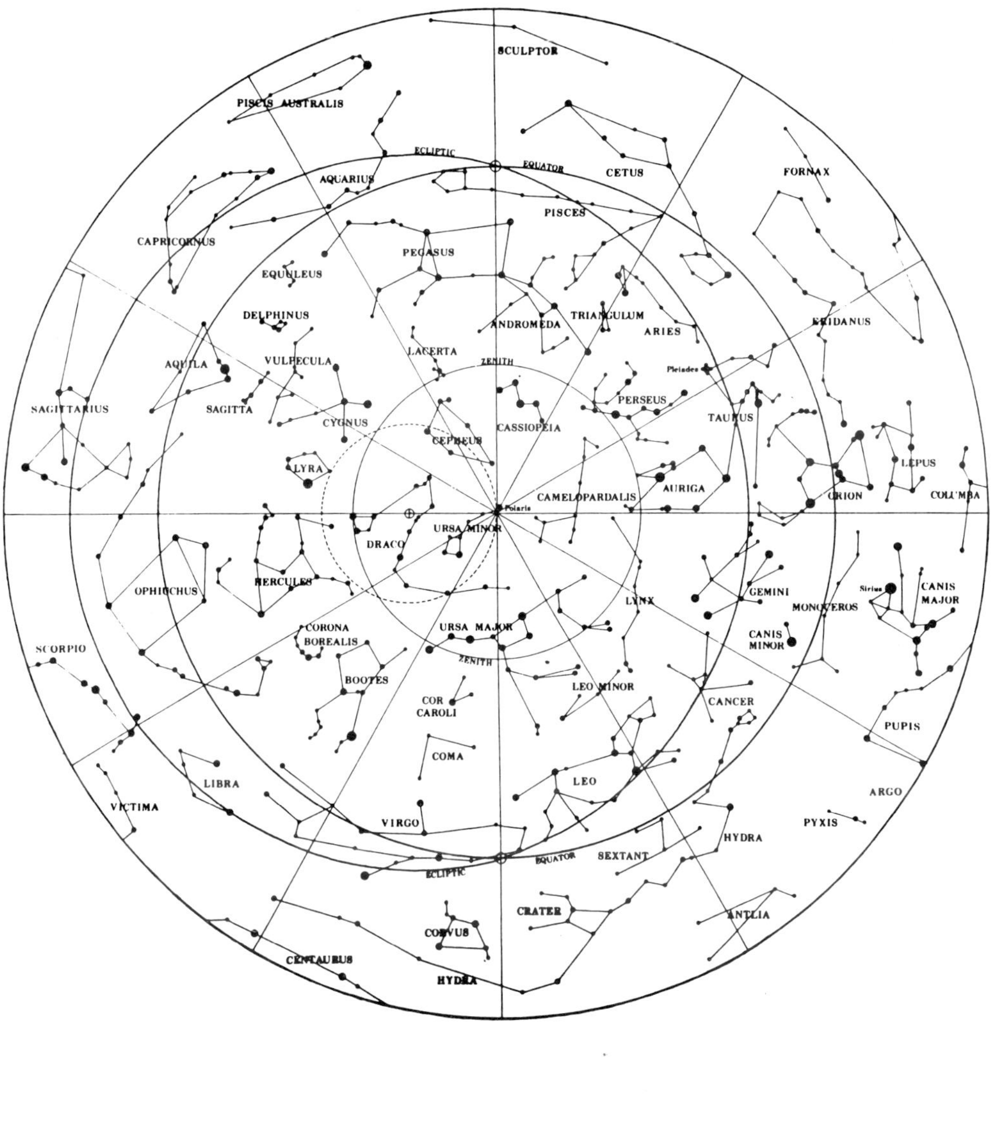

Diagram J: CELESTIAL MAP

(12) WILLIAM CECIL, 1st Baron Burghley, aged 65 years (b.1520-d.1598) - oil painting (c. 1585) attributed to Marcus Gheeraerts.

Sir William Cecil was appointed Secretary of State in 1558, created Baron Burghley in 1571, and appointed Lord Treasurer in 1572. With his appointment as Lord Treasurer, Burghley was created a Knight of the Garter. In this portrait he is depicted in his full Garter robes, wearing a particularly elaborate Greater George, and holding his white staff of office in his right hand. His principal country houses were Theobalds in Hertfordshire and Burghley in Northamptonshire, whilst his palatial London house was Sir Thomas Palmer's house, built on the site of the rectory of St. Clement Danes, which Burghley greatly extended and embellished.

(13) ROBERT CECIL, 1st Earl of Salisbury, aged 39 years (b.1563-d.1612) - oil painting (1602) by
Critz.

HAM, Principal Secretary of State (b.1532-d.1590) - line drawing by Simon
nd's *Herwologia* (1620).

influential merchant middle-classes, who were far more interested in accumulating private wealth and living the quiet life than seeking glory and fame. This polarisation can perhaps be seen and understood as an expression of the cabbalistic Twin Pillars, *Fire* and *Water*, or Force and Form - the former providing the inspiring, stimulating, illuminating driving force of life, the latter contributing the stabilising, enfolding, enabling, form-making aspect of life. Queen Elizabeth had to listen to and take account of both factions, which, were it not for the insinuous machinations of the young Robert Cecil that slowly poisoned her mind, she managed remarkably well.

Holding the middle ground, as it were (although particularly linked with the Dudleys), were the Bacons, the Walsinghams and the Sydneys. In addition there was one particularly remarkable man (other than Francis Bacon) whom we have so far only fleetingly mentioned,[20] but whose influence on the whole course of Elizabethan civilization was immense. This man was known as Dr. John Dee.

(15) DR. JOHN DEE (b.1527-d.1608) - oil painting, date and artist unknown.

John Dee: the Magus

John Dee was born on 13th July 1527, the son of Rowland Dee, an official at the court of Henry VIII. He was of Welsh descent, and Dee always claimed that he was in fact related to the Tudors.[1] Furthermore, Dee was an earnest exponent of the ancient Tudor descent and its imperial (or Arthurian) right to kingship of the country. These are two interesting and important points, putting Dee firmly in the Tudor camp. Recent cipher discoveries amongst Francis Bacon's works would indicate that one should consider John Dee as John '**D**', and think about what that is signifying.[2] Dr. John Dee was a master exponent of the art of symbolism, dissimulation and disguise - and this may have been inherited in part.

In 1542, aged fifteen, John Dee was sent to St. John's College, Cambridge, by his father. He studied with a group of scholars who promoted the new learning for which St. John's became noted in the latter half of the 16th century. Here Dee encountered Sir John Cheke, who actually dared to foster Greek and mathematics as well as the more acceptable subjects of orthodox Humanism. Dee threw himself into learning all he could, studying sixteen hours every day. In 1546 Trinity College was founded, and Dee's ability was rewarded by his being given a fellowship as an under-reader of Greek at the new college. By this time he was immersed not only in Greek and mathematics, but also in the sciences (so-called), especially mechanics. He arranged a mechanical 'flying' device for use in a play performed on stage at Trinity, and so astonished the audience that they attributed it to "magic" and called him a "conjuror" - an unfair ascription which unfortunately stuck in the general public's mind. Here he also began to study astrology, which included the science of astronomy.

Apparently dissatisfied with the mathematical and scientific education available in England, Dee made the first of his many journeys abroad in May 1547. He went to the Low Countries, where he formed close friendships with Mercator, Frisius, Myricaeus and Gogava. Gemma Frisius, one of the most renowned geographers of the period, helped to teach Dee navigation; whilst Gerard Mercator, the famous cartographer, assisted him with a knowledge of map-making. Armed with navigational instruments that had never been seen before in England, and a unique knowledge, Dee returned to Trinity College for a brief residence. Then, in 1548, having received his Master of Arts, he left Cambridge and enrolled in Louvain, a centre of Neoplatonic and Cabbalistic studies, and the place where Agrippa had died eleven years previously.

When Dee arrived in Louvain he had already been studying Hermeticism (including what is called alchemy) and Cabbala for several years; and he clearly went there to forward these studies, as the English universities were inimical to these subjects. At Louvain he also studied civil law, "for recreation." Many well-known people came to visit him, including Sir William Pickering (afterwards English ambassador to France) who studied under Dee. In 1550, after graduating as a Doctor at Louvain, Dee made a short visit to Antwerp, then to Brussels, returned to Louvain, and finally went to Paris. Here he

(16) JOHN DEE'S COAT OF ARMS - as depicted in a wood cut at the end of his *Letter Apologeticall* (1599).

became famous as a teacher, at the remarkably young age of twenty-three, expounding cabbalistic, hermetic and magical theory, particularly relating to numbers which, it was believed, was the one science that linked all three worlds of existence - *i.e.* (i) the lower, elemental or natural world, (ii) the mathematical, middle or celestial world, and (iii) the supercelestial, religious or intellectual world - a science which he hints at in his 'Mathematicall Preface' to *Euclide* that was published in England in 1570. Some of the best minds of Europe sought him out.

Dee was offered a King's Readership at Paris University, but he refused this honour and chose instead to return to England in 1551. With him he carried a storehouse of ideas and knowledge in geography, navigation, geometry, arithmetic, numerology, mechanics, astrology, alchemy, Greek, Hebrew, Cabbala, Hermeticism, and Vitruvian concepts on architecture. All in all, he was fitted to become a Renaissance *magus. Arriving in England,* he was brought to the notice of William Cecil by Sir John Cheke, who in turn introduced Dee to Edward VI. In return for presenting the king with two treatises, Edward rewarded Dee with an annual pension of 100 crowns. (These he converted into ownership of the rectories of Upton-upon-Severn and of Long Leadenham in 1553, from which he received an annual income of £80 for the next thirty years.) He also started a correspondence with the young Princess Elizabeth, and cast her horoscope. Under the name of 'Benger' he became the princess' auditor.[3]

Dee became closely associated with the proponents of religious reform at court. One very important figure whom Dee met - a Renaissance *magus* of the highest order, who profoundly influenced Dee - was Geronimo Cardano. Born in Milan in 1501, Cardano became a remarkable philosopher, physician, scientist and mathematician. Leonardo da Vinci was a friend of his in earlier years. He is probably best known nowadays as a pioneer in the popularisation of science, and as the author of the world's first treatise on the Theory of Probability.[4] His literary output was prolific, consisitng of 131 books published in his lifetime, and 111 manuscripts unpublished at the date of his death in 1576.[5] His writings embraced all the major ideas of his time, and contained the rudiments and ideas pertaining to almost every scientific and philosophical doctrine developed during the 17th century. Gabriel Naudé (1600-1653) writes of Cardano:-

> "Not only was he beyond dispute an outstanding physician, he was also probably the first and only man to distinguish himself in all sciences at once. He is one of nature's illustrations of what man is capable of achieving. Nothing of significance was unknown to him in philosophy, medicine, astronomy, mathematics, history, metaphysics or the social sciences, or in other more remote areas of knowledge. He, too, erred, of course, that is only human; it is marvelous, though, how seldom he erred."[6]

Cardano also wrote on astrology, physics, chess, immortality, dialectics, gems, colours, poisons, air and water, dreams, music, morals, wisdom, and much more besides, including cryptography. Of all this, it is his stenographic system of cryptography, known as 'the Cardano grille' method, that has handed his name down to posterity. This is an almost perfect cipher system which Dee learnt and taught to his pupils, and which Walsingham used in his Intelligence Service. This 'Cardano grille' system was later adapted and improved upon by Anthony and Francis Bacon, and used in their printed works.[7] Cardano visited England in 1552, staying at John Cheke's house, and there Dee went to meet and learn from the great man.

At the end of February 1552, he entered the service of the Earl of Pembroke (a member of the Herbert family). Also during this time, Dee became attached to the Duke of Northumberland's household, as tutor to the Lord Protector's family. (Robert Dudley was at that time aged twenty.) Most of Northumberland's children were by then married - John to Anne Seymour, Ambrose to Anne Whorwood, Henry to Margaret Audley, Robert to Amy Robsart, Mary to Sir Henry Sidney - but Dee clearly provided what we would nowadays call 'adult education', in the most privileged sense. Northumberland wished his children to have the best scientific education available, and he himself was particularly interested in navigation. Available to Dee and the Dudley family was Northumberland's great library. Both Robert and Mary (Lady Sydney) became renowned as being both scholars and patrons of learning. Not surprisingly, Robert Dudley gained the reputation of being a champion of the sciences, for which he was rebuked (in 1564) by the more Humanistic and orthodox Roger Asham, who had also been Robert's teacher. Similarly the Princess Elizabeth, whose private tutor and mentor since February 1548 had been Roger Asham, was also being influenced by Dee.

In 1553, at the request of the Duchess of Northumberland, Dee wrote two further treatises.[8] But then came the dramatic events of Summer 1553, when the young king died and the Duke of Northumberland attempted to put Lady Jane Grey on the throne, illegally. By August he and his sons were prisoners in the Tower of London, their lands and titles forfeit, and Mary Tudor firmly on the throne of England. Following Wyatt's rebellion in February 1554, the Princess Elizabeth was also imprisoned in the Tower. Expecting to be executed at any time, she and Robert there pledged their troth and underwent a form of marriage conducted by an itinerant priest. In May 1554 Elizabeth was removed from the Tower and taken under guard to Woodstock. But, as an act of goodwill, on the occasion of Prince Philip of Spain's visit to England in the summer for his marriage to Queen Mary, Elizabeth was released from house arrest and was reinstated at court. As a further act of good faith, the Dudley brothers were freed from the Tower in October 1554, by request of Philip (who acceded to the throne of Spain as Philip II on 25th October), after their mother had beseeched Philip for her sons' release. Except for John, who died within a few days of his release, and Guildford, who had been executed, they took service in Philip's army, campaigning in France. Henry was killed in the fighting, but Ambrose and Robert were home by the Spring of 1557 - Robert with high commendations, as a result of which he, his brother Ambrose and his three sisters were "restored in blood".

Meanwhile John Dee had undergone a harrowing experience. Some time after Mary I came to the throne of England, Dee was requested to cast Mary's natal horoscope, together with that of her husband's. This he did, and then he performed the same service for the Princess Elizabeth, whilst she was at Woodstock. But a Spanish agent, named Prideaux, and another man called George Ferrys, falsely accused Dee of attempting to enchant the Queen by calculating the nativities of the King of Spain, the Queen of England and the Princess Elizabeth. Accusations of sorcery were easily believed and stuck hard in those days of ignorance, superstition and religious dogma. Accordingly, on the 28th May 1555, Dee was arrested by orders of the Privy Council and taken to Hampton Court to face charges. He managed to clear himself of the principal charge of treason, but was then sent to the Star Chamber in the custody of the Catholic Bishop of London, to be tried on ecclesiastical charges. He remained a prisoner until the 29th August 1555, when he was released without penalty. Although exonerated, from that time on Dee was in constant fear of his life.

On the 15th January 1556 John Dee wrote a supplication to Queen Mary, imploring "the Recovery and Preservation of Ancient Writers and Monuments."[9] Dee was acutely worried at the continuing destruction of the old libraries in England, that had begun with the dissolution of the monasteries in Henry VIII's reign, and which was compounded firstly by the Edwardian reforms and the rejection by the universities of the old philosophical and scientific traditions (considered to be papistical and evil), and secondly by the actions of Mary's commissioners who sought to ensure that no books prejudicial to the Roman Catholic religion remained in the university libraries. Couching his words carefully, Dee tried to interest Mary in the foundation of a national or royal library of books and manuscripts. He suggested that, if the work could be put in hand with "speedy diligence," a great deal might still be saved from the dispersion and destruction that was taking place everywhere. He himself offered to travel abroad if necessary, to obtain copies of all the important manuscripts in such great libraries as the Vatican in Rome and St. Mark's in Venice, if the Queen would defray the costs of the journey.

Dee was not the first to propose the founding of a royal library. The antiquary, John Leland, in 1536, had proposed a similar scheme to Henry VIII, *via* Thomas Cromwell, for the same reasons. Leland at least had partial success in persuading the government to preserve some manuscripts in the royal library that would otherwise have been lost; but mostly he had to collect as much as he could, with his own finance, and form his own library. Queen Mary rejected Dee's ideas as being too costly, and so Dee did the same as Leland: he started his own collection. It was not too difficult at that time, as whole libraries could be bought for almost nothing; and so Dee acquired the habit of buying books and manuscripts at home and abroad whenever his finances permitted, whilst sometimes he would receive them as presents. Eventually his library was to become the largest, most famous and most comprehensive in the land.

On November 17th, 1558, Queen Mary died and Elizabeth Tudor succeeded to the throne of England. The new Queen sent her beloved Robert Dudley to Dr. John Dee, to consult him as to the best time for her coronation. Dee advised that the coronation should take place on 15th January 1559, which it accordingly did.

By the end of 1562 Dee had returned to the continent on another visit, which he proclaimed to Cecil was hugely successful. He visited Antwerp and the printer, William Silvius (who later published Dee's *Monas Hieroglyphica*), and whilst there he was able to purchase, along with other books, a manuscript copy of Trithemius' immensely important work on cryptography, cabbalistic numerology and magic, *Steganographia.* Leaving Antwerp, Dee went on to Zurich, then to Italy, in search of the type of knowledge such as could be found in the *Steganographia.* In Italy he stayed some time at the court of the Duke of Urbino, renowned as a centre for the study of the exact sciences. In collaboration with Commandinus, the court mathematician, he published a mathematical text, *De superficierum divisionibus,* which Dee had discovered.[10]

From Urbino, Dee went to Rome, then on to Pressburg in Hungary, where he seems to have attended the coronation of Maximilian of Hapsburg as Hungary's king. During that period of time Dee had managed to write his *Monas Hieroglyphica,* a Hermetic treatise which he said summed up his philosophy and embodied universal wisdom - a work which he had been contemplating for seven years and which he wrote in seventeen days (including a long letter of dedication to Maximilian). Having completed it, Dee returned to Antwerp, to supervise the printing of this treatise, which was published on 31st March 1564. The book reached England before he did, and it would appear from a letter written by Dee to Queen Elizabeth (in 1592) that Elizabeth, in Dee's absence, had commended the book and defended it against the attacks of university graduates and other gentlemen who "dispraised it, because they understood it not."[11] By the 14th June 1564, Dee had

(17) TITLEPAGE TO *MONAS HIEROGLYPHICA* (1564).

Inscription: *"Qui non intelligit, aut taceat, aut discat."*

Translation: "Who does not understand, should either learn or be silent."

returned to England, to the court at Greenwich, where the Queen showed great interest in the *Monas Hieroglyphica,* desiring to have its secrets explained; after which she did "comfort" and "encourage" Dee in all his studies "philosophicall and mathematicall, &c."[12] More than this, the Queen gave Dee her personal protection, which was certainly needed in those times of religious bigotry, superstition and intolerance. Dee writes that the Queen "promised unto me great security against any of her kingdome, that would by reason of any my rare studies and philosophicall exercises, unduly seeke my overthrow."[13]

After Dee had returned to England from the continent, he settled at Mortlake, in a rambling old house belonging to his mother, that was situated between Mortlake church and the river Thames. Here he established a private centre of learning and set up his library, adding further rooms and buildings from time to time, and acquiring adjacent properties, so as to house not only his growing library but also his many collections. What Dee established and equipped was a centre of universal learning, an Academy, to which many scholars continuously resorted; for he placed his library and other collections at the service of English scholars and scientists, and it was frequently used. As Francis Bacon was later to declare about himself to Burghley, Dee had taken all knowledge as his province, and was helping others to do the same.

By 1583 Dee's library numbered "neere 4,000: the fourth part of which were written bookes."[14] It was the largest and most comprehensive library in Elizabethan England, rivalling the best on the continent (except for Jacques de Thou's colossal library of 8,000 printed books and 1,000 manuscripts). The next largest collection in England was that of Lord Lumley, numbering about 3,000 separate works by 1609 (although most of it was collected by 1596). When one considers that a library of more than a few hundred volumes was considered large for the 16th century, and that the University Library at Cambridge only had about 451 books and manuscripts in 1582, and Corpus Christi College at Oxford had about 379 volumes in 1589, then it puts Dee's efforts and library into perspective. The Queen, with her Privy Council and other lords and ladies, made a special visit to see Dee's library on the 10th March 1575.

Dee's library was indeed comprehensive, and he despised no aspect of knowledge. But one of the most significant facts that we can perceive, from a study of Dee's literary collection and his known interests and work, is the major part that poetry, drama, architecture and occult mathematics played in the whole scheme. This cannot be stressed enough. Occult mathematics is really the scientific basis of Cabbala, which links numbers with letters (and hence geometry and numerology with language and culture), and with levels of consciousness and being. Every letter of the alphabet has its own numerical value, which is itself a rational expression of a particular idea, power and purpose. Moreover, all number is linked together in one comprehensive and geometric whole that is perfectly ordered and harmonious, just as all ideas are essentially organised parts of one Divine Idea. The same is true for the alphabet, whose letters, properly (*i.e.* mathematically) ordered, make comprehensible words, and whose words form the language that expresses and communicates ideas. Not for nothing was said:-

> "In the beginning was the Word, and the Word was with God, and the Word was God.
> The same was in the beginning with God.
> All things were made by Him; and without Him was not anything made that was made.
> In Him was life; and the life was the light of men."[15]

Occult mathematics is the rational basis of all *Gnosis,* or Gnostic teaching, of whatever language, culture or religious teaching. It is the foundation that underlies and holds together, for instance, the Hebrew, Greek and Latin Bibles. After John Dee's and Francis Bacon's work was completed, this was also true of the English Bible. Moreover, occult mathematics forms the basic law or principle of all sacred architecture the world over, from time immemorial; which architecture is itself a deliberate 'copy' of the whole cosmos or universe, of which man is a particular living symbol.

When one realises that all desire or e-motion is a movement or vibration, and that all vibration is a 'sound' (or 'word'), and that each sound embodies a thought that gives rise to an idea (or 'light') perceived in the mind, and that ideas result in and govern all actions, then we are beginning to comprehend truth. Furthermore, when we know that perfect desire or love is God's 'Word', which is life itself, and that its perceived radiance is 'Light', and that the ideas of God are known occultly as 'spirits' or 'angels', then we will understand why Dee was associated with "conjuring" or invoking spirits through the use of mathematics and language. Those who did not understand what Dee was doing charged him with conjuring demons; whilst Dee himself was always at great pains to explain that he was communicating with angels.

Mathematics is thereby understood to be the rationale of life, which itself is divine Love expressed in a myriad of ways and forms. As such, mathematics is the rationale of language and of all thought, and is the key to the working of intelligence. This is precisely how Dee understood it and used it, as had other great sages before him. One of its uses, in the realm of practical intelligence, was in the form of cryptology and cryptography. It could be used, in language (written and spoken) and in pictures (drawn on paper and portrayed dramatically), to both reveal and conceal truth all at the same time. Unless the listener or spectator had the key, the truth would remain hidden; but, possessing the key, the truth would unfold itself.[16]

> "And when he had said these things, he cried, 'He that hath ears to hear, let him hear.'
> And his disciples asked him, saying, 'What might this parable be?'
> And he said, 'Unto you it is given to know the mysteries of the kingdom of God: but to others in parables; that seeing they might not see, and hearing they might not understand.'"[17]

> "And he turned him unto his disciples, and said privately, 'Blessed are the eyes which see the things that ye see: for I tell you, that many prophets and kings have desired to see those things which ye see, and have not seen them; and to hear those things which ye hear, and have not heard them.'"[18]

Dee, and those directly associated with him, used cryptographic methods in all that they published, and in their associated programme to usher in a Golden Age. Dee believed in the royal dignity and importance of the ancient Tudor line as being the genuine Arthurian royal blood line, and he knew the mystery and promoted the importance of the British Isles in their world role. Undoubtedly he also knew that the Renaissance period (from 1440 to 1620) corresponded to the zodiacal period of Virgo in the cycle of the Piscean Age, and that the Elizabethan period in world history was thus exceedingly important.[19] This led him to actively promulgate 'Elizabethan Imperialism' - not an imperialism of lust and greed, but one of glory and fame, achieved through courage, virtue and wisdom. His own motto was "Nothing is useful unless it is honest." He saw Britain performing its occult role openly, as the Virgin Heart of the world - a role that he and others saw personified in the

(18) QUEEN ELIZABETH I, as 'Astraea', the Heavenly Queen - engraving by F. Delaram, from drawing by Nicholas Hilliard.

English Queen. For only when the Wise Virgin, Astraea, comes again on earth, and Saturn rules once more, will there be another Golden Age.

Astraea is a synonym for Justice, which (in Renaissance thinking) corresponded to the cabbalistic principle, *Geburah,* meaning 'Judgment'. Until there is Justice on earth, which means true judgment or discernment, Saturn cannot rule. Another term for what Astraea represents is 'Righteousness', which means purity of mind and clarity of thought. In other words, it is man's ignorance and his selfish thoughts that have banished Astraea from the world; and all thoughts result in deeds. Hence, human vice and the evils of this world are due to a lack of justice or righteousness: "Father, forgive them for they know not what they do."[20]

Balancing Justice is Mercy - the compassion or clemency that every true king or queen should have. ("The quality of a king is mercy.") True justice cannot develop without mercy, which is an outflowing of generous love. Hence Justice goes hand in hand with Mercy, and together they make it possible for Saturn to reign over a Golden Age. Saturn is another name for the cabbalistic principle *Daath,* meaning 'Knowledge'. Saturn, or *Satan,* is the Doorkeeper of heaven, who holds the keys of the door that leads to the Divine Mystery. Cabbalistic Justice is not just Judgment, but the perfect blend of both Judgment and Mercy, or Perception and Compassion. The Knowledge referred to is knowledge of truth - not just intellectual knowledge, which is only partial illumination, but the final and full knowledge (or illumination)[21] that comes from the experience of having put the intellectual knowledge into practice: in other words, combining Mercury and Mars.[22]

Also at the heart of Dee's library was an almost complete collection of *prisci theologia,* which were clearly of the highest importance to him. The *prisci theologi* were divinely inspired pre-Christian authors, and included Hermes Trismegistus (also known as Mercurius), Zoroaster, Orpheus, Pythagoras, Plato, Philolaus and other Greek philosophers. Their writings carried sublime truths, and demonstrated that the great truths were understood by sages of many cultures and epochs, and that the same truths underlie all religions. They showed, for instance, a handing on of the Wisdom Tradition from Ancient Egypt to the Greek philosophers as well as to Moses, and from thence into Christianity and the Renaissance. Considered together with other notable works, and with Dee's own writings, this collection points out Dee as being what might be called a 'religious Hermetist', which is precisely what Bacon's Great Instauration is founded upon.

> "Religious Hermeticism - the whole cult of the *prisca theologia* - encouraged religious toleration and inspired a new and liberal approach to religion in the sixteenth century. Religious Hermeticists like Dee thought that a religion of the world, one of love and unity, could be developed through the rediscovered *prisca theologia,* and they envisioned a healing of the breach in Christendom. Religious Hermeticism did away with dogmatic theology, stressed the mystical unity of all religions and the oneness of God, and emphasised pristine Christianity and man's innate knowledge of the Divinity; the writings of the *prisci theologi* were used to support these ideas."[23]

> "Dee's nationalism was strong but, like his religion, was based on broader concepts. He had a mystical vision of Britain as the new leader in reuniting Christian Europe and re-establishing the new golden age of civilization for the benefit of the entire world."[24]

Dee was by no means alone in being aware of these ancient knowledges and what they meant for those present times, and he was far from being the only proponent of them, but he beçame the principal *magus* of the movement in early Elizabethan England. The whole Tudor lineage, with its entourage, embodied this knowledge and purpose, guarded it and bore it from one generation to the next; and every so often the bloodline produced someone who could understand and interpret what it all meant better than the others.

(Looked at another way, all mankind shares in the same 'royal' blood, as descendants of Adam and Eve, the first 'son-daughter of God'; and also because of the continual and healthy intermixing of bloodlines - the 'sharing of the holy grail'. But the real Holy Grail is a description of the sainted or illumined souls of all ages, who are pure vessels of love-light, and who together form the Communion of Saints or Brotherhood of Light. They are the true Grail line or family, who are in a real communion of love and consciousness with each other and all of life. But the natural world has its legitimate echo or symbolic manifestation of the more spiritual reality; hence the physical 'royal' bloodline has its genuine role to play in the theatre of this world. And what is man, but an actor on his stage?)

In 1570 there was published the first English edition of *Euclides Elementes,* translated from the Greek by Sir Henry Billingsley, an alderman and later Lord Mayor of London. To the translation Dee had added his important and influential 'Mathematicall Preface', together with annotations throughout the folio volume, and introductions to the various books. In this preface Dee manages to outline the whole state of science as it then existed. He calls for the recognition of mathematics as a key to all knowledge, and advocates the application of mathematical principles for the definite use and help of all mankind. Although pointing out the utilitarian uses to which mathematics can be put, Dee's Preface also points to something even more profound and useful to the soul of man - a means by which the soul of man may reach joy and perfection.

> "But well you may perceive by *Euclides Elementes,* that more ample is our Science, then to measure Plaines: and nothyng lesse therin is tought (of purpose) then how to measure Land. An other name, therefore, must nedes be had, for our Mathematicall Science of Magnitudes: which regardeth neither clod, nor turff: neither hill, nor gale: neither earth nor heaven: but is absolute *Megethologia:* not creping on ground, and dasseling the eye, with pole perche, rod or lyne: but liftyng the hart above the heavens, by invisible lines, and immortall beames: meteth with the reflexions, of the light incomprehensible: and so procureth Joye, and perfection unspeakable."[25]

Another name for this 'Mathematicall Science of Magnitudes', as described by Dee (and based largely on Plato), would be Free and Accepted Masonry - as indeed this cabbalistic science later came to be called, under the guidance of Francis Bacon. But in Dee's work lie the germs of what has now become a world-wide philosophical organisation with philanthropic aims, practising charity. Later, when Dee published his *General and Rare Memorials pertayning to the Perfect Arte of Navigation* in 1577, this 'Freemasonic' connection becomes clearer. Dee's Preface to Euclid and his Memorials form a definite blueprint for Modern Freemasonry, which Francis Bacon and his group later formed into a practical working example of what Dee may have had in mind. In many ways it is nothing new, but a revival of something very ancient - for Freemasonry is said to be the oldest science and art, stemming from the first man, Adam, who taught it to Cain.

Mortlake formed a good centre for Dee's 'Academy', for several important reasons.[26] One of these reasons is that it was close to the river Thames, which formed the main arterial route for trade and transport between London, Greenwich, Westminster, Richmond,

Hampton Court, Windsor, *etc.*. There was thus relatively easy access between the principal residences of Queen Elizabeth, where the court was usually to be found - *i.e.* the Palace of Placentia at Greenwich, York Palace (later known as Whitehall), Richmond Palace, Hampton Court Palace, Nonsuch Palace, Windsor Castle. Dee was often at court, where he enjoyed almost universal esteem, and likewise many courtiers were to be found at Dee's house at all times. Sir Francis Walsingham, Adrian Gilbert and Edward Dyer, for instance, constantly visited Dee, as also did the rest of the group of scholar-poets that became known as the English Areopagus, together with the Syon house group. Dee also received many visitors from abroad at his Mortlake home.

Within easy reach were the residences of many of Dee's closest friends, such as Syon House, home of Henry Percy, ninth Earl of Northumberland (later to be known as 'the Wizard Earl' for his scientific experiments), which was used as a centre of learning by the 'Syon House group'. Barn Elms, the country seat of Sir Francis Walsingham, was not far away, a centre of the intelligence network that Francis Walsingham and his cousin Thomas ran. The great houses that led off the Strand, between Westminster and the City of London, were not far away and readily accessible by river, such as Leicester House, Arundel House and York House. York House was the London home of Sir Nicholas Bacon, Lord Keeper of the Great Seal, until his death, and a hub of communication. Arundel House was the London home of Henry FitzAlan, Earl of Arundel, in which he had assembled a magnificent library and collections of marble statues, busts, and other treasures from Greece and Rome which, like Dee's collections, he made available to scholars. Arundel was probably collecting books even before Dee. He took over Arundel House from Thomas Seymour, Lord High Admiral, upon Seymour's fall from power and execution in 1549. Leicester House was next door, the London home of Robert Dudley, Earl of Leicester, who also had a library and made it a centre of poets and intellectuals. Leicester had acquired this house from Lord Paget in 1569. These houses, each a centre of learning with valuable libraries and other collections, together formed a flourishing 'university' of far greater significance than those institutions which bore that name. It is this informal 'university' which laid out and influenced the whole course of Elizabethan civilization, and created the English Renaissance and Scientific Reformation.

(19) SIR PHILIP SYDNEY, aged 23/24 years (b.1554-d.1586) - engraving, artist unknown, from oil painting (c.1577).

Sir Philip Sydney was the son of Sir Henry Sidney and Mary (*née* Dudley). He was thus the nephew of Ambrose Dudley, Earl of Warwick, and Robert Dudley, Earl of Leicester. Robert Sidney, Viscount Lisle, and Thomas Sidney were his brothers. Mary Sidney, who married Henry Herbert, 2nd Earl of Pembroke, was his renowned and beloved sister. Francis Dudley/Tudor (adopted Bacon) and Robert Dudley/Tudor (adopted Devereux), Earl of Essex, were his natural cousins. He was knighted in 1582. In 1583 he married Frances Walsingham, daughter of Sir Francis Walsingham, and the same year was appointed Governor of Flushing under Leicester's command. He died from a wound to the thigh sustained in battle at Zutphen in the Netherlands.

The English Areopagus and Arcadia

Although there is no direct documentary reference to Francis Bacon having met or known Dr. John Dee before 1582 (a record in Dee's diary for that year states, "Aug. 11th, Mr. Bacon and Mr. Phillips of the court cam."), there is a wealth of other evidence to pointedly suggest that Francis must have known Dee intimately, and as a teacher, even before he returned home from France in 1579 and joined the Dudley-Sydney circle. Indeed, considering the company that Francis was raised in - the very court circles that held Dee in high regard and paid him to tutor their children - plus the fact that Francis was a noted child genius and received special attention and instruction from the best teachers available, it would be surprising if Francis had not received some instruction or influence from Dee before he (Francis) was banished to France in 1575. The question is really whether Francis was influenced by Dee before his great vision in 1573-4[1] or immediately after. Certainly by 1575 Francis was fired by the whole Tudor Arthurian theme and seems to have played a major role in the Kenilworth and Woodstock Entertainments in which the hermit, Hemetes, could easily have been modelled in part upon Dee.

These lavish Entertainments, which burst upon the Elizabethan scene in the Summer of 1575, set the scene and laid out the symbolic imagery that was the original basis of all Bacon's work, and which was developed in all its poetic, mathematical and mystical imagery by him under the pseudonym of 'Immerito' (later masked by Edmund Spenser).[2] This original basis is an Arthurian theme, and the Entertainments deliberately set out to revive the Order of the Round Table, with its Ancient British and Druidic associations, and at the same time to introduce the Faerie Queen, who is Astraea or Gloriana in another form - the Virgin Queen and British Minerva. The Entertainments also introduced the deeply mystical symbolism of Christianity and its Hermetic origins - the theme of the Good Shepherd (*i.e.* the Divine Pymander), in the form of the shepherd-knight and Hermetes (*i.e.* Hermes), the hermit or wise man. With the Orphic or Greek imagery added, these Entertainments set the scene for the development of the Arthurian and Arcadian imagery of the 'Areopagus' of English poets, notably in the principal literary works of Philip Sydney's *Arcadia* and Immerito's *Shepheards Calender* and *Faerie Queen,* which permeated and influenced the whole development of Elizabethan pageantry, culture, learning, politics and national belief.

It is almost impossible to think of these Entertainments having occurred without Dee's involvement in one way or another, as they were concerned with so much that he himself was fostering; and many of those who took part were both fellow-scholars and pupils of Dee. The sources and influences which helped create the Entertainments were various and varied, but they were all interwoven to produce the first organised effort on a grand scale in Elizabethan England to promote the 'Arcadian' theme of Utopia, a Golden Age, employing poetry and all the creative (or imaginative) arts channelled into high drama.

Compared with what followed in later years, these particular Entertainments were the product of youth, with a certain naïvité and immaturity about them. But the principal author (or authors) did take the themes that such *magi* as Dee were proclaiming (and which had been used by the earlier Tudor kings, Henry VII and Henry VIII), plus the poetic and artistic talents that others offered, and through sheer brilliance and poetic gift wove them into a *Magnificence.*

The Kenilworth and Woodstock Entertainments were the prototypes for the vitally important and influential grand Accession Day Tournaments, which seem to have been instituted thereafter and which continued to the end of Elizabeth's reign. It is possible to see that all these later Accession Day Tournaments, plus some other tournaments held at other times (notably the May Tournaments), were a continuation of the theme that was introduced at the Woodstock Entertainment, and that they not only developed and elaborated the initial theme but unfolded the story (or romance) step by step in true Arthurian or Bardic fashion.

There were two principal stars of these tournaments who were involved from the beginning - Sir Henry Lee and Sir Philip Sydney. Lee featured as the Queen's Champion from the time of the 1575 Woodstock Entertainment until his retirement in 1590, during which period of time the whole theme of the knight who eventually retires as the hermit was developed - Lee beginning as the knight who is taught by a hermit and then in the end retiring as a hermit himself. Sir Philip Sydney featured as the Champion of Protestantism and Chivalry for the general populace, his first (known) appearance as a knight-champion being at the Accession Day Tournament of 1577; but his career was cut short by his untimely death on the battlefield in 1586. After Sydney's death his place was taken up by Robert, Earl of Essex.

The Areopagus of English poets first appeared on the scene openly with the publication of the *Shepheards Kalendar* on the 5th December 1579. It was published anonymously, and dedicated on the titlepage to Francis' cousin and friend, Sir Philip Sydney, who (at least to all outward appearances) was presented to the public eye as the leader of the group (although 'Immerito' always stood as the chief or greatest of the poets). After the dedication there follows a cryptic little poem entitled "To His Booke" and signed "Immerito." Also included was a letter from 'E.K.', preceding the twelve *Eclogues* (poems covering the twelve months of the year), addressed "To the most excellent and learned both orator and poete, Mayster Gabriell Harvey, his verie special and singular good friend E.K. commendeth the good lyking of this his labour, and the patronage of the new Poete." The letter is dated "from my lodging at London thys 10 of Aprill, 1579". This 'E.K.' adulates the "new Poete" and proclaims that he has added commentaries (or "glosses") to the Eclogues. The commentaries, like the introductions to the poems in each Eclogue, show such an intimacy with the poet's meaning and sources of information that it seems clear that 'E.K' (the commentator) and 'Immerito' (the poet-author) are one and the same. Francis Bacon makes this quite clear in his Biliteral cipher when he declares not only that he wrote the works published under Spenser's name, but also that 'E.K.' stands for himself as he hoped he would be - "England's King."

> "E.K. will be found to be nothing less than the letters signifying the future sov'raigne, or **England's King**..........The present Queen, purely selfish in all that doth in a sort make for proper though tardie recognition of that true prerogative of royal blood, doth most boldly and constantly oppose with her arguments the puny effort in our cause which hath most disproved ability to uphold our true and rightful (but at this present time, very little seen or only

TO THE MOST EXCELLENT and learned, both Oratour and Poet, maſter *Gabriel Haruey*, his verie ſpeciall and ſingular good friend, *E. K.* commendeth the good liking of this his good labour, and the patronage of the new Poet.

V*Ncouth, vnkiſt*, ſaide the old famous Poet *Chaucer*: whom for his excellencie and wonderfull skill in making, his ſcholler *Lidgate*, a woorthy ſcholler of ſo excellent a maſter, calleth the loadſtarre of our language: and whom our *Colin Clout* in his Eglogue calleth *Tytirus*, the God of Shepheards; comparing him to the worthineſs of the Roman *Tytirus, Virgil*. Which prouerbe, mine owne good friend M. *Haruey*, as in that good old poet, it ſerued well *Pindarus* purpoſe, for the bolſtering of his bawdie brocage, ſo very wel taketh place in this our new Poet, who for that he is vncouth (as ſaid *Chaucer*) is vnkiſt; and vnknown to moſt men, is regarded but of a fewe. But I doubt not, ſo ſoone as his name ſhall come into the knowledge of men, and his worthineſſe be ſounded in the trumpe of Fame, but that he ſhall be not onely kiſt, but alſo beloued of all, embraced of the moſt, and wondred at of the beſt. No leſſe, I thinke, deſerueth his wittineſſe in deuiſing, his pithineſſe in vttering, his complaint of loue ſo louely, his diſcourſes of pleaſure ſo pleaſantly, his paſtorall rudeneſſe, his morall wiſeneſſe, his due obſeruing of *Decorum* euerie where, in perſonages, in ſeaſons, in matter, in ſpeech, and generally, in all ſeemelie ſimplicitie of handling his matters, and framing his words: the which of many things that in him be ſtrange, I know will ſeeme the ſtrangeſt; the wordes themſelues beeing ſo ancient, the knitting of them ſo ſhort and intricate, and the whole period and compaſſe of his ſpeech ſo delightſome for the roundneſſe, and ſo graue for the ſtrangeneſſe. And firſt of the words to ſpeake, I grant they be ſomething hard, and of moſt men vnvſed, yet both Engliſh, and alſo vſed of moſt excellent Authours, and moſt famous poets. In whom, when as this our poet hath beene much trauailed and throughly read, how could it be (as that worthy Oratour ſaid) but that walking in the Sunne, although for other cauſe hee walked, yet needes hee muſt be ſunne-burnt; and hauing the ſound of thoſe ancient poets ſtill ringing in his eares, hee mought needs in ſinging, hit out ſome of their tunes. But whether hee vſeth them by ſuch caſualtie and cuſtome, or of ſet purpoſe

A 2. and

(20) FIRST PAGE, LETTER TO GABRIEL HARVEY, signed 'E.K.' (dated 1579), written as a foreword to *The Shepheards Calender,* and printed in the Spenser 1st Folio (1611). The "new Poet" ('Immerito') is one and the same as 'E.K.' - *i.e.* Francis Bacon.

partly guessed) claim to royal power. In event of death of Her Majesty, - who bore in honourable wedlock Robert, now known as son to Walter Devereux, as well as him who speaketh to the yet unknown aidant decypherer that will open the doors of the sepulchre to break in sunder the bonds and cerements of a marvellous history, - we the eldest born, by the divine right of a law of God made binding on man, inherit sceptre and throne."[3]

"As I have often said, and as you well know by this time, you have poems and prose works on divers themes in all such various styles as are put before the world as Greene's, as Shakespeare's, Burton's, as Peele's, Spenser's, as Marlowe's, as Jonson's dramas or my own long devised and but well begun labour, - than which none hath a better object, - for I varied my style to suit different men, since no two show the same taste and like imagination, and all doth contain the great Cipher I constantly teach....."[4]

"Next write a comedy, a quaint device for making known the men that do give, lend, sell, or in any other way have put me in possession of their names. These I have used as disguises that my name might not be seen attached to any poem, stage-play, or any of the light works of this day......... The title of the comedy is, *Seven Wise Men of the West.* Actor's names: Robert [Burton], Christopher [Marlowe], William [Shakespeare], another Robert [Greene], George [Peele], Edmund [Spenser] and Francis [Bacon]."[5]

These 'seven wise men' or masks were the principal masks (*personas*) that Francis used - 'Bacon' being one of them, as Francis was really a Tudor or Dudley - but they were not the only ones. Francis also adopted other pseudonyms, sometimes being responsible for the authorship of the whole work in question, sometimes including his own writings within a body of writings by another, and sometimes inserting the creations of others into his own. But he always seems to have signed or indicated his own works cryptically, according to consistent rules; and throughout his principal literary works run the Biliteral and Word ciphers which tell the story of his life, and other things besides.

The 'seven wise men' were used as a deliberate cabbalistic scheme, the key to which was only published fifty-three years after Francis' death to the world as 'Bacon'. This key was the *Abcedarium Naturae* (or "Alphabet of Nature" - "a Metaphysical Piece") which was published by Archbishop Thomas Tennison in 1679.[6] Because this cabbalistic and cryptographic scheme is present in *The Faerie Queene,* it is quite clear that Francis had this particular project planned out before 1580, which is when he first submitted early proofs of *The Faerie Queene* to the criticism of his old tutor, Gabriel Harvey (as known from letters supposedly written between Harvey and 'Immerito', dated April 1580 and printed in 1580.)

It was the published Harvey-Immerito letters that first fully publicised not only England's "new Poet" but also the fact that a group of poet-scholars had deliberately banded together to create an English Renaissance in poetry, language, the arts generally and scientific-mystical teaching. Concerning this, a manuscript letter-book and diary, belonging to Gabriel Harvey and containing his correspondence with 'Immerito' (written 1578-1580) is fortunately still in existence. The whole of the correspondence was published by the Camden Society. Mr. Scott, who was responsible for the editing of the correspondence for the Camden Society, noted that in the portions containing copies or drafts of letters to 'Immerito', much had been torn out. Henry Seymour, in his article *The History or Mystery of Edmund Spenser,* says:-

> "Mr. Parker Woodward, in his *Tudor Problems*, refers to this [the missing portions] and cites a number of particulars from the Harvey letters in an effort to identify 'Immerito'; and when these are pieced together they appear to fit Francis Bacon, rather than Edmund Spenser, to the smallest detail. There can be no doubt that Harvey shared Bacon's confidence to some extent in those early days; and if we presume that Bacon was then masquerading as 'Immerito', it is plain to see that Harvey was in the position of a spiritual father to Bacon, to whom he gave much sound advice in all that related to literary and poetical forms whilst he was yet a youth."[7]

The letters are dated October 1579 and April 1580, and were published in 1580.[8] In the letters, the "new Poet" is referred to under the names of 'Benevelo' and 'Immerito'. The first letter from 'Immerito', written from "Leycester House, this 5 October, 1579", begins, "To the Worshipfull his very singular good friend, Maister G.H.", which carries a Freemasonic allusion pointing to the probability that Gabriel Harvey was a Worshipful Master in a Cabbalistic or Freemasonic lodge - one that in all likelihood was associated with (or supervised by) Dr. John Dee and two other 'Grand Principals'. The same applies elsewhere to Philip Sydney. The fellow poets that are identified in these letters as belonging to the group, besides 'Immerito', Gabriel Harvey and Phillip Sydney, are Edward Dyer, Thomas Drant, a "Maister Preston" and a "Maister Still".[9] Fulke Greville is also known to have belonged to this company of talented friends.[10] As a group, 'Immerito' refers to them as "αρειωπαγω" - *i.e.* the *Areopagitœ*, or judges of the Areopagus. This title conveys succinctly what the group was all about.

The Areopagus was the 'hill of Mars', located near Athens, on which was situated the great law-court of Athens. Lempriere's *Classical Dictionary* has this to say about the Areopagites:-

> "The precise time at which this celebrated seat of justice was instituted is unknown.......The number of judges that composed this august assembly is not known.......The most worthy and religious of the Athenians were admitted as members, and such archons as had discharged their duties with care and faithfulness. If any of them were convicted of immorality, or they were seen sitting at a tavern, they were expelled from the assembly; though the dignity of a judge of the Areopagus always was for life. The Areopagites took cognizance of murders, impiety, and immoral behaviour; and particularly of idleness, which they deemed the cause of all vice. They watched over the due execution of the laws, and had the management of the public treasury; they had also the liberty of rewarding the virtuous, and of inflicting severe punishment upon such as blasphemed against the gods, or slighted the celebration of the holy mysteries. They always sat in the open air, because they took cognizance of murder; and by the laws of Athens, it was not permitted for the murderer and accuser to be both under the same roof. This custom might also have originated from an idea that the persons of the judges were sacred, and they were consequently afraid of contracting pollution, by conversing in the same house with men who had been guilty of shedding innocent blood. Opposite to them were erected two stones, on which the accuser and the accused were seated; and on two columns near them were engraved the laws, according to the tenor of which they were to pass judgment.[11] Their decision was delivered secretly; and two urns, one of brass and the other of wood, were placed to receive the pebbles, which, according as they were black or white, decreed condemnation or absolution. They always heard causes and passed sentence in the night, that

(21) PALLAS ATHENA - emblem from titlepage to *Nova Atlantis* (1643) by Francis Bacon.

Inscription: *"Et flore et fructu."*

Translation: "With both flower and fruit."

Pallas Athena is Goddess of Intelligence and Enlightenment, by means of whom Wisdom is born and made manifest in the world. According to ancient mythology, Apollo, God of Wisdom, is her spouse, and Esclepius, the great Healer-Teacher (a Hermes or Christ figure), is their son. Mount Parnassus, the Mount of Inspiration and Enlightenment, is their mutual home.

Athena is the Goddess Fame, Patroness and Leader of the Seven Liberal Arts and Sciences, Goddess of peace and all learning, and hence the Protectress of the heroes - the brave, valorous and righteous - who seek the 'fame' of enlightenment. As the great Goddess of Intelligence, she is the upholder of justice and virtue. She is the Goddess of Poetry, in the sense of Poetry embracing all other things, and is known as the Tenth Muse, Leader of the other Nine.

Athena's principal symbols - as shown in these illustrations - are the olive tree (signifying enlightenment, peace, concord, prosperity), the owl (signifying perception and sagacity), a spear of light (signifying the power of bestowing illumination upon others), a golden helmet (akin to the halo, signifying illumination, which is salvation, and which bestows 'invisibility' or protection upon its wearer), a mirror-like shield, sometimes containing the impress or image of the Gorgon Medusa's head (signifying the clear, reflective mind and its understanding), a goatskin breastplate, sometimes fringed with snakes and concealing a bag containing the head of Medusa (signifying righteousness and the protection of pure love), and an ægis or tunic, sometimes a cloak (worn by maidens as a symbol of chastity and virtue, as well as concealment).

(22) PALLAS ATHENA - emblem from titlepage to *Scripta in Naturali et Universali Philosophia* (1653).

Inscription: *"Ne extra oleas."*

Translation: "Nothing but the olive."

they might not be prepossessed in favour of the plaintiff or of the defendant by seeing them. Whatever causes were pleaded before them, were to be divested of all oratory and elaborate speaking, lest eloquence should charm their ears, and corrupt their judgment. Hence arose the most just and impartial decisions; and their sentence was deemed sacred and inviolable, and the plaintiff and defendant departed equally convinced of its justice."[12]

Cabbalists will know that Mars stands for the cabbalistic principle of *Gevurah,* 'Judgment', on the Tree of Life, and so the appropriateness of the Athenian designation is exact. As mentioned before, Judgment means perception, discernment, and relates to the intellectual faculty. It is also referred to as Justice and Righteousness which Astraea, the Faerie Queen, symbolises. It belongs to and is an attribute of the Pillar of Form ruled over by Divine Intelligence, which the Greeks personified as Pallas Athena. The story of the birth of Athena from the head of Zeus is a parable of the release of intelligence from the head (or reflective part of the mind) as a burst of understanding. Athena springs forth from Zeus' head brandishing or shaking her spear, hence her name *Pallas Athena,* which means 'Shaker of the Spear'. Athena's spear signifies the light of illumination, which indeed lights up the face and shines from the head as a golden crown or halo. This crown of light is symbolised by Athena's glorious helmet, which remains invisible to those with eyes that cannot see what is before them. ("Behold Good!" is the cry of Hemetes the Heremyte; but not everyone is able to see good. People tend to be blind or to look for other things!)[13]

The whole of Athens was dedicated to this principle of Intelligence, from which the city derives its name. Athens contained the magnificent Parthenon on the top of the Acropolis, dedicated to Athena and housing the renowned gold and ivory statue of the goddess, one of the masterpieces of Phidias. At the foot of the Acropolis was the Athenum, also sacred to Athena, "where the poets, philosophers and rhetoricians regularly declaimed and repeated their compositions. It was open for the use of all the professors of the liberal arts."[14] Then there was the Areopagus.

Athens was also known as Astu, one of the eyes of Greece - the learned city, the school of the world, the common patroness of Greece. It was said to have been founded *c.* 1556 BC by Cecrops, as an Egyptian colony. It played a major part in the so-called Eleusinian Mysteries. Athens to the east, Mount Parnassus (on the side of which lies Delphi) to the west, and Mount Olympus to the north, form a sacred triangle in Ancient Greece, not unrelated to certain cabbalistic principles. They 'oversee' the country of the Peloponnesus, the southern peninsular of Greece which contains, as its heartland, Arcadia.

The especial god of Arcadia was Pan, who presided over the shepherds, cattle-herders and hunters of this beautiful and mountainous country. The Arcadians considered themselves to be the most ancient people in Greece, and more ancient than the moon. They were passionately fond of music and poetry and pastoral song, and became renowned for this and their learning. They were also warriors. Acorns formed a large part of their diet, as the country was once well planted with majestic oaks.

Arcadia derives its name from Arcas, the son of Zeus and the beautiful Callisto. Callisto was given the form of a bear by Zeus, represented in the sky by *Ursa Major,* the Great Bear. When Arcas died, Zeus made him into a constellation and set him beside his mother. In Arcas we have *Ursa Minor,* the Little Bear.[15] The first star of the seven stars of the Little Bear is the present North Pole Star, and it carries the name of Arcas. It signifies the crown 'jewel' of the world and its heavens. *Ursa Minor* was also known as *Phoenice,* or

(23) TITLEPAGE TO *HYMNE TO ELIZABETH*, *Sovereign of England, France and Ireland* (1586) - manuscript written and illustrated by 'George de la Mothe'. The picture shows an emblem of a phoenix (or eagle), between two cabbalistic pillars, rising from the mundane world to the spiritual sun.

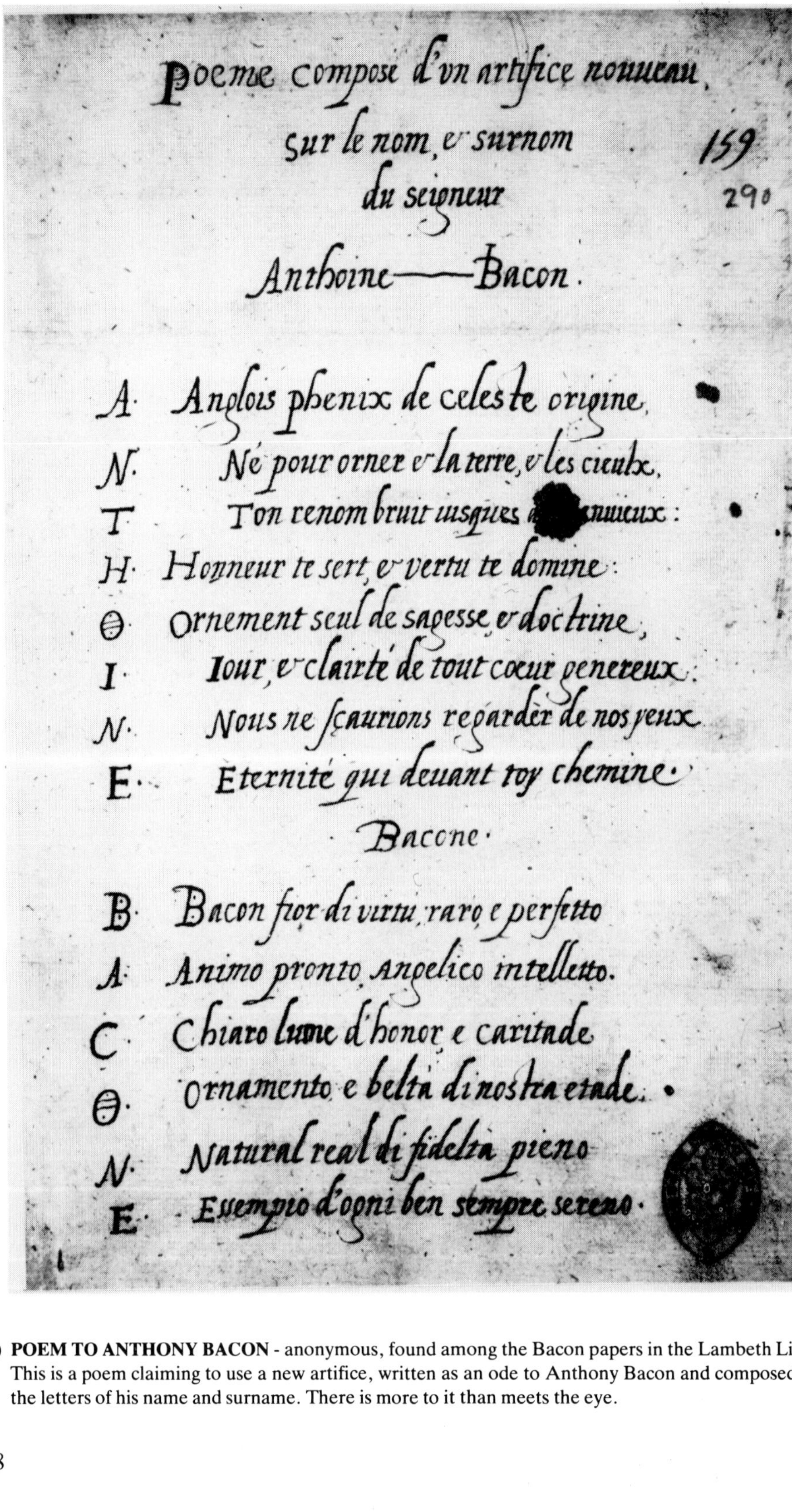

Poeme composé d'vn artifice nouueau,

sur le nom, & surnom 159

du seigneur 290

Anthoine——Bacon.

A· Anglois phenix de celeste origine,
N· Né pour orner & la terre, & les cieulx,
T· Ton renom bruit iusques [illegible]:
H· Honneur te sert, & vertu te domine:
O· Ornement seul de sagesse, & doctrine,
I· Iour, & clairté de tout coeur genereux:
N· Nous ne sçaurions regarder de nos yeux
E· Eternité qui deuant toy chemine.

Bacone·

B· Bacon fior di virtu, raro e perfetto
A· Animo pronto, Angelico intelletto.
C· Chiaro lume d'honor e caritade
O· Ornamento e belta di nostra etade.
N· Natural real di fidelta pieno
E· Essempio d'ogni ben sempre sereno·

(24) POEM TO ANTHONY BACON - anonymous, found among the Bacon papers in the Lambeth Library. This is a poem claiming to use a new artifice, written as an ode to Anthony Bacon and composed upon the letters of his name and surname. There is more to it than meets the eye.

Ursa Phoenicia. The Arcadians were known as 'the Bear Race'; they were also (in their most ancient sense) the original Phoenicians.[16] Furthermore, the stars of *Ursa Minor* were those principally used in all navigation by land as well as by sea, to direct the course of the traveller. Thus the Arcadian or Bear Race were known as the guides or pathfinders of mankind, leading and lighting the way like Mercury or Hermes. The 'Bear' is also known as an 'Ark' or 'Boat'; and the Arcadians were the 'People of the Ark' or the 'Arch Race' - 'the People of the Covenant'.[17] As Phoenicians, they were renowned sailors and navigators, as well as teachers and healers.

With all this in mind we can begin to see the real significance of the Areopagus of English poets, the chief of whom were Dudleys or Tudors, and their symbolism of Arcadia intermixed with the symbolism of Britain. When we also come to realise that the Ancient British derived their ancestry from the Trojans, who came to 'the Fortunate Isles' from the Ægean (the 'Summerland' of the Trojan Empire, which once included Greece, Thracia and Asia Minor as a mountainous 'ring' around the Ægean Sea) in the 2nd millenium BC, we can begin to see the whole picture with far more meaning. Then add to this the fact that the Trojans claimed to be but another, later wave of settlers to Britain from the Ægean area, and were related as *Cymry* to the previous 'giant' inhabitants of Britain - the megalithic builders - who had been led from Ægaea to these 'Fortunate Isles' by Albion in about the 4th millenium BC.[18] Finally, take into account that the Cymry relate, in the stories concerning their origins, that they originated from the 'Blessed Land' (later called the 'Fortunate Isles') in the first place; but that, when the Great Flood occurred, their ancestors (who were known simply as the 'holy man' and 'holy woman') built an ark and preserved the race. When the flood subsided and the 'ark' came to rest, the Cymri settled in Ægaea (which was once an extensive land surrounded by a ring of mountains, until the Black Sea burst through the Bosphorus and formed the Ægean Sea and its many islands); but whenever conditions became suitable, or when they were forced to, they made successive migrations back to their original homeland.

Immerito

The *Shepheard's Calender* is first mentioned in 'Laneham's Letter' (as "the Sheperdz kalender") written by Francis shortly after the Kenilworth Entertainment in 1575 by request of his patron, the Earl of Leicester. The letter was published soon afterwards, and masked by the pseudonym of 'Robert Laneham', an obscure or fictitious personality who was described as 'Keeper of the Council Chamber Door'.

'**Laneham**' is an interesting name, invented probably at the same time as 'Burnham' was created as a pseudonym for Anthony Bacon.[1] It is composed of the words 'lane' and 'ham', the former being a variant of the word 'lone' and the latter suggesting, by inference, 'Bacon'. Several times Francis alludes to his mother, the Queen, as an "alone" Queen, meaning not only someone who is set apart by reason of her birth, but also by reason of her excellency and virginity (purity):-

> "I pray God she may not be too much alone, but it is a Name of excellency and virginity."[2]

Here Francis appears to use it for himself, both because he is heir to the throne by birth and also because of his own 'virginity' and excellence, which sets him apart from others. In a rough draft of a light play, which still exists written in Francis Bacon's handwriting, couched in the form of a petition signed and presented by Philautia (goddess of Self-love) to Queen Elizabeth, he refers again to the Queen as 'the alone Queen'. This rough draft also confirms Francis' association with the Muse of Genius, labelled Pallas Athena, the 'Shake-Spear':-

> "Excellent Queen, Making report to Pallas, upon whom Philautia depends, of my last audience with your Majesty and of the Opposition I found........she, putting her Shield before her eyes as she useth when she studieth to resolve....said....'this shall you do....and the alone Queen - so she ever terms your Majesty, will see that she hath Philautia's first offer. If she reject it, it will be received elsewhere to her disadvantage.....' *etc.*. Philautia."[3]

Philautia might refer to the author himself, who considered himself as Solomon II and who knew the import of Solomon's words, "All is vanity"; or else this play-petition may have been written for and on behalf of another person, represented by Philautia, for presentation to the Queen. It certainly fits in with Francis' continual plea to his royal mother to give him financial support and recognition in his philanthropic ventures, which he was performing to a large extent on her behalf by building up her public image and thereby the perceived greatness of her kingdom. But Philautia could also have been an allusion to the Earl of Leicester, as Philautus (the man) was said to have been married to Camilla, the virgin queen of the Volsci and priestess of Diana (goddess of the Moon and of the Hunt), and Camilla can certainly be related to Queen Elizabeth in this sense.

'Laneham's Letter' records some of the Kenilworth festivities in a jocular manner, and mentions that "the Sheperdz kalender" was enacted (or recited) there.[4] This would indicate that the original ideas and version of this extraordinary poetic work was first foisted onto a select audience in 1575, when Francis was aged 14, but that it was probably worked up into the final masterly version during the months immediately after his return in 1579. Its publication at the end of 1579 is said to have been due to the request of Gabriel Harvey; but quite clearly it was part of a carefully laid design, and marked the deliberate start of what later scholars regard as the beginning of the English Renaissance.

Francis Bacon had been translating the classics and writing poetry and other literary works from a very early age. He was, it must be realised, a noted child genius and wonder of his day. Just as many of our greatest musical geniuses are known to have begun composing as well as playing beautiful music before the age of 10, so also did Francis Bacon in the sphere of poetry and literature. In terms of the so-called 'Spenser' works, Francis had translated into English six epigrams (*i.e.* poetical pieces in the style of sonnets) of French Petrarchist verses and at least eleven sonnets of Joachim du Bellay by the time he was 8 years old; which epigrams and sonnets appear as *Visions of Petrarch and Visions of Bellay* that were published together under Spenser's name in 1591. These poetic translations were originally published in 1569 under the title of *A Theatre wherein be represented as well the Miseries and Calamities that follow the Voluptuous Worldlings as also the great Joyes and Plesures which the Faithful do enjoy. An Argument both Profitable and Delectable to all that sincerely loue the Word of God. Deuised by S. John Vander Noodt.*[5] This was an English translation of *Het Theatre* that was first published in 1568 by John Daye (printer of Foxe's *Acts and Monuments,* and of Sir Henry Billingsley's and John Dee's *Euclide*). The English translation contains a dedication to Queen Elizabeth I which provides an "early formulation of the Astraea theme in relation to the Queen"[6]:-

> "The Kingdom of Saturn and the Golden world is come again, and the Virgin *Astraea* is descended from heaven to build her seat in this your most happy country of England."

The Petrarchian theme and influence on the young Francis Bacon was of some importance, as well as the reformative, anti-papal tone of the *Theatre of Worldlings.* Petrarch is most famed for his poem, *I Trionfi,* which is wrtiten in the form of a series of Roman triumphs, each triumph allegorizing some virtue and the whole forming a moral sequence. First comes the Triumph of Love, then of Chastity, of Fame, of Death, and finally the Triumph of Eternity.

> "This remarkable poem was extremely well known, and its influence was spread through its popularity as a subject for the visual arts. The triumphs were displayed in countless paintings and tapestries and frequently also they were enacted in court and public pageantry."[7]

The sequence is vitally important, and is based on the law of life: *i.e.* desire-thought-action, which, when raised to perfection, becomes love-understanding-service. These are the three initiatory steps or degrees of the Lesser Mysteries of Purification, which lead to the truly Christian or 'resurrected' life. In Petrarch's terminology, which is also adopted by Francis Bacon, 'chastity' signifies understanding, which is only attained when the mind is pure, clean, virtuous, and uncorrupted by vile or tempting thoughts. 'Fame' signifies service, which is charity, which is the only lasting and true fame that a man can ever hope for or achieve. (This 'fame' is the *fama* of the later Rosicrucian manifestos, and the "fame, that all hunt after in their lives" of *Love's Labour's Lost.*) Such fame, or charity, leads to 'death', for it involves a sacrifice or offering of the personal will in order to do God's will; and such death leads to resurrection, which is 'eternity'. This 'eternity' is knowledge of God, or knowledge of Truth. It is full illumination, and the crown of man's achievement.

As Frances Yates points out, it is indeed significant that, "at about the same time that Spenser was writing his *Shepherd's Calender,* John Dee was exercising his mathematical, astronomical and astrological knowledge on the project of the reform of the calendar."[8] **Time** is of vital importance to mankind, and the figure of Old Father Time is identical with that of **Pan**, who is also the same as Saturn, or Satan. Moreover, Pan is equated with man himself, who falls as Lucifer from heaven into the 'abyss' of matter in order to climb up to the top of the heavenly Mount of the Congregation - the Mount of Knowledge or Light - in order to know God by having discerned the difference between good and evil, and having redeemed (or spiritualised) all matter. Pan, or Time, is the god of Arcadia. The rotation of the two 'Bears' about the northern pole of the heavens not only gives direction to travellers, but also provides a celestial clock by which to tell the passage of time. The emblem of Pan was widely used on all the Baconian works.

The Harvey-Immerito letters mention other works as being either commenced or already written by 'Immerito' at that time; *viz. Dreames,*[9] *Legends, Court of Cupide, The English Poet, The Dying Pelican, Stemmata Dudleiana,*[10] *Slomber, Nine English Comedies,*[11] *The Epithalamium Thamesis,*[12] and the *Faerie Queene*. Of these, only the *Faerie Queene* is definitely known to have been published, and then not until 1590 when Part I, containing the first three books out of a projected twelve was printed.[13] Part II, containing the next three books, was published in 1596. *Prosopopoia, or Mother Hubbard's Tale,* which, like *The Shepheard's Calender,* was based on Chaucer (whom 'Immerito' refers to as his master), was also written in "the raw conceipt of my youth". He published this latter work together with *The Ruines of Time, The Teares of the Muses, Virgils Gnat, The Ruines of Rome (by Bellay), Muiopotmos or The Tale of the Butterflie, Visions of the Worlds Vanitie, Bellayes Visions* and *Petrarches Visions,* in 1591, in a volume entitled *Complaints.*

'Edmund Spenser' is not mentioned in any published work as the name of the poet, England's Virgil, until 1589, shortly before the first part of the *Faerie Queene* was published. This allusion to Spenser comes in the preface written by Nashe for Greene's *Menaphon*:-

> "Divine Master Spenser, the miracle of wit to bandie line for line, in honor of England, gainst Spaine, France, Italie, and all the worlde."

According to the evidence of the writings themselves, to Francis's own ciphers, and to clear hints made by Gabriel Harvey, both Nashe and Greene were additional masks for his authorship.[14] The preface to *Menaphon* thus carefully prepared the way for publication of the *Faerie Queene* and others of his poetic works under Spenser's name, whilst Spenser the man (but not the poet) was busy building a career and a home in Ireland.

Until 1590, only the pseudonyms 'Immerito', 'Benevelo', 'E.K.' and, in the context of the *Shepherd's Calender,* 'Colin Cloute', were used to represent the author in the publications later masked by Spenser.

'**Immerito**' means 'the blameless one' or 'spotless one' - one who is innocent but made to bear the weight of blame or sin of others. For Francis it had two distinct allusions: one to the circumstances of his birth, with his royal mother ever blaming him and his arrival into this world as the cause of her woes, and the other to the more esoteric idea of the perfect

(25) PAN, OR TIME - emblem from titlepage to Francis Bacon's *New Atlantis* (1626).

Inscription: *"Tempore patet occulta Veritas."*

Translation: *"Hidden Truth brought forth by Time."*

The Great God, PAN, is also known as Old Father Time - 'Old' in the sense of Ancient (as in the 'Ancient of Days'), 'Father' in the sense of Spirit as opposed to Matter, and 'Time' in the sense of All-Time or Eternity, in which is to be found all divisions and sub-divisions of relative time. 'Pan' means 'Universal' - *i.e.* the One or All. As Francis Bacon explains in his essay on Pan (see 'Pan, or Nature', *The Wisdom of the Ancients*), Pan signifies Universal Nature - the offspring of the Word of God (*i.e.* Spirit) and Matter.

The symbolical form of Pan was invented so as to suggest the gradual transformation and transmutation of nature from the lowest earthly forms to the highest celestial forms - the highest form of life being celestial or angelic man, the Christ man. The highest form transcends and contains all the lower ones, and hence Pan is pre-eminently a symbol of Man who, as Adam, was intended to be the Lord of Nature, and who may still (and is intended to) attain that end. Pan is the Archetype for the whole evolution and crowning achievement of the human soul. In other words, Pan signifies the Spirit of Man, whilst the Bride (or Truth) represents the human soul in whom Truth is revealed when 'she' attains perfection and virtuous mastery over all nature. Bacon reveals that what is now known as the 'Church of Christ' was once known as the 'Brotherhood of Pan'.

Pan is also known as Saturn or Satan, meaning the 'Tester' or 'Adversary'. Properly understood, this signifies Intelligence, which tests the Spirit to see whether it is good (*i.e.* God) or not. This is the function of the mind, or intellect, which is directly associated with what primarily Man is - a 'thinker'. Through thinking man comes to understand and eventually, through thought put into action, to know truth - to know God. Hence Pan, or Saturn, or Satan, also represents Knowledge, or Enlightenment. Pan carries the sickle and hourglass, symbols of death and time, for it only through the changing experiences of nature, which we call mortality, that the human soul is perfected and true knowledge is obtained.

(26) PAN, OR TIME - emblem from Philip Sydney's *The Countess of Pembroke's Arcadia* (1625).

and royal sacrifice - the 'lamb of God' - which he believed in as an ideal and enacted out throughout his life to make it real and manifest. The Good Shepherd becomes a shepherd by first being the perfect sheep who is gathered into the sheepfold. Then he may go out of the sheepfold as a shepherd to gather in other sheep.[15]

'**Benevelo**' would appear to be a deliberate play on the Latin word *benevole,* or *benevolus,* meaning 'kind', 'obliging', 'well-disposed', and the pseudonym obviously goes hand in hand with the meaning of 'Immerito'.

'**E.K.**', as mentioned before, stands for 'England's King'; or rather, England's future king by right of birth.

As for '**Colin Cloute**', the author explains that, "Now I think no man doubteth but by Colin is meant the Authour selfe...,"[16] and "Under which name (Colin Clout) this poet secretly shadoweth himselfe."[17] In the Epistle addressed to Gabriel Harvey, signed 'E.K.' and dated 10th April 1579, is written the following:-

> "As for Colin, under whose person the Author selfe is shadowed......
>
> "And also appeareth by the basenesse of the name, wherein it seemeth he chose rather to unfold great matter of argument covertly then, professing it, not suffice thereto accordingly.
>
> "Which moved him rather in Æglogues then otherwise to write, doubting perhaps his habilitie, which he little needed, or mynding to furnish our tongue with this kinde, wherin it faulteth; or following the example of the best and most auncient Poetes, which devised this kind of wryting, being both so base for the matter, and homely for the manner, at the first to trye theyr habilities; and as young birdes, that be newly crept out of the nest, by little first to prove theyr tender wyngs, before they make a greater flyght. So flew Theocritus, as you may perceive he was all ready fully fledged. So flew Virgile, as not yet well feelig his winges. So flew Mantuane, as not being full somd. So Petrarque. So Boccace. So Marot, Sanazarus, and also divers other excellent both Italian and French Poetes, whose foting this Author every where followeth; yet so as few, but they be well sented, can trace him out. So finally flyeth this our new Poete as a birde whose principals be scarce growen out, but yet as one that in time shall be able to keepe wing with the best.
>
> "Now, as touching the generall dryft and purpose of his Æglogues, I mind not to say much, him selfe labouring to conceale it. Onely this appeareth, that his unstayed yougth had long wandred in the common Labyrinth of Love, in which time to mitigate and allay the heate of his passion. or els to warne (as he sayth) the young shepheards, s. his equalls and companions, of his unfortunate folly, he compiled these xij Æclogues, which, for that they be proportioned to the state of the xij monethes, he termeth the *Shepheards Calender,* applying an olde name to a new worke."[18]

In the Harvey-Immerito letters, Gabriel Harvey refers to Immerito as "You a gentleman, a courtier, a youth." From all this one can see that 'Colin Clout' represents the author, and that this author is a youth who has recently been involved in a passionate love affair that turned out to be "an unfortunate folly". That this love-affair was with Marguerite of Navarre, which Francis ciphers in this work as Rosalinde, we will see later. The author is

IANVARIE·

Aegloga prima.

ARGVMENT.

IN this firſt Aeglogue, *Colin Cloute*, a ſhepheards boy, complayneth himſelfe of his vnfortunate loue, beeing but newly (as it ſeemeth) enamoured of a Countrey Laſſe called *Roſalind*: with which ſtrong affection being verie ſore trauelled, he compareth his carefull caſe to the ſad ſeaſon of the yeere, to the froſty ground, to the frozen trees, and to his owne winter-beaten flocke. And laſtly, finding himſelfe robbed of all former pleaſance and delight, he breaketh his Pipe in peeces, and caſteth himſelfe to the ground.

COLIN CLOVT.

A Shepheards boy (no better doe him call)
VVhen winters waſtfull ſpight was almoſt ſpent,
All in a Sunne-ſhine day, as did befall,
Led forth his flocke, that had been long ypent,
So faint they woxe, and feeble in the fold,
That now vnnethes their feet could them vphold.

All as the ſheepe, ſuch was the ſhepheards looke,
For pale and wan he was, (alaſſe the while!)
May ſeeme he lou'd, or elſe ſome care he tooke:
VVell couth he tune his Pipe, and frame his ſtile.
Tho to a hill his fainting flocke he led,
And thus him plaind, the while his ſheepe there fed.

Ye gods of loue, that pitty louers paine,
(If any gods the paine of louers pitty:)
Looke from aboue, where you in ioyes remayne,
And bow your eares vnto my dolefull ditty.
And PAN thou ſhepheards God, that once did loue,
Pitty the paines, that thou thy ſelfe didſt proue.

Thou barren ground whō winters wrath hath waſted,
Art made a mirrour, to behold my plight:
Whilom thy freſh Spring flowr'd, and after haſted
Thy Sommer proud, with Daffadillies dight.
And now is come thy winters ſtormy ſtate.
Thy mantle mard, wherein thou maskedſt late.

Such rage as winters, raigneth in my heart,
My life blood freezing, with vnkindly cold:
Such ſtormy ſtoures doe breed my balefull ſmart:
As if my yeeres were waſte, and woxen old,
And yet alaſſe, but now my ſpring be gonne,
And yet alaſſe, it is already donne,

You naked trees, whoſe ſhady leaues are loſt,
Wherein the birds were wont to build their bowre,
And now are cloath'd with moſſe and hoary froſt,
In ſtead of bloſſoms, wherewith your buds did flowre,
I ſee your teares that from your boughes doe raine,
VVhoſe drops in drerie yſicles remaine.

Alſo

(27) **FIRST PAGE TO 'JANUARY', *THE SHEPHEARDS CALENDER*** - Spenser 1st Folio (1611).

also described as being a gentleman and a courtier. None of these descriptions fit the man, Edmund Spenser, who was never a courtier and who certainly was not in his youth at the time that the letters were written.[19] But they certainly do fit Francis Bacon, who was then aged 18, who was a gentleman and a courtier, who had just returned from the court of France where he had been deeply but hopelessly in love with Marguerite de Valois, who was fired with the purpose of enriching the English language and creating an English Renaissance equal to, if not surpassing, anything to be found on the continent, and who had certainly had the experience and studied the necesssary authors to make the 'Spenser' works (and all his other works) possible.

A further help to an identification of the true author, from the glosses and poems alone, is given in the April Æclogue. In the sixth verse we read:-

> "Colin, thou kenst, the Southern shepheard's boye."

Referring to this sentence, the glosse states:-

> "*Colin thou kenst,* knowest. Seemeth hereby that Colin perteyneth to some Southern noble man, and perhaps in Surrye or Kent, the rather bicause he so often nameth the Kentish downes....."[20]

In the September Æclogue a character called 'Roffynn' is mentioned, who is the father of Colin Clout. ("Colin Clout, I wene, be his selfe boye.") In the glosse it says:-

> "*Roffy,* the name of a shephearde in Marot his Æclogue of Robin and the Kinge. Whom he here commendeth for greate care and wise governaunce of his flock."[21]

In these poems and glosses we are told that Colin was the son of (*i.e* "perteyneth to") a 'southern shepherd', that this 'southern shepherd' was a nobleman who was associated with Surrey or Kent, that this nobleman was commended for the great care and wise governing of his 'flock', that he was shadowed under the name of 'Roffy', and that Roffy was associated with Robin and the King (or Sovereign). 'Robin' was the nickname by which the Queen (often referred to as 'King') called her husband and lover, Robert Dudley. In this we have subtle hints pointing out the author's father, the Earl of Leicester, who indeed spent a large part of his time in the Queen's favourite palaces south of the Thames, and in the houses of his friends and relatives directly associated with (or part of) the Areopagus - *viz.* Penshurst in Kent (home of the Sydneys), and Mortlake, Nonsuch, Richmond, *etc.,* in Surrey. Confirming the relationship of 'Robin' to the Queen, the October Æclogue says:-

> "There may thy Muse display her fluttryng wing,
> And stretch her selfe at large from East to West;
> Whither thou list in fayre Elisa rest,
> Or, if thee please in bigger notes to sing,
> Advaunce the worthy whome shee loveth best,
> That first the white beare to the stake did bring."[22]

The glosse tells us not only that by 'Elisa' is meant "our most gratious soveraugn", the Queen, but also:-

THE SHEPHEARDS CALENDER.

CONTAINING
TVVELVE ÆGLOGVES, PRO
PORTIONABLE TO THE TWELVE
MONETHS.

ENTITVLED,
To the Noble and vertuous Gentleman, most
*worthy of all titles, both of learning and chi-
ualrie, Master* Philip Sidney.

AT LONDON,
Printed by *H. L.* for *Mathew Lownes,* and are to
be sold at the signe of the Bishops head in
Paules Church-yard. 1611.

(28) TITLEPAGE TO *THE SHEPHEARDS CALENDER* - Spenser 1st Folio (1611).

"*The Worthy,* he meaneth (as I guesse) the most honorable and renowned the Erle of Leycester, whom by his cognisance[23] (although the same be also proper to other)[24] rather then by his name he bewrayeth, being not likely that the names of worldly princes be known to country clowne."[25]

In his *Prothalamium* Francis sings of swans whom he saw in a vision, floating down the river Thames, and alludes to his royal Tudor ancestry which is, as he says, a house of ancient fame, as no other:-

"At length they all to mery London came,
To mery London, my most kindly nurse,
That to me gave this lifes first native sourse,
Though from another place I take my name,
An house of auncient fame."

Right at the start of the *Shepheards Calender,* in the Argument that commences the January Æclogue, is the plain statement that Colin Clout (the author) is still suffering from his unrequited love, and that Rosalinde in the poem shadows (*i.e.* veils) the name of the one he loves:-

"In this fyrst Æclogue Colin Cloute, a shepheardes boy, complaineth him of his unfortunate love, being but newly (as semeth) enamoured of a countrie lasse called Rosalinde....."[26]

The glosse tells us that:-

"*Colin Cloute,* is a name not greatly used, and yet I have sene a Poesie of M. Skeltons under that title. But indeede the word Colin is French, and used of the French Poete Marot (if he be worthy of the name of a Poëte) in a certain Æglogue. Under which name this Poete secretly shadoweth himself, as sometimes did Virgil under the name of Tityrus, thinking it much fitter then such Latine names, for the great unlikelyhoode of the language.

"*Rosalinde,* is also a feigned name, which, being well ordered, wil bewray the very name of hys love and mistresse, whom by that name he coloureth. So as Ovide shadoweth hys love under the name of Corynna, which of some is supposed to be Julia, the emperor Augustus his daughter, and wyfe to Agrippa.......And this generally hath been the a common custome of counterfeicting the names of secret Personages."[27]

In the 7th verse of the April Æclogue is added further information:-

"But now from me hys [Colin's] madding mynd is starte,
And woes the Widdowes daughter of the glenne;
So now fayre Rosalind hath bredde hys smart,
So now his frend is changed for a frenne."[28]

The glosse tells us:-

"*The Widowes,* He calleth Rosalind the Widowes daughter of the glenne, that is, of a country Hamlet or borough, which I think is rather sayde to colour and concele the person, then simply spoken. For it is well knowen, even in spighte

THE SHEPHEARDS CALENDER.

CONTAINING
TWELVE ÆGLOGVES, PROPORTIONABLE TO THE TWELVE MONETHS.

ENTITVLED,
To the Noble and Vertuous Gentleman, moſt worthy of all titles, both of Learning and Chiualrie,
Maſter Philip Sidney.

LONDON,
Printed by *Bar: Alſop* for *Iohn Harriſon the elder*, and are to bee
ſolde at his ſhop at the ſigne of the golden Anker
in Pater Noſter Row, 1617.

(29) TITLEPAGE TO *THE SHEPHEARDS CALENDER* - Spenser 2nd Folio (1617).

> of Colin and Hobbinoll, that shee is a Gentlewoman of no meane house, nor endewed with anye vulgare and common gifts, both of nature and manners: but such indeede, as neede nether Colin be ashamed to have her made knowne by his verses, nor Hobbinol be greved, that so she should be commended to immortalitie for her rare and singular vertues.....
>
> "*Frenne,* a straunger. The word, I thinke, was first poetically put, and afterwarde used in common custome of speach for forene."[29]

From this we gather that the author's love, shadowed under the name 'Rosalinde' which, "being well ordered," will betray the true name of his love and mistress, is the Widow's daughter of the glen, a gentlewoman of no mean house, and a foreigner.

The phrase "being well ordered" points to an anagram of the name '**Rosalinde**' which, in a reordering of the letters, will reveal the true name. A straight-forward and meaningful anagram of 'Rosalinde' is '**Daisie lorn**'. Daisy, in French, is *Marguerite.* 'Lorn' means 'lost'. Thus 'Daisie lorn' means 'Lost Marguerite', which is an extremely good description of the author's love, whom he has lost. That this is Francis Bacon's love, Marguerite de Valois, is confirmed by the statement that she is the Widow's daughter of the glen, for Valois is not only a district (hence a "borough") of Northern France, but *de Valois* infers 'of the Valley' which is the same as 'of the Glen'. Furthermore, Marguerite de Valois was the daughter of Catharine de Medici, widow of Henri II of France - not just any widow, but the Queen Mother. As such, not only was Marguerite "the Widowes daughter", but she was also "of no meane house" and a foreigner. To add even further detail, Marguerite de Valois was the wife of Henri de Navarre, and thus the Queen of Navarre, even whilst being the object of Francis' love and (for a short time) his lover. This condition explains the pointed comparison with Julia, daughter of Augustus and wife to Agrippa.

Five separate quarto editions of *The Shepheards Calender* were issued, in 1579, 1581, 1586, 1591 and 1597 respectively - all of them published anonymously, even whilst the *Faerie Queene* and other poetic works were being published under Spenser's name from 1590 onwards. Then, when the first folio edition of 'Spenser's' works was published in 1611, *The Shepheards Calender* was included with the other works attributed to Spenser, but without any explanation whatsoever and no apparent signature on its titlepage. That there was confusion (deliberate?) concerning the authorship can be seen as follows.

George Whetstone, a contemporary poet, ascribed the authorship of *The Shepheards Calender* to Sir Philip Sydney in his poem entitled *Sir Philip Sydney, his Life, his valiant Death, and true Virtues* (which was published in 1587, immediately after Sydney's funeral), even though the *Shepheards Calender* is dedicated to Sir Philip himself:-

> "The last Sheppards Calender, the reputed work of S. Phil. Sydney - a work of deep learning, judgment and witte, disguised in Shep. Rules."[30]

William Webbe, in his *Discourse of English Poetrie,* published in 1586, four years before Spenser's name was used on any titlepage, has this to say:-

> "This place have I purposely reserved for one who if not only, yet in my judgment principally deserveth the tytle of the rightest English Poet, that ever I read: that is the Author of the Sheepeheardes Kalender, intituled to the woorthy gentleman Master Philip Sydney, whether it was Master Sp. or what rare Scholler in Pembrooke Hall soever, because himself and his freendes, for

TO HIS BOOKE.

Goe, little Booke: thy ſelfe preſent,
As child whoſe parent is vnkent,
To him that is the preſident
Of nobleneſſe and chiualrie:
And if that Enuy barke at thee,
As ſure it will, for ſuccour flee
Vnder the ſhadow of his wing.
And, asked who thee forth did bring,
A ſhepheards ſwaine ſay did thee ſing,
All as his ſtraying flocke he fedde;
And when his honor hath thee redde,
Craue pardon for thy hardy-head.
But if that any aske thy name,
Say thou wert baſe begot with blame:
For why thereof thou takeſt ſhame.
And when thou art paſt ieopardie,
Come tell me what was ſaid of mee,
And I will ſend more after thee.

Immeritò.

(30) PREFACE TO *THE SHEPHEARDS CALENDER* - Spenser 1st Folio (1611).

what respect I knowe not, would not reveale it, I force not greatly to sette downe."[31]

The *Arte of English Poesie,* published anonymously in 1589 (and later ascribed to 'George Puttenham'), when speaking of the various poets who have arisen in England, and when describing their various characteristics, says:-

> "For Eclogue and pastoral Poesie, Sir Philip Sydney, and Master Challenner, and that other Gentleman who wrote the late *Shepherd's Calender.*"[32]

After the titlepage of *The Shepheard's Calender,* and preceding the Epistle to Harvey, the General Argument and the main poem itself, is a little eighteen-line verse signed 'Immerito', the first seven lines of which are as follows:-

> "Goe, little Booke: thy selfe present,
> As child whose parent is unkent,
> To him that is the president
> Of nobleness and chivalrie:
> And if that Envy barke at thee,
> As sure it will, for succour flee
> Under the shadow of his wing....."

'Unkent' means 'unknown', 'unguessed at'; hence the statement is that the book should present itself to the "president of nobleness and chivalry" with its authorship concealed. The comparison of the book with a child whose parentage is concealed, is a case history drawn from life - Francis' own life. Then follows the distinctly Rosicrucian phrase, "Under the shadow of his wing" - a phrase taken from *Psalms 57* and *91,*[33] and used to mark the Rosicrucian manifestos of the 1610's; and which became, in its Latin form, the well-known motto of the Rosicrucian work ("***Sub Umbra Alarum Tuarum Jehovah***").[34] The motto appeared, at the same time as the manifestos, in an emblem dedicated to Francis Bacon: *viz.* in *Mirror of Modestie* (1618). It also appeared in an associated emblem in Jacob de Bruck's *Book of Emblems* (1616). Both these emblems depict Jehovah as an eagle with protective wings outstretched: cryptically alluding to St. John (whose symbol is an eagle), whose name is derived from the same sacred vowels - I.O.A. - as is *JIHOVAH.*

I.O.A., as in Iona, also means 'Dove' - the Holy Ghost that brings divine Light into the world of material manifestation. Furthermore, as the symbol of the Dove is derived from the Hebraic word *Dôhveh,* meaning rest, strength, security, plenteousness, peace' (which is the description of *I.O.A.*), and as this word is directly associated with *Dohver,* a 'sheepfold', and with *Dôhv,* a 'bear', *I.O.A.* is represented by all these symbols - eagle, dove, sheepfold, bear. *Ursa Major* and *Ursa Minor* are the celestial signs, or symbols, of *I.O.A.*; thus the eagle in the emblems, or *Jehovah* in the motto, can be translated as the Great or Little Bear, referring again to Arcadia and the Arcadians, to the Ancient British race, and in particular to the Dudleys or Tudors.

Following on from what was previously said in the F.B.R.T. Journal I/4, *The Great Vision*, this should make it distinctly clear that the so-called 'Rosicrucian' work that was declared publicly in the opening decades of the 17th century is one and the same with the Baconian work, which Francis - as Fr. C.R.C., or Francis Rosicross[35] - referred to as the Great Instauration, or the General and Universal Reformation of the whole wide world, or the Renewal of all Arts and Sciences. It also becomes clear that the "president of nobleness and chivalry" is none other than God, the Most High, the Almighty and Heavenly Father;

TO HIS BOOKE.

Goe little Booke: thy selfe present,
As child whose parent is vnkent,
To him that is the president
Of noblenesse and chiualrie:
And if that Enuie barke at thee,
As sure it will, for succour flee
Vnder the shadow of his wing.
And, asked who thee forth did bring,
A Shepheards swaine say did thee sing,
All as his straying flocke hee fed:
And when his Honour hath thee read,
Craue pardon for thy hardy head.
But if that any aske thy name,
Say, thou wert base begot with blame:
For why thereof thou takest shame.
And when thou art past ieopardie,
Come tell me what was sayde of mee,
And I will send more after thee.

Immerito.

(31) PREFACE TO *THE SHEPHEARDS CALENDER* - Spenser 2nd Folio (1617).

EMBLEME 6.

NEuer ſhould any thinke himſelfe ſo ſure
Of friends aſſiſtance,that he dares procure
New enemies: for vnprouok'd they will
Spring out of forg'd,or cauſeleſſe malice ſtill.
Elſe, why ſhould this poore creature be purſu'd,
Too ſimple to offend,a beaſt ſo rude.
Therefore prouide (for malice danger brings)
Houſe-roome to find vnder an *Eagles* wings.
You are this *Eagle*,whcih ore-ſhades the *ſheepe*
Purſu'de by *humane wolues*, and ſafe doth keepe
The poore mans honeſt,though might-wronged cauſe,
From being cruſhed by oppreſſions pawes.
Faire Port you are, where euery *Goodneſſe* findes
Safe ſhelter from ſwolne *Greatneſſe*, ſtubborne winds
Eager to drench it: but that feareleſſe reſt
Dwels in your harbour,to all good diſtreſt.
I bid not you prouide,you are compleate,
The good for to protect,or bad defeate.

C 2

(32) ***"SUB UMBRA ALARUM TUARUM"*** ("Beneath the Shadow of Thy Wings") - Rosicrucian emblem assigned to Sir Francis Bacon, from the emblem book, *Mirror of Modesty* (1618). Note that Francis Bacon is likened to the eagle (*c.f.* phoenix) which protects the sheep, like a shepherd, from the human wolves.

and that the book (*The Shepheards Calender*) is, therefore, dedicated and offered to God. Moreover, through the comparison of the book with his own state, Francis was also alluding to his own personal dedication and offering to God - which offering is summed up in his pseudonym, 'Immerito'.

Thus Francis Tudor stamped his first major printed work, marking the actual commencement (at the age of 19) of his planned Rosicrucian work and of the English Renaissance, which he had been designing and testing out in previous years at home and abroad. Part of this work was the enriching of the English language with additional words, majestic sounds and rhythms, and numerical (or cabbalistic) values; so that the English language could become a major sacred language, and a powerful and adaptable vehicle for the holding and communication of knowledge - a tool of divine consciousness. The English Renaissance and the English language did not just happen: they were created. Chaucer began the process; Bacon completed it. Each had help, divine and human, but each was the genius principally responsible.

Francis Bacon devoted every opportunity to study language, customs and habits of every kind of person available. Not only did he read widely, but he also travelled extensively, at home and abroad, as a student and as a messenger in Walsingham's and the Queen's service. In addition, he had the unique opportunity to observe and listen to people from all over England, and from other lands, both in his foster-father's house and at the very 'earthy' Elizabethan court, where a wonderful mixture of characters and dialects was to be found and capable of being studied at leisure - especially in his role as 'Keeper of the door to the Council Chamber'. Right from the start, in the Epistle to the *Shepheards Calender,* 'E.K.' states that his purpose is to first renew and then afterwards to expand and beautify the English language:-

> "...for in my opinion it is one special prayse of many, whych are dew to this Poete, that he hath laboured to restore, as to their rightfull heritage, such good and naturall English words, as hath ben long time out of use, and almost cleane disherited. Which is the onely cause, that our Mother tonge, which truely of it self is both ful enough for prose, and stately enough for verse, hath long time ben counted most bare and barrein of both. Which default whenas some endevoured to salve and recure, they patched up the holes with peces and rags of other languages, borrowing here of the French, there of the Italian, every where of the Latine; not weighing how il those tongues accorde with themselves, but much worse with ours: So now they have made our English tongue a gallimaufray, or hodgepodge of al other speches."[36]

Not only was it words that were the especial concern of Francis, but also the structure and quality of sentences, the rhythms of language, the rules of verse, and so on.

> "Now, for the knitting of sentences, whych they call the joynts and members therof, and for al the compasse of the speach, it is round without roughnesse, and learned without hardnes, such indeede as may be perceived of the leaste, undertsoode of the moste, but judged onely of the learned. For what in most English wryters useth to be loose, and as it were ungyrt, in this Authour is well grounded, finely framed, and strongly trussed up together."[37]

It would appear that Francis Bacon, freshly returned from his sojourn abroad in France, went to Leicester House and enrolled the help of his cousin, Philip Sydney, and their like-minded friends, in carrying out Francis' great scheme. The task of the specially chosen scholar-poets - the Areopagites - was to begin work on the English language, and find ways of using it as a powerful tool for good, "to fashion a gentleman or noble person in vertuous and gentle discipline,"[38] whilst at the same time teaching the Mysteries of life and bringing people to the point of self-knowledge and then divine knowledge. Other scholar-friends provided help in researching, experimenting, teaching and collecting information of all kinds - for the basis of the Pyramid of Philosophy (and thus of the Great Instauration) is the gathering together of a 'History' of all that exists and has been experienced or observed in life, even to the minutest or most common detail - and taking "all knowledge" for their province. Without a great language, none of this would be possible. For patronage he had the help of his father, Leicester (for a few years only), and of his cousins, the Herberts and Sidneys. For guidance he had the help of Dee, and of his Muse, Pallas Athena.

(33) SPENSER MEMORIAL (restored), Poets Corner, Westminster Abbey, restored by private subscription 1778.

(34) SPENSER MEMORIAL (original), Poets Corner, Westminster Abbey, erected by Anne Clifford in 1619, 20 years after Spenser's death - engraving from the 1679 edition of *The Faerie Queene.* The accompanying explanation infers that Robert, Earl of Essex, first provided this "tomb" for "Spencer's learned reliques".

Spenser

The question is, of course, why did Francis Bacon choose Edmund Spenser as a mask to his name, when he was already using pseudonyms anyway? The answer, I think, lies in (a) the need at the time to create even greater security against discovery, by using the names of contemporary personages rather than relying wholly on symbolic pseudonyms that might give the game away too readily, and (b) the intention to lay a fascinating treasure-trail, with all the necessary clues to aid discovery by future generations who are able to notice and prepared to question anomalies - the clearest anomalies occurring in matching the life and qualities of the supposed authors to the works their names are attached to.

Little is actually known about the life of Edmund Spenser. His biography has, like Will Shaxpere's, been constructed mainly out of inferences drawn from the poems ascribed to him and, when the external sources of information present difficulties, they are discarded for what is accepted arbitrarily as 'internal evidence'. The date and place of his birth are not accurately known, nor of his death. The original monument erected to Spenser's memory in Westminster Abbey (1619)[1] states that he was born in 1510 and died in 1596. The restored monument (of 1778) gives different dates - "born in 1553 and died in 1598." The Globe edition (1879) of the 'Spenser' poems says that more modern scholars consider the birthdate of Spenser to have been 1552, deriving their date from an analysis of one of his sonnets, *viz.* sonnet 60, which, they say, is fairly well ascertained as having been composed in the year 1593.[2]

Leaving aside the internal evidence of the poetic works themselves - which is applicable only to the true author and his friends, and may not agree (and indeed does not agree) with documentary facts concerning the life of Spenser - we have the following information, not all of it irrefutable. The names of Edmund Spenser's parents are unknown, although his father is supposed to have been John Spenser, a journeyman tailor in London. He received an elementary education at the Merchant Taylor's School as a "poor scholar". Through the influence of a benefactor he was admitted as a 'sizar', or serving-scholar,[3] to Pembroke Hall, Cambridge, in 1569, where he is again reported as in receipt of relief. (If born in 1510, he would have been 59 years old. If born in 1553 be would have been 16 years old.) He is said to have been industrious and remained at Cambridge until 1576, when he obtained his M.A.. (If true, he may have had a chance to meet Francis and Anthony Bacon, who were at Cambridge 1573-1575.) At Cambridge he met Gabriel Harvey, the friend and tutor of the Bacon brothers in rhetoric and poetry. (Harvey had become a fellow of Pembroke Hall in 1570.) Harvey is said to have found Spenser a post in the household of the Earl of Leicester after he left Cambridge in 1576, in which household he met Philip Sydney. (Sydney had returned from his continental tour in 1575.) Alternatively, Spenser is said to have been introduced to Philip Sydney by Gabriel Harvey, and in turn Philip Sydney introduced Spenser to Leicester. Dr. Fulke, Master of Pembroke College at that time, was chaplain to the Earl of Leicester, and so it is conceivable that he might have obtained a position for Spenser as a clerk in the employment of Leicester. From inferences in writings connected with 'Immerito', it would appear that Spenser was for some time in the north of England (although this might apply to the real author), and by 1579 in the households of Sir Henry Sidney in Kent and the Earl of Leicester in London. There is also a suggestion that Spenser was employed by Leicester as a messenger bearing letters to France in April 1579; also that he was similarly sent to Spain and Rome later that year (November/December), returning to England in April 1580. Whilst he was absent, Francis Bacon published the *Shepheards Kalendar* (5th December 1579).

The first incontestable fact in Spenser's life was his appointment as private secretary to Lord Grey of Wilton in July 1580. Lord Grey had been appointed Lord Lieutenant of Ireland, and he and Spenser went to Ireland forthwith. There are allusions to Spenser having been in Ireland before (which might help to explain this appointment), which could have happened on or soon after he left Cambridge if he entered Sir Henry Sidney's service then. Sir Henry, as Lord Deputy of Ireland at that time, was in Ireland when Spenser left Cambridge.[4] It is postulated that Spenser may have gone to Ireland with Philip Sydney in 1576,[5] perhaps together with Walter Devereux, 1st Earl of Essex, who was sent to Ireland in August/September 1576 as Earl Marshal. (A possible marriage was at that time considered between Philip Sydney and Essex's daughter Penelope, whom he loved - the 'Stella' of his sonnets - but nothing came of this as Essex, who greatly desired that match, died of dysentry soon after his arrival in Ireland, on September 22nd, 1576, before the match could be arranged. It was soon after this that Leicester married Essex's widow, 'Lettice'.)

(In the *View of the Present State of Ireland,* the author, referring either to himself or to Spenser under the pseudonym of 'Irenaeus', states that he was present at the hanging of O'Brien at Limerick, in July 1577. This reference could be to either Edmund Spenser or to Francis Bacon. Certainly we know that Francis, whilst in France, undertook a secret and dangerous journey sometime between July and September 1577, and that he returned from it "safe, sound and in good health, and worthy of" Sir Nicholas Bacon's favour.[6] In *The Great Vision*[7] I suggested that Francis' mission might have been as a special ambassador to his friend, Henri de Navarre, to sue for peace on behalf of the French King. But it is equally likely that Francis may have been sent from France to Ireland as a messenger and for the purpose of gathering intelligence for Walsingham and the Queen.)

Lord Grey stayed in Ireland for two years, during which time Spenser remained his Secretary and obtained rewards for his services. When Lord Grey left Ireland, Spenser chose to stay on, and did so more or less to the end of his life. In 1581, whilst still secretary to Lord Grey, he was appointed Clerk of Degrees and Recognizances in the Irish Court of Chancery in Dublin, a post which he held for seven years (1581-1587), at the end of which time he purchased the Clerkship to the Council of Munster (June 1588) from Ludovick Bryskett.[8] He secured a lease of Abbey lands at Enniscorthy, County Wexford, in 1581, and of New Abbey, County Clare, in 1582. In June 1586 Spenser became an 'undertaker', one of those who undertook to resettle with Englishmen whatever Irish land that was confiscated from the Irish rebels. He was granted 3028 acres of land in County Cork, with a home at Kilcolman Castle (formally made over to him in 1588). In 1589 Sir Walter Raleigh (the principal of the undertakers) visited him there.

Spenser maintained himself at Kilcolman until 1598, when the undertakers were involved in general ruin. During this time he was sued for improperly appropriating some lands (1589-1593). In 1593 he assigned his Clerkship of Munster Council to R. Curteys. On the 30th September 1598, the Privy Council recommended to the Irish Government that Spenser be appointed Sheriff of Cork. But, before the letter arrived, the Munster rising

(35) SIR WALTER RALEIGH (b.1552?-d.1618) - miniature painted (c. 1585) by Nicholas Hilliard.

took place, which resulted in the abandonment of the settlement and the death of many English settlers. Ben Jonson is reported as saying that "Spenser's goods were robbed by the Irish, and his house and little child burnt, he and his wife escaped, and after died for want of bread in King Street; he refused 20 pieces sent him by my lord Essex, and said he was sure he had no time to spend them." Camden wrote that Spenser "had scarcely secured the means of retirement and leisure to write when he was ejected by the rebels, spoiled of his goods and returned to England in poverty, where he died immediately afterwards, and was interred at Westminster near to Chaucer, at the charge of the Earl of Essex, his hearse being attended by poets, and mournful elegies, with the pens that wrote them, being thrown into his grave."

Spenser must have been in Ireland up to at least the 9th December 1598, as Norreys, the President of Munster, "wrote home a despatch about the state of the province. This despatch was sent to England by Spenser....."[9] The report made by Ben Jonson of Spenser's extreme poverty is not entirely accurate, for Spenser is reported to have been in receipt of a small pension[10] from the Queen. Except for a petition to the Queen which Spenser supposedly presented on his arrival in England in late December 1598, and for which service he was paid, this despatch was the last original document which remains about Spenser. Thus he would appear to have remained in Ireland for more than two months after his home was sacked by the rebels. It appears that he died in a tavern in King Street, Westminster, shortly after his arrival in England, on January 16th, 1599. His widow married again sometime before 1603.

Whilst living in Ireland, Spenser is conjectured to have visited England twice, although all the evidence for these visits is based on the poems, their dedications and their registrations. His first visit is deduced to have been from the end of 1589 to some time early in 1591; whilst his second visit hypothetically took place during the whole of 1596 and part of 1597. During the first supposed visit the *Faerie Queene* was entered in the Stationers Company register (1st December 1589); the letter to Raleigh expounding the author's intention in the *Faerie Queene* was written and signed 'Edm. Spenser', dated "23. Januarie, 1589." (*i.e.* 1590 modern style). Shortly afterwards the first three books of the *Faerie Queene* were published by William Ponsonby, together with the letter to Raleigh, six commendatory poems by friends of the poet, and seventeen sonnets addressed by the poet to various illustrious personages (*e.g.* Sir Christopher Hatton, Lord Burghley, the Earl of Essex, Lord Charles Howard, Lord Grey of Wilton, Lord Buckhurst, Sir Francis Walsingham, Sir John Norris, Knight, Lord President of Munster, Sir Walter Raleigh, the Countess of Pembroke, and others - all friends of or people closely connected with Francis Bacon, but barely known, if at all, by Spenser). Then Spenser is supposed to have returned to his home in Ireland in time to write *Colin Clouts Come Home Again,* whose dedication is dated "from my house at Kilcolman, the 27 of December, 1591." Yet the dedication to *Daphnaida* is dated "London this first of Januarie 1591" (*i.e.* 1592 new style) - a practical impossibility in those days if Spenser had been the true author and if the dating from Kilcolman had not been a blind.

Whilst Spenser was back in Ireland, *Complaints,* containing nine poems, was registered for publication on 29th December 1590 and published in 1591. In it were mentioned other works by the author: *viz. Ecclesiastes, The Seven Psalms, Canticum Canticorum, A Sennight Slumber, The State of Lovers,* the *Dying Pelican, The Howers of the Lord, and The Sacrifice of a Sinner.* The elegy, *Daphnaida,* was also published that year. The *Amoretti* and *Epithalamium* were registered on the 19th November 1594, and published in 1595. *Colin Clouts Come Home Again* was republished in 1595, with several additions made to the original version.

Complaints.
Containing ſundrie ſmall Poemes of the Worlds Vanitie.
VVhereof the next Page maketh mention.
By Ed. Sp.
Caroli Bernard
LONDON.
Imprinted for *VVilliam Ponſonbie*, dwelling in Paules Churchyard at the ſigne of *the Biſhops head.*
1591.

(36) TITLEPAGE TO *COMPLAINTS* (1591) by Ed. Sp.

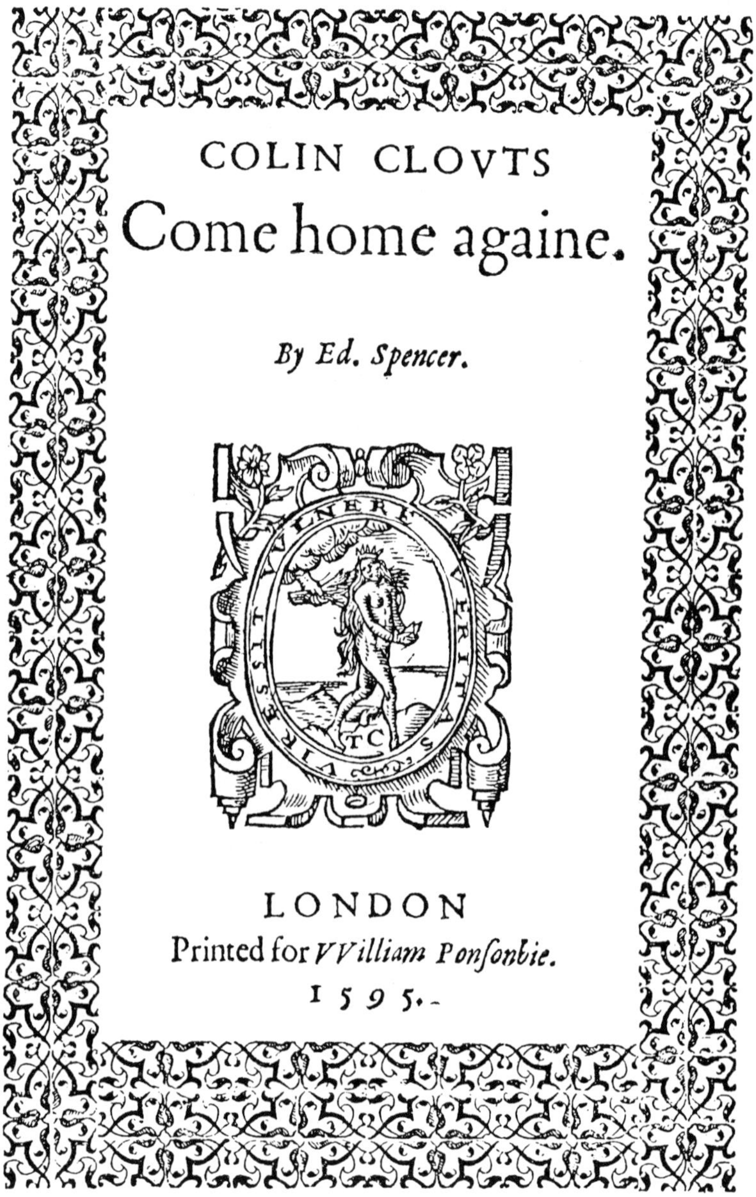

COLIN CLOVTS
Come home againe.

By Ed. Spencer.

LONDON
Printed for *VVilliam Ponſonbie.*
1 5 9 5.

(37) TITLEPAGE TO *COLIN CLOUTS COME HOME AGAIN* (1595) by Ed. Spencer.

Spenser is supposed to have returned to England again at the end of 1595. Part II of the *Faerie Queene* (the 4th, 5th and 6th books) was registered on the 20th January 1596 and published later that year. The *View of the Present State of Ireland,* which Spenser is supposed to have brought over with him from Ireland (and which probably was written by Spenser), was not registered until April 1598, and then only conditionally.[11] But Spenser was back in Ireland by then. During this second visit to England, *Hymns to Heavenly Love and Heavenly Beauty,* and the *Prothalamium* - the so-called last works of 'Spenser' - were supposedly written (although two of the *Hymns* had been in circulation for some years, and were composed in the author's "youth"). The dedication to the *Hymns* is dated from Greenwich, September 1st, 1596. The beautiful *Hymns* are an exceptionally clear statement of the poet's Neoplatonism, abounding in references to Plato and to Platonic philosophy, dealing with heavenly love and the inner beauty of the heart, and with the (Hermetic) ascent and descent through the spheres of the universe.

Until the publication of the first three books of the *Faerie Queene* in 1590, Spenser's name was not associated with the 'Immerito' works. There are so many anomalies between the life of Spenser and the allusions and statements in the poems, that they are mostly irreconcilable. But, when related to Francis Bacon, the content of the poems, together with their manner and dates of publication, fit like a glove to a hand. The illustrious people mentioned in the poems and letters, with whom 'Spenser' had deep and long-lasting friendships or aquaintance - most of them noblemen and ladies at court - were part of Francis Bacon's life, but not of Edmund Spenser's. For instance, Spenser could hardly have known Robert Devereux, the 2nd Earl of Essex, with whom the poet boasts a particularly long and special friendship; but Francis Bacon did - of course, for Essex was his blood-brother, and they had much to do with each other!

Harvey's manuscript 'letter-book' is very helpful, and gives numerous references bearing upon the identity of 'Immerito'. For instance:-

(i) 'Immerito' was "a Hertfordshire gentleman". This fits Francis Bacon, whose foster-father's country seat, Gorhambury, was in Hertfordshire. It does not fit Edmund Spenser, who is supposed to be a Londoner.

(ii) Harvey referred to 'Immerito' as "Illustrious Anglo-francitalorum." This certainly would seem to refer to an Englishman who had lived in France and had a love for France as well as his own country. This definitely applies to Francis Bacon, but not to Spenser.

(iii) Harvey addressed 'Immerito' with much deference and politeness: *viz.* "Magnifico Signor Benevolo," "Your delicate Mastershipp," "My younge Italianate Seignoir and French Monsieur," "Good-natured and worshipful young gentleman," "....take my leave of your Excellencies feet and betake your gracious Mastershipp," *etc...* This deference simply would not have been the attitude for Harvey to take towards an ex-sizar of his College, in which he was a tutor; not even to an employee of the Earl of Leicester. But it is an attitude that would have been consistent with the position of a young gentleman such as Francis Bacon, particularly if Harvey had known of Francis' real parentage.

(iv) 'Immerito' is identified as a lawyer, a gentleman, a courtier and a youth. This was true of Francis Bacon, but not true of Edmund Spenser:-

> "So trew a gallant in the Court, so toward a lawyer and so witty a gentleman."

(38) ROBERT DEVEREUX, 2nd Earl of Essex, aged 21/22 years (b.1566-ex.1601) - miniature painted *(c. 1588) by Nicholas Hilliard.*

"We are yet to take instructions and advertisements at your lawiers and courtiers hands, that are continually better trayned and more lively experienced therein than we University men are."

"So honest a youth."

"Good lord, you a gentleman, a courtier and youth."

(v) 'Immerito' wrote his letters sometimes from Leicester House, sometimes from the Court at Westminster. He had private audience with the Queen, which would have been a highly improbable occurrence for an ex-sizar and clerk, but, as we know, a normal occurrence for Francis Bacon:-

"Your desire to hear of my late being with her Majesty must die in itself."

(vi) 'Immerito' wrote his poems and their accompanying letters for the nobility, most of whom he knew well (but Spenser had no chance of knowing the majority of them, both because he was in Ireland most of his life and also because he was not a member of the aloof and privileged aristocracy); and he wrote them for purposes other than for "gaine and commodity":-

"First I was minded for awhile to have intermitted the uttering of my writings leaste by over much cloying of their noble ears I should gather a contempt of myself or else seem rather for gaine and commoditie to doe it."

However, a variety of ciphers and other clues help to tell the true story and authorship. For instance:-

The *Faerie Queene,* Part I (1590), is titlepaged to 'Ed. Spenser'. The *Complaints* (1591) is titlepaged to 'Ed. Sp.' This pointed difference in the two abbreviated versions of the author's pseudonym on the two consecutive titlepages clearly suggests the presence of cipher signatures which underlie the pseudonyms, wherein each confirms the other. '**Ed. Spenser**' renders, in the Simple English Cabbala, **100** (*i.e.* E.D.S.P.E.N.S.E.R = 5.4.18.15.5.13.18.5.17 = 100).[12] **100** is the cipher for '**Francis Bacon**' in the Simple Cabbala (*i.e.* F.R.A.N.C.I.S + B.A.C.O.N = 67 + 33 = 100). Instead of repeating this same cryptic signature in the following year, Francis Bacon contrived to indicate his authorship in a similar but not identical vein, so as to confirm that the cipher and signatures are intended; hence '**Ed. Sp.**' renders, in the same Simple Cabbala, **9** + **33** (*i.e.* E.D + S.P = 5.4 + 18.15 = 9 + 33). **9** + **33** stands for '**I, Bacon**' (*i.e.* I = 9; BACON = 33).

This particular method of enciphering signatures was used by Francis Bacon over and over again throughout his life, and forms a signal and a lead in to a great deal of interesting, but otherwise concealed, information. It should be noted, for instance, that 33 is (to the nearest whole number) one third of 100, and 67 is two-thirds of 100, *etc.*. Similarly, it is valuable to note that 100 is called a 'Century', from the Latin *Centuria,* which word Bacon uses to denote each chapter in his books. The word 'chapter' is derived from *chapiter,* which means 'capital' or 'head', and it is not by chance that the words 'century' and 'chapter' are intimately associated in this way. The Roman numeral for 100 is the letter **C**, and, as explained elsewhere in this book, the "**C**-men" (a play on the words 'Seamen', 'See-men' and 'Semen') are the Rosicrucian Cabbalists, who are the great Navigators or Phoenicians. They are also the top intelligencers, for Intelligence is both their game and also their Muse (*i.e.* Pallas Athena).

The works of 'Spenser' and of Sir Philip Sidney are cabbalistic and magical: magical, in the sense of using poetry and music to control passions, exalt the mind and bring all things into perfect order and harmony - as did Orpheus, David, Typander, Amphion, Pythagorus and Timotheus; and cabbalistic, in that their works are based upon cabbalistic knowledge and principles, such as geometry and numerology coupled with the power of imagination and symbolic images. By means of numbers, correctly used, divine principles are discovered and angelic forces invoked (*i.e.* attuned to). Angels are spirits, or intelligencies, which are purely and simply qualities or aspects of Universal Intelligence. This is what Dee was trying to do, only Dee was not gifted as a poet - but his special pupils were.

The totality of Universal Intelligence is summed up in the number 100, which is 10 x 10 (or 10 squared). 10 is the number of Divine Wisdom, which is composed of ten Principles of life (or light) - neither more nor less, for God is the Three and the Seven. To square the Wisdom is to understand it, hence 10 squared is the cipher of Holy (*i.e.* complete) Intelligence, and hence the 'magic square' of Universal Knowledge. Furthermore, **C** is the third letter of the trinity **ABC**, and denotes the Third Person of the Holy Trinity, which is the Holy Intelligence or Ghost - Immerito's Muse:-

"Lo, I the man, whose Muse whilom did mask,
As time her taught, in lowely Shepheards weeds,
Am now enforc't a far unfitter task,
For trumpets stern to change my oaten reeds,
And sing of Knights and Ladies gentle deeds;
Whose praises having slept in silence long,
Mee, all too meane, the sacred Muse areeds
To blazon 'broad, amongst her learned throng:
Fierce warres, and faithful loves, shall moralize my song.

"Help, then, O holy Virgin, chiefe of nine,
Thy weaker novice to performe thy will:
Lay forth out of thine everlasting scrine
The antique rolles, which there lie hidden still,
Of Faerie Knights, and fairest *Tanaquill,*
Whom that most noble Briton Prince so long
Sought through the world, and suffered so much ill,
That I must rue his undeserved wrong:
O! help thou my weake wit, and sharpen my dull tongue.

"And thou most dreaded impe of highest *Love,*
Fair Venus sonne, that with thy cruell dart
At that good Knight so cunningly didst rove,
That glorious fire it kindled in his hart,
Lay now thy deadly Heben bowe apart,
And with thy mother milde come to mine ayde:
Come both, and with you bring triumphant *Mart,*
In loves and gentle jollities arrayd,
After his murdrous spoiles and bloody rage allayd.

"And with them eke, O Goddess heavenly bright,
Mirrour of grace and Majestie divine,
Great Lady of the greatest Isle, whose light
Like Phoebus lampe throughout the world doth shine,

Shed thy fair beames into my feeble eyne,
And raise my thoughts, too humble, and too vile,
To think of that true glorious type of thine,
The argument of mine afflicted stile;
The which to heare, vouchsafe, O dearest dread a-while."[13]

(39) ST. GEORGE AND THE DRAGON - woodblock emblem from the 1590 edition of *The Faerie Queene,* dedicated to Queen Elizabeth.

(40) MARY HERBERT (*née* SYDNEY), Countess of Pembroke, aged 57 years (b.1561-d.1621) — engraving (1618) by Simon van de Pass.

Mary was born the daughter of Sir Henry Sidney and Mary (*née* Dudley). She was sister to Philip, Robert and Thomas Sidney, and the niece of Robert Dudley, Earl of Leicester and Ambrose Dudley, Earl of Warwick. Francis Dudley/Tudor (adopted Bacon) and Robert Dudley/Tudor (adopted Devereux) were her natural cousins. William Herbert, 3rd Earl of Pembroke, and his brother Philip Herbert, Earl of Montgomery (and later 4th Earl of Pembroke) were her two sons, who were immortalised in the 1st Shakespeare Folio as "the Most Noble and Incomparable Paire of Brethren".

(41) HENRY HERBERT, 2nd Earl of Pembroke (b.1534-d.1601) - line drawing by Simon Passe and engraved for Holland's *Herwologia* (1620).

Henry Herbert succeeded his father as Earl of Pembroke in 1570. He married Mary Sydney (his third wife) in 1577. In 1586 he succeeded his father-in-law, Sir Henry Sidney, as President of Wales. His family home was at Wilton House near Salisbury in Wiltshire, which became a haven for the *Areopagitæ* and other scholar poets and scientists whom he and his wife patronised. From 1592-1600 he maintained a company of actors (Pembroke's Men), who included Shakespeare plays in their repertoire.

(42) SIR PHILIP SYDNEY, aged 23/24 years (b.1554-d.1586) - oil painting (c. 1577) by unknown artist. (See illustration no. 19.)

Euphues

During August 1579 a certain matter exploded like a time-bomb in Leicester's household, which affected many people connected with him. In January Jean de Simier, the close friend of and Master of the Wardrobe to the Duc d'Anjou, had been sent over to England by the Duc in order to 'court' the Queen on his behalf and to prepare for his own visit to England. Elizabeth had renewed her pretended courtship of this younger brother of the French King in the summer of 1578, for political advantages, and was still 'playing the game'. Simier was well liked by the ladies at the English Court, and he began to make such an impression upon the Queen that Leicester became jealous and then extremely anxious that perhaps a royal marriage might become a reality if his own secret and bigamous marriage with Lettice Knollys became known. If this happened, Leicester would inevitably lose all his power and status with the Queen that he still enjoyed. He might even be ruined. As a result Leicester became intolerant of and exceedingly rude to Simier, doing his utmost to ruin the courtship; and it was this very reaction which brought about his own undoing.

In April discussions were begun seriously on arranging for Anjou to visit the Queen in England, plus tentative negotiations for marriage terms. These discussions with the Queen and her councillors took place at Leicester's house at Wanstead. Subsequently Elizabeth signed the necessary passport for Anjou to enter the country later that summer. Leicester's reaction was to retreat to his own house and retire to bed in a sulk, feigning illness.

Whilst the Earl of Leicester was absent from the Court (which was then at Greenwich) in this great sulk, an assassination attempt was made on Simier. Rightfully or wrongfully, Simier believed that Leicester was behind this attempt on his life, and so he counter-attacked. Early in July he had discovered Leicester's secret. So, what Simier did was to go straight to the Queen and tell her about Leicester's marriage with Lettice, giving the Queen sufficient evidence that it had taken place. The eruption was colossal! Elizabeth was deeply hurt, both in her heart and in her pride. In a burst of hysteria and fury, she ordered Leicester to be sent to the Tower. This severe and rather unwise action was somewhat modified by the counsel of the Queen's friend and Lord Chamberlain, the Earl of Sussex, who pleaded that such a course might do her and the country irreparable damage. Leicester was instead banished from her presence and required to place himself under house-arrest in the Tower Mireflore - a little tower built by Humphrey of Gloucester on a mound in Greenwich Park. Whilst incarcerated there it was given out that he was ill and taking physic, and as a result could see noone.

After a few days Elizabeth relented somewhat, and Leicester, although still banished from the Queen's presence, was allowed to retire to his home at Wanstead. As for Lettice, she was banished from Court for evermore, and the whole of Leicester's household lived for a while under a heavy cloud.

This action of Leicester, in marrying again bigamously and the new marriage becoming publicly known (whilst his prior marriage to the Queen remained a well-kept secret), deeply affected the future of Francis and his blood-brother, Robert (Devereux). It effectively placed the Queen in a moral dilemma and political straitjacket, for from that moment on any choice still left to her of whether to declare her marriage to Leicester or not, and hence whether to declare her natural-born sons as her rightful heirs, was seemingly removed from her. Although she had seen to it that the law of the land had been adjusted to allow her to declare any natural-born child of hers (whether born in wedlock or not) heir to the throne, yet her inner conditioning coupled with the outer image being built of her as the Virgin Queen probably made this choice a nightmare to her. To the end of her life she kept her secret, unable to find the strength to declare the truth openly and so choosing instead to maintain the image of an unmarried, virgin maid. Probably the other choice was too much for her to face alone. No wonder Francis referred to his mother as "the alone Queen", and no wonder he and Robert bewailed their fate, never being quite sure of their royal mother's intentions towards them and so never being able to build their lives on a firm social and political footing during her reign. This is possibly why Francis, as 'Immerito', named Leicester as the causer of his care and pain. (See *Virgils Gnat.*)

As for Francis' cousin, Mary Herbert (Lady Pembroke, *née* Sidney), she found the situation so distressing that she retired to Penshurst Place, her parent's home in Kent, relinquishing her apartments at Court. Later in the summer she went to live at her husband's great country house of Wilton, in Wiltshire, where she became pregnant with her first child.

Eventually the Queen's wrath subsided and she gave out that Leicester was no longer forbidden to appear at Court in his usual place, provided that he appeared publicly as a man humbled. But it took him a long time to swallow his own pride and recover from the hurt he had given to the Queen, his first and still rightful wife. In fact Leicester never fully recovered his standing with Elizabeth, which is hardly surprising, although the Queen appeared to forgive him and to continue loving him.

Anjou's visit duly took place, from the 17th to 29th August 1579. The Queen attended him with gaiety and fervour, appearing enchanted with this lively and fascinating French prince; so much so that when Anjou left England a marriage between him and the Queen did indeed seem possible. But it was not a popular match with a large section of the English government and people. Leicester, Sir Philip Sydney, Sir Francis Walsingham, their families and friends, and indeed the majority of parliament, the mass of the clergy and the Protestant population of England who had not forgotten St. Bartholomew's Eve, the loss of Calais and the treachery of the Houses of Guise and Valois, were very much against such a marriage taking place. Many preached openly against the marriage.

Philip Sydney, who was held in high esteem by Elizabeth as well as by the country, took the risk of writing plainly to the Queen about this issue, on behalf of all English Protestants and at the request of Leicester and Walsingham. Elizabeth is said to have wept when she read Sydney's letter. He for his part diplomatically retired from Court for a whole year, probably first going to live at his parent's home, Penshurst. Then, in March 1580, he went to stay for the best part of a year with his sister, at her husband's great house in Wiltshire. It was at Wilton, in the company of his sister Mary, that Philip Sydney began to write his epic, *The Countess of Pembroke's Arcadia.* Also, with the help of his sister, he began to translate the *Psalms* into English.[1] It is probably from that year onwards that both Mary and Wilton House became famed for patronage of poets and men of learning. At that time

VIRGILS
GNAT.

LONG SINCE DEDICATED
To the moſt noble and excellent Lord, the Earle
of Leiceſter, deceaſed.

(*.*)

VVRongd, yet not daring to expreſſe my paine,
To you (great Lord) the cauſer of my care,
In clowdie teares my caſe I thus complaine
Vnto your ſelfe, that onely priuie are:
    But if that any *Oedipus* vnware,
Shall chaunce, through power of ſome diuining ſpright,
To read the ſecret of this riddle rare,
And knowe the purport of my euill plight,
    Let him be pleaſed with his owne inſight,
Ne further ſeeke to gloſe vpon the text:
For griefe enough it is to grieued wight
To feele his fault, and not be further vext.
    But what-ſo by my ſelfe may not be ſhowne,
    May by this *Gnats* complaint be eaſily knowne.

(43) DEDICATION PAGE TO *VIRGILS GNAT* - Spenser 2nd Folio (1617).

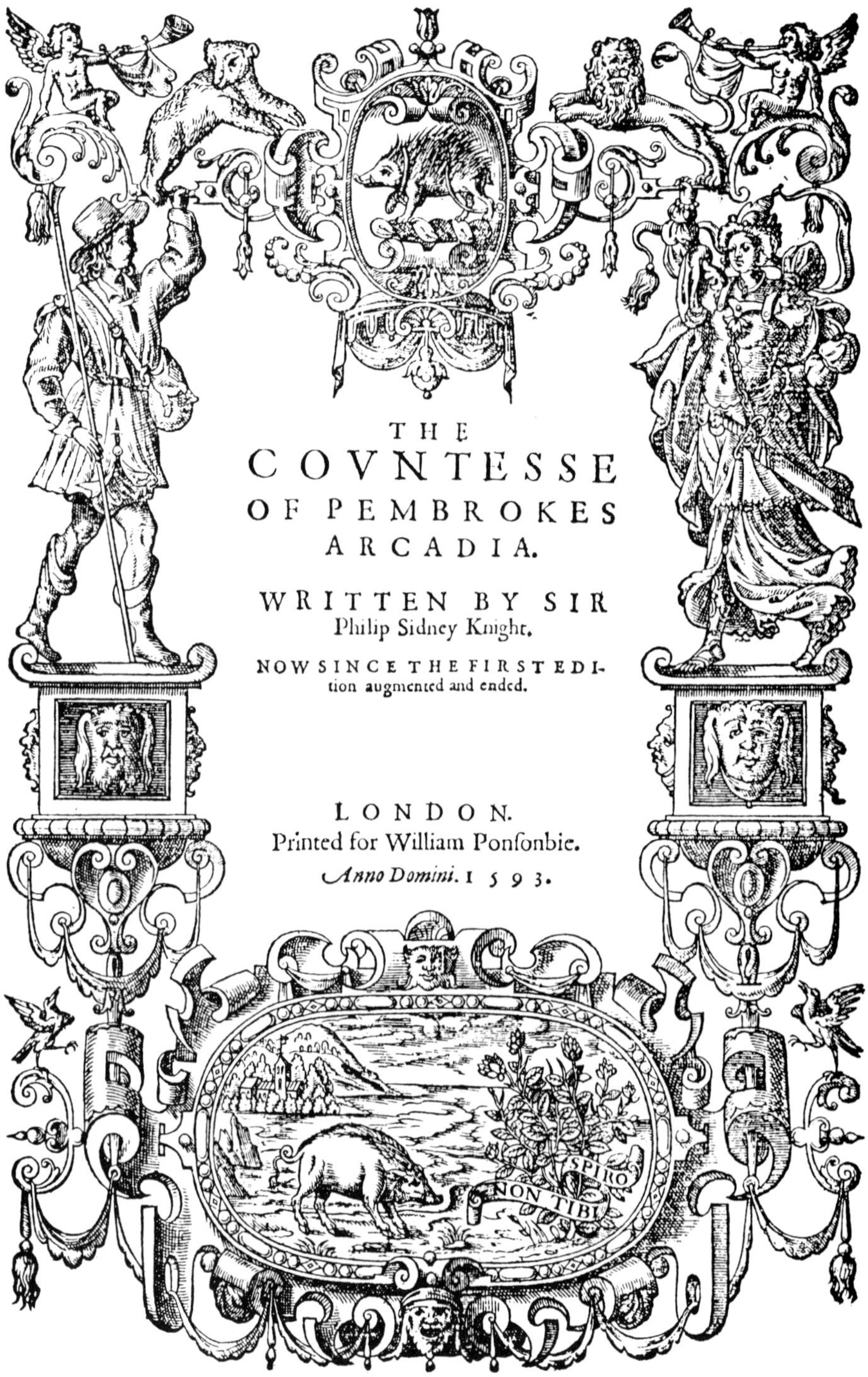
THE
COVNTESSE
OF PEMBROKES
ARCADIA.

WRITTEN BY SIR
Philip Sidney Knight.

NOW SINCE THE FIRST EDI-
tion augmented and ended.

LONDON.
Printed for William Ponſonbie.
Anno Domini. 1593.

SPIRO NON TIBI

(44) TITLEPAGE TO *THE COUNTESS OF PEMBROKES ARCADIA* (1593) by Sir Philip Sydney.

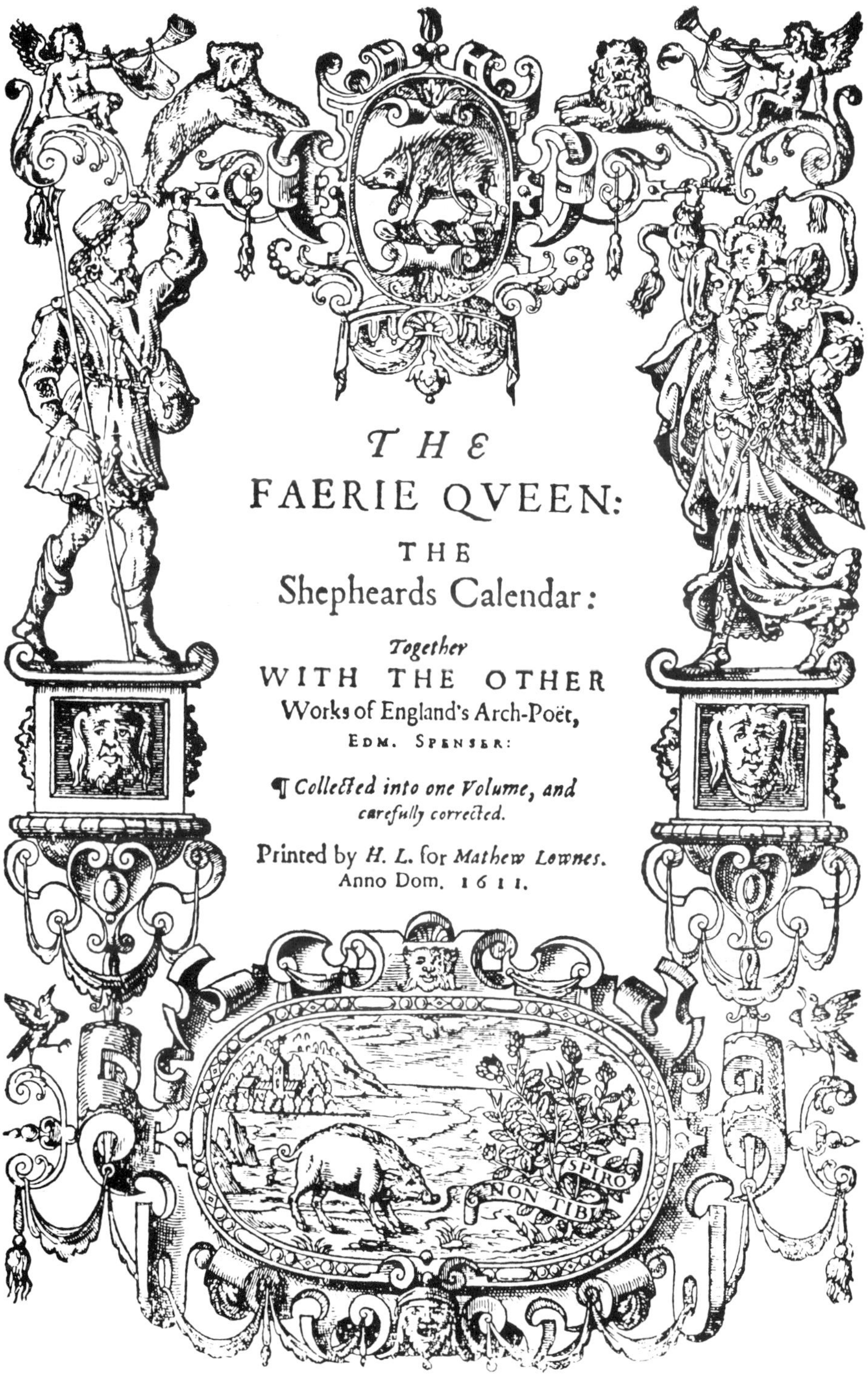

THE

FAERIE QVEEN:

THE

Shepheards Calendar:

Together

WITH THE OTHER

Works of England's Arch-Poët,

EDM. SPENSER:

¶ *Collected into one Volume, and carefully corrected.*

Printed by *H. L.* for *Mathew Lownes.*

Anno Dom. 1611.

SPIRO NON TIBI

(45) TITLEPAGE TO *THE FAERIE QUEENE, etc.* - Spenser 1st Folio (1611).

Wilton boasted a splendid library, which included much Italian literature. Aubrey reports in his *Brief Lives*:-

> "In her time Wilton House was like a college, there were so many learned and ingenious persons. She was the greatest patroness of wit and learning of any lady in her time. She was a great chemist and spent yearly a great deal in that study.....
>
> "This curious seat of Wilton and the adjacent countryside is an Arcadian place and a paradise. Sir Philip Sydney was much here....."[2]

Philip Sidney arrived at Wilton in time for the birth of his sister's first child. William Herbert was born on 8th April 1580 and, fortunately, the Queen agreed to act as godmother to the child, with Leicester and Warwick, great-uncles to the child, being the godfathers. (Anne, Countess of Warwick, stood proxy for the Queen, and Philip Sidney for Leicester, at the baptism.) William grew up to become the 3rd Earl of Pembroke and Lord Chamberlain to James I; whilst his brother Philip, born four years later, became Earl of Montgomery. These two brothers, sons of Francis' dearly loved cousin, Mary, were children with whom Francis had a lot to do during their growing up years. They were the "most noble and incomparable paire of brethren" to whom Francis dedicated his 1st Folio of Shakespeare Plays in 1623.

As to the royal marriage, the Queen's Council, although mostly hostile to the proposals, refused to advise the Queen one way or another, wishing to displease neither the Queen nor the country. Elizabeth was not amused by this lack of positive recommendation one way or the other, and for several weeks she stormed at them for not giving her definite support of the marriage - for without such support it would have been far too dangerous for her to proceed further. Eventually, in November of that year, she sent a quorum of her councillors (excluding Leicester) to talk marriage terms with Simier; but she made Simier agree, on Anjou's behalf, to a moratorium of two months in the marriage treaty, so that during that time she could try to win popular support for the match. But by then, in effect, she had already decided to proceed no further, or at least to procrastinate. Not until 1582 did she revive her affair with her 'Frog' (as she called Anjou), and then only because the Catholic powers had again become more menacing.

Before she knew of Leicester's secret marriage with Lettice, Elizabeth probably never intended to marry Franois, Duc d'Anjou, but only to play him along for the sake of politics; and she was probably advised in this way. But it would seem that the discovery of her husband's bigamous marriage very nearly threw her into the arms of Anjou out of spite, or despair, or both. The situation was clearly dangerous, and could have led to a bloody and horrendous civil war if the marriage with Anjou had taken place. If that had occurred, the English Renaissance would never have happened; for, as Francis so rightly said, "the arts flourish only in times of peace."

Whilst all this was going on Francis not only completed and saw to the publication of *The Shepheards Calendar,* but he also published some other related works that he had composed before he had gone to France, whilst he was there and after he had returned. These were printed under other pseudonyms. The first of these publications was *Euphues Anatomy of Wit,* which he referred to as "my first counterfeite" (*i.e.* dissembling),[3] and which came out in 1579. This was followed the next year by *The Shepheards Calendar* and *Euphues and his England,* the latter containing a clever commentary upon the English court as the author found it on his return from France, and concluding with high praise of the Queen. *Euphues Anatomy of Wit* (1579) was published anonymously.[4] *Euphues and his England* (1580) was titlepaged to John Lyly, another mask.[5]

Euphues Anatomy of Wit had been registered and licensed for publication on December 2nd, 1578. In *Eupheus and his England* the author states that the *Anatomy of Wit* was mainly autobiographical and that "it was sent to a nobleman to nurse, and was hatched in the hard winter with the Alcyon." He also writes, at the end of the *Anatomy,* "I have now finished the first part of *Euphues,* whom now I left ready to cross the seas to England." From this we can infer that, having written the first part of the work in France, Francis gave it to a nobleman friend to look after and take to England. There, in due course, it was registered on behalf of the author in the winter of 1578, whilst Francis was still enjoying his life in France before it was rudely interrupted in the spring of the following year, and before the Leicester storm blew up. When Francis returned to England he completed the second part of the work and saw both parts through the presses, one after the other.

According to *The Schoolmaster,* written by Francis' and the Queen's old tutor, Sir Roger Ascham, '**Euphues**' means "he that is apt by goodness of wit and appliable by readiness of will to learning." The name has an affinity with Orpheus, the founder of the Orphic Mysteries, and of the Society of the Rose (instituted in order to create a high understanding between people in which art would flourish naturally and war would become impossible). It also has an affinity with Euphorbus, a Trojan hero of whom Pythagorus claimed to be a reincarnation. As if to confirm these allusions, Euphues is said to have had his cell on Mount Silexedra. *Silex* means 'a hard stone', 'crag', 'rock', 'cliff', and *Silexedra* means 'a stone Seat' - hence Mount Silexedra refers to such a place as the Acropolis of Athens, or Mount Parnassus, sacred to Apollo, Athena, Dionysus and the Muses. It should be compared with *Silentes,* 'the Silent Ones', the name given to the Pythagoreans; and with Silenus, the tutor and attendant of Dionysus.

Euphuism was the ebullient but learned style of writing of the young Francis Bacon. He used it in his early works, and it forms the principal style of his Word Cipher. It became fashionable in literature and polite conversation in 16th century England. That is to say, this difficult and high-flown style of writing (which is very much the style of a young but highly vivacious and educated young man, who is well aware of his own worth) was adopted by a number of writers whose works appeared shortly after publication of the *Anatomy of Wit* and *Euphues and his England.* These authors (as one might expect, since Euphuism is so difficult to emulate) were in fact one writer, Francis Bacon, writing under various masks - Lyly, Greene, Lodge, Watson, Gosson and Rich. In this way he released his own writings anonymously, without suspicion, and in a manner which made it seem as if there was a huge literary outburst by many talented people. The effect of this was, as intended, to make England (if not all Europe) excited and interested in what was going on and in the ideas that were being broadcast.

As with Edmund Spenser, the biographical details of John Lyly's life are not altogether certain, particularly his earlier years. Much has been attributed to him from the internal evidence of the 'Lyly' works themselves, and therefore may be misleading or, indeed, irreconcilable (as with the statement that the first part of *Euphues Anatomy of Wit* was written abroad, when the person, John Lyly, did not travel abroad). Lyly's date of birth is not known. From records at Magdalene College, Oxford, the Lyly in question is thought to have been one John Lylie who had matriculated there in 1571.[6] (However, the surname, 'Lylie,' in its various forms of spelling, is fairly common at the college.) This particular Lylie was, when matriculated, described as "plebeii filius." His family is described as coming from Hampshire, his father as being a diocesan official in Canterbury, and his grandfather as being a high master of St. Paul's School, Canterbury. In 1574 he petitioned Lord Burghley to be made a fellow. In 1575 he took his M.A., and in 1584 owed to the bursars of the college £1.3s.10d. for his share of the college provisions, "pro communis et batellis."[7] But other biographers say that Lyly stayed at Oxford until 1573 when, having failed to obtain a fellowship, he went to London, living in the Savoy. After leaving Oxford he is thought to have entered the service of Lord Delawarr. By 1580 Lyly was serving Edward de Vere, the Earl of Oxford. Soon afterwards he took up the management of a company of boy actors under Oxford's name - a combination of St. Paul's Boys and the Children of the Chapel Royal. In 1583 Lyly married, and from about June 1583 Lyly leased the so-called Blackfriar's Theatre - a 60 by 20 feet rectangular room in the dissolved Blackfriars monastery, on the edge of the Thames, where the various properties for use in the Court revels and interludes had been kept from the time of Edward VI.[8] There the boys began to rehearse the 'Lyly' plays, which were then performed at Court.[9] (Later, in 1596, the Blackfriars Theatre was built.) In July 1584 Lyly was in prison for debt. All the 'Lyly' plays, except *The Woman in the Moon,* were presented by Paul's boys, either alone or with the Chapel children. After the inhibition of Paul's boys in 1590, Lyly's fortunes rapidly declined. He is thought to be the same John Lillie who became M.P. for Hendon in 1589, Aylesbury in 1593, Appleby in 1597, and Aylesbury in 1597. It is also assumed that he was hoping for the reversion of the Mastership of the Revels whenever Edmund Tylney (appointed 1579) should retire. But Tylney held the post until 1610, and the reversion was granted to George Buck in 1597 in anticipation of Tylney's future retirement. Lyly wrote two petitions to the Queen and Robert Cecil, supposedly in 1598 and 1601, bitterly complaining of having wasted ten years in hanging about the court in hope of preferment. "John Lyllie, gent," was buried on November 30th, 1606, in the grounds of St. Bartholomew's.

Under the mask of 'Lyly', Francis wrote two major novels (the *Euphues* series) and eight dramas (or comedies). The dates of writing of the plays remain conjectural, although it is thought that they were probably all written before 1595. The following is a list of the 'Lyly' works:-

	TITLE	WRITTEN	PUBLISHED	TYPE
1.	*Euphues Anatomy of Wit*	1576-1578	1579	novel
2.	*Euphues and his England*	1579-1580	1580	novel
3.	*Campaspe*	1583	1584	play
4.	*Sapho and Phao*	1583	1584	play
5.	*Gallathea*	1584	1592	play
6.	*Endimion*	1586-1587	1591	play
7.	*Love's Metamorphosis*	1589	1601	play
8.	*Midas*	1589	1592	play
9.	*Mother Bombie*	1589-1590	1594	play
10.	*The Woman in the Moon*	1594-1595	1597	play
11.	*Pappe with an Hatchet*	1589	1589	pamphlet

Of these, all but the last are in prose. The plays were all what are known as Court Comedies, and were performed before the Queen and her Court. Of the above, Nos. 1, 3, 4, 5, 6, 8, 9 and 11 were first printed without any author's name to them. The second edition of No.1 and the first edition of No.2 were printed as by 'John Lyly, Master of Arts.' No.11 was attributed to 'Lyly' by Gabriel Harvey, and Nos. 7 and 10 were first printed as by 'Lyly' in the years 1597 and 1601. The pamphlet is attributed to 'Lyly' because of style, and is one of a set of pamphlets written by 'Lyly' and 'Nash' in reply to the Martin Marprelate pamphlets of 1589-1590.[10]

The *Euphues* novels established the style of Euphueism as the "new English." Edward Blount, 'Lyly's' first editor, reported in 1632 (in the *Sixe Court Comedies*), "that Beautie in Court, which could not Parley, *Euphueisme,* was as little regarded; as shee which now there, speakes not in French." Most of the plays - sophisticated and witty comedies - are derived from classical history or legend, and the author shows familiarity with passages from Ovid, Virgil, and Cicero. These comedies, although they included lyrics, were the first to be written in prose and paved the way for the later 'Marlowe' and 'Shakespeare' productions. *Campaspe* was derived from Pliny's Natural History, *Sapho* and *Gallathea* from Ovid. At least four of the comedies were carefully composed eulogies of the Queen's virtue, beauty, chastity and wisdom. *Euphues and his England* concludes with a similar panegyric, as does the anonymous play, *The Arraignment of Paris* (a pastoral performed before the Queen by the same children, sometime between 1580-1584). The Ephœbus passages of *Euphues Anatomy of Wit* were almost entirely translated from Plutarch on the subject of Education, whilst, besides using Plutarch, large passages are borrowed from the Spanish writer Guevara. Not only is the writing ebullient, but it is filled with seemingly unnecessary puns. (But, in fact, the use of puns is one particular way in which to introduce and teach new English words and phrases, and to make them stick in the minds of people.) These same puns also "bubble up" in 'Spenser,' 'Nash,' 'Kyd,' 'Greene,' 'Peele' and 'Shakespeare.'[11]

> "Anyone carefully reading the works attributed to 'Lyly' will find in them evidence of wide and copious reading combined with an exceptional memory. The author was familiar with Pliny, Plutarch, Plato, Ovid, Aristotle, Cicero, Pythagorus, and Anaxagoras. He had studied his Bible, and thought much upon religion. David and Solomon were favourite lives. He wrote of Tymon of Athens, Diogenes, the Labyrinth of Crete, of Apollo, Pan, Proteus, Orpheus, Venus, Vulcan, and other gods of ancient mythology. He had read of Homer and the Trojans, of Dido, Titus, Cæsar, Cleopatra, Cornelia, Tarquin and Lucretia, Troilus and Cressida, Damon and Pithias, Hero and Leander, and the fable of the Phoenix. He was familiar with falconry and hunting. He affirmed, 'Philosophy, Physic, Divinity, shall be my study.' "[12]

> " 'Lyly,' like Bacon, appreciated Atalanta, Orpheus, Vulcan, and many more of the ancient myths. In his epilogue to *Sapho,* he refers to the Labyrinth of conceits, and wishes everyone a thread to lead them out of it. Bacon in later life entitled one of his papers *Filum Labyrinthi.*"[13]

> "In all 'Lyly's' work we have many examples of that triform construction of sentences common to Bacon, Greene and Shakespeare. Here is one: 'Virtue shall subdue affections, wisdom lust, friendship beauty.' "[14]

Such triform statements carry such deep wisdom, and are carefully constructed statements made in true Bardic tradition (usually referred to as 'triads'). In the 'Lyly' works, the plays in particular, we can see the beginnings of the great art that became, when polished to perfection, the art of 'Shakespeare.'

It is not certain why John Lyly was chosen as a mask: whether it was because he was a man already existing in a suitable position to be used as such in 1580, or whether Francis invented the name 'John Lily' or 'Lyly' as a symbolic pseudonym - the lily being a symbol of the Virgin, and of virtue, *etc.*, which the 'Lyly' works were promoting and foisting as an attribute upon the Queen of England. 'John' is, as discussed before, a mystery name for man himself, and particularly in the context of man as 'mind' or 'soul'. 'John Lily' then takes on the meaning of 'pure mind'. 'John' is a name which Francis especially associates with himself, as a 'Dove' or bearer (*i.e.* harbinger) of Light. Francis referred to himself as "the herald of the new age." It is possible that the man, John Lyly, did not exist as the manager of the boy actors and lessee of the Blackfriars rooms, but that this was simply Francis himself using an invented name and rehearsing the boys; and that the man John Lyly, whoever he might have been, was only associated with the 'Lyly' works later. On the other hand, it seems more probable that the man really did exist as a manager of the boy actors, and that he cooperated with Francis and lent his name right from the start.

'Euphues' is principally associated with 'Lyly', but there is also a further association with another scholar and actor-director named Thomas Lodge. In 1588[15] there was published a pastoral romance entitled *Rosalynde: Euphues golden legacie,* "fetched from the Canaries by T. L. Gent." T. L. (or T. L. G.) is supposedly Thomas Lodge, whose name is signed as such at the end of the dedicatory letter to Lord Hunsdon, the Lord Chamberlain. *Rosalynde,* itself seemingly derived from Chaucer's *Tale of Gamelyn,* formed the basis of the Shakespeare play, *As You Like It,* the latter play being thought to have been written and first acted in 1599-1600, and then first published in the Shakespeare Folio of 1623. In the dedication to *Rosalynde,* Thomas Lodge claims that he wrote the book on the open seas, whilst voyaging with Captain Clarke to the islands of Tercaras and the Canaries, "to beguile the time with labour". The story is prefaced with an extraordinary "Scedule", which describes *Rosalynde* as being "Euphues Testament, the tenour of his Legacie, the Token of his Love". This prefatory letter or "Scedule" is written to 'Philautus', whom we have met before (see chapter 4, 'Immerito') in another play in Francis Bacon's own handwriting, as a probable pseudonym for the Earl of Leicester. It mentions that Philautus "hast sons by Camilla" - Camilla (the priestess of Diana) being a pseudonym for Queen Elizabeth - and exhorts Philautus and Camilla to leave their sons virtue and glory, rather than earthly estates. Seen in the light of Francis Bacon's authorship or guiding light, it is a further example of Francis trying to ensure that his royal mother should really understand that his great desire in life was purely philanthropic and that he had no great wish for worldly riches (beyond what was needed for the philanthropic work) but that he did desire the honours of his parents - that is, to be publicly acknowledged as their son.

The story of *Rosalynde* is one concerning the meaning of true love and wisdom, and the name itself is the same as the Rosalinde of the *Shepheards Calendar,* being the lady (*i.e.* Marguerite de Navarre) whom Colin Clout (*i.e.* Francis Bacon) was greatly in love with. (See chapter 4, 'Immerito'.) *Rosalynde,* and its dramatic sequel *As You Like It,* is a wonderful story concerning love, its wisdom and its powers of transformation. It is an example of one of the finest esoteric or Mystery stories ever invented, and contains a perfect Cabbala. It is truly a "golden legacie" left mankind by this great poet-teacher of Love:-

> THE SCEDULE ANNEXED TO EUPHUES TESTAMENT, THE TENOUR OF HIS LEGACIE, THE TOKEN OF HIS LOVE.
>
> The vehemency of my sicknes (Philautus) hath made mee doubtfull of life, yet must I die in counsailing thee like Socrates, because I love thee. Thou hast sons by Camilla, as I heare, who being yong in yeres have green thoghts, and nobly born have great minds: bend them in their youth like the willow, least thou bewayle them in their age for their wilfulnes. I have bequeathed them a golden legacie, because I greatly love thee. Let them read it as Archelaus did Cassender, to profit by it; and in reading let them meditate, for I have approved it the best methode. They shall find love anatomized by Euphues

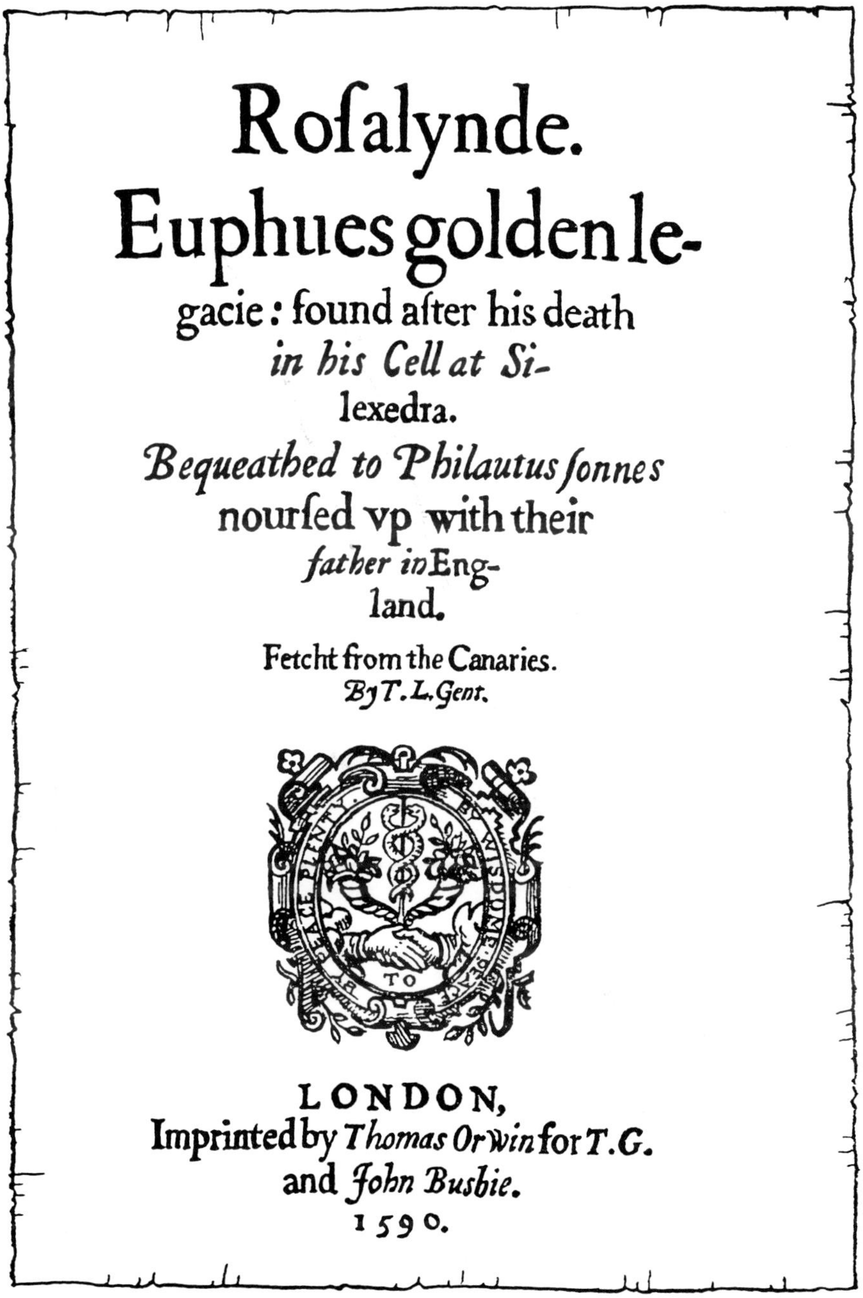

Roſalynde.

Euphues golden legacie : found after his death *in his Cell at Silexedra.*

Bequeathed to Philautus ſonnes nourſed vp with their *father in* England.

Fetcht from the Canaries.
By T. L. Gent.

LONDON,
Imprinted by *Thomas Orwin* for *T. G.* and *John Busbie.*
1590.

(46) TITLEPAGE TO *ROSALYNDE* (1590) by T. L. ('Thomas Lodge'), the forerunner of the Shakespeare Play, *As You Like It.*

(47) QUEEN ELIZABETH AS 'DIANA' - oil painting (c. 1560) attributed to Frans Floris.

> with as lively colours as in Appelles table: roses to whip him when he is wanton, reasons to withstand him when he is wilie. Here may they read that vertue is the king of labours, opinion the mistres of fooles; that vanitie is the pride of nature, and contention the overthrow of families: here is elloborus, bitter in taste, but beneficial in triall.[16] I have nothing to sende thee and Camilla but this counsel, that in stead of worldly goods you leave your sons vertue and glorie; for better were they to bee partakers of your honours then lords of your mannors. I feele death that summoneth me to my grave, and my soule desirous of his God. Farewell, Philautus, and let the tenor of my counsaile be applyed to thy childrens comfort.
>
> *Euphues dying to live.*[17]
>
> If any man find this scrowle, send it to Philautus in England.[18]

Thomas Lodge (1558-1625) was, supposedly, the son of Sir Thomas Lodge (Lord Mayor of London, 1563). He was educated at Merchant Taylor's School, and afterwards at Trinity College, Oxford (1573-7), under Sir Edward Hobby, at the same time as two sons of Lord Hunsdon were at the university. Having gained a degree in arts, he left for London where he entered Lincoln's Inn as a law student (1578-?). Around this time he joined a theatrical company and became an actor and dramatic author. In 1581 he wrote *Defense of Stage-Plays* in response to Stephen Gosson's *School of Abuse.* His *Defense* was suppressed by the authorities, which fact he explains in a later publication, *Alarum Against Usurers* (1584), and he himself had to report regularly to the Privy Council. In 1588 he went on a voyage to the Canaries, and it is this voyage which is linked with the writing of *Rosalynde,* which was published that same year. In 1591 he went on another voyage, this time to South America. About 1596 he became a convert to Roman Catholicism, and went to live on the continent of Europe. There he took up the study of medicine, taking his degree in medicine at Avignon. From 1602 to the end of his life he practised as a physician. He returned to England in 1602-3, where he eventually died, in London, in 1625.

The principal works ascribed to Lodge (but which are probably by Francis Bacon) are all novels, euphuistic in style, the best known of which are *Rosalynde* (1588/90) and *A Margarite of America* (1596). But also printed under Lodge's name (and probably written by him) was *Scylla's Metamorphosis* (1589), which is said to be the earliest romantic poem on a classical subject, and *The Wounds of Civil War* (*c.* 1588), which is a tragedy on the lives of Marius and Sulla, and which was acted by the Admiral's Men. This tragedy is the only known dramatic composition that he supposedly wrote on his own, although he collaborated with Robert Greene in the morality play, *A Looking Glass for London and England* (*c.* 1593). In similar satirical vein he wrote *A Fig for Momus* (1595) and *Wit's Miserie and the World's Madness* (1596). He also translated the works of Josephus (1609) and Seneca (1614), wrote a *Treatise on the Plague* (1603) which was then raging in London, and published a collection of lyric poems, songs and sonnets under the title of *Phillis* (1593).

It appears highly probable that, like other good scholar-writers or "university wits",[19] Thomas Lodge was employed by Francis Bacon as one of the "good pens" in the Bacon scrivenry at Twickenham Lodge, having first of all been 'discovered' in the Inns of Court. These "good pens" assisted Francis with translations of all kinds, with the writing of poetry, prose and drama, with the producing of their group's plays and publications, with the creation of new words for the English language, with the develoment of the language itself, with cipher and letter-writing, with dictation and copying, with proof-reading and the supervision of printing, with illustrations, with research, with ideas, with stories, with reports and records, each according to his or her ability. Some also served as messengers, as explorers, and as actors to train the less educated players. Where appropriate, some of these "good pens" lent their names to works that were written either wholly or at least principally by Francis Bacon. Others of the 'Rose Cross Fraternity' did the same. Often Francis would simply improve upon writings produced by one or other of the "good pens", inserting mystery teachings and, like a Midas, turning all to gold, before the work was published under that person's name. At other times Francis would suggest the whole story and, except for supervision, leave it to a "good pen" to write it up, and then put the finishing touches to it at the end. Francis was, however, famed for his dictation, and his "good pens", trained in Francis' shorthand method, would always carry with them a note book and pen so as to catch the words as they flowed from Francis' constantly alert and creative mind. He even had one or two chosen "good pens" sit with him through the night so that, when he awoke, they should be able to record the ideas and inventions that had come to him in his dreams, before they were lost in the heat of the day. Certain of his "good pens" would walk with him whenever he took exercise in the gardens (which he loved to do, meditating on the beauties of nature), and he would continue to dictate to them even then, whilst he was relaxing, so that every thought was caught fresh as it flew into his mind. At meal times everyone had their note-books at hand, so as not to miss anything of note in the conversation at the table. If possible, nothing was wasted; and, through Francis' supervision and masterly touch, everything was woven into a fabric of light - as the English Renaissance.

Under the mask of a very different and dissolute character, Robert Greene,[20] Francis published *Mamillia* (entered on the Stationer's Register without an author's name in 1580, when Greene was in Cambridge, and published in 1583), followed quickly in 1584 by *Morando, Arbasto, Myrrour of Modestie, Carde of Fancie,* and *Debate between Folly and Love.* Also from 'Greene's' pen came *Planetomachia,* in which the author is described as a "Student in Physicke," and which is dedicated to the Earl of Leicester. The motto, "*Omne tullit punctum qui miscuit utile dulci*", which Francis, as 'Immerito', quoted in one of his 1580 letters to Harvey, was habitually placed on the title-pages of books ascribed to Robert Greene.[21] Later, still under Greene's pseudonym, were published the better known *Menaphon* (1589), republished as *Greene's Arcadia,* and *Pandosto* (1588), the source for the plot of the Shakespeare play, *Winter's Tale.* Several plays, published in the 1590's, are ascribed to Greene, including *Selimus* (1594), *Orlando Furioso* (1594), *Looking Glass for London and England* (*c.* 1593), *Friar Bacon and Friar Bungay* (1594), *Alphonsus, King of Arragon* (1597), *James VI of Scotland* (1598), and *Pinner of Wakefield* (1599). These plays were all printed anonymously after Green's death, and nearly all of them are forerunners of or source material for Shakespeare plays.

Like Marlowe, Greene is considered to be the 'father' of Shakespeare, and often it is difficult to separate the style of either Greene or Marlowe from the style used in the Shakespeare plays - so much so, that several of the Shakespeare plays are thought by Shakespeare scholars to have been jointly written with either Greene or Marlowe. But the earlier style of 'Greene' was close to the style of 'Lyly'; hence Harvey referred to 'Greene' as "the Ape of Euphues." The simple answer is that they were all the works and masks of one man, one author, one genius, who deliberately set out to create a Renaissance of universal learning and art, and a language worthy of the future, whilst himself remaining concealed. The development of his dramatic art, which has never been bettered and which reached its peak of greatness with the 'Shakespeare' plays, can be followed in the sequence of works produced under the mask of these other names.

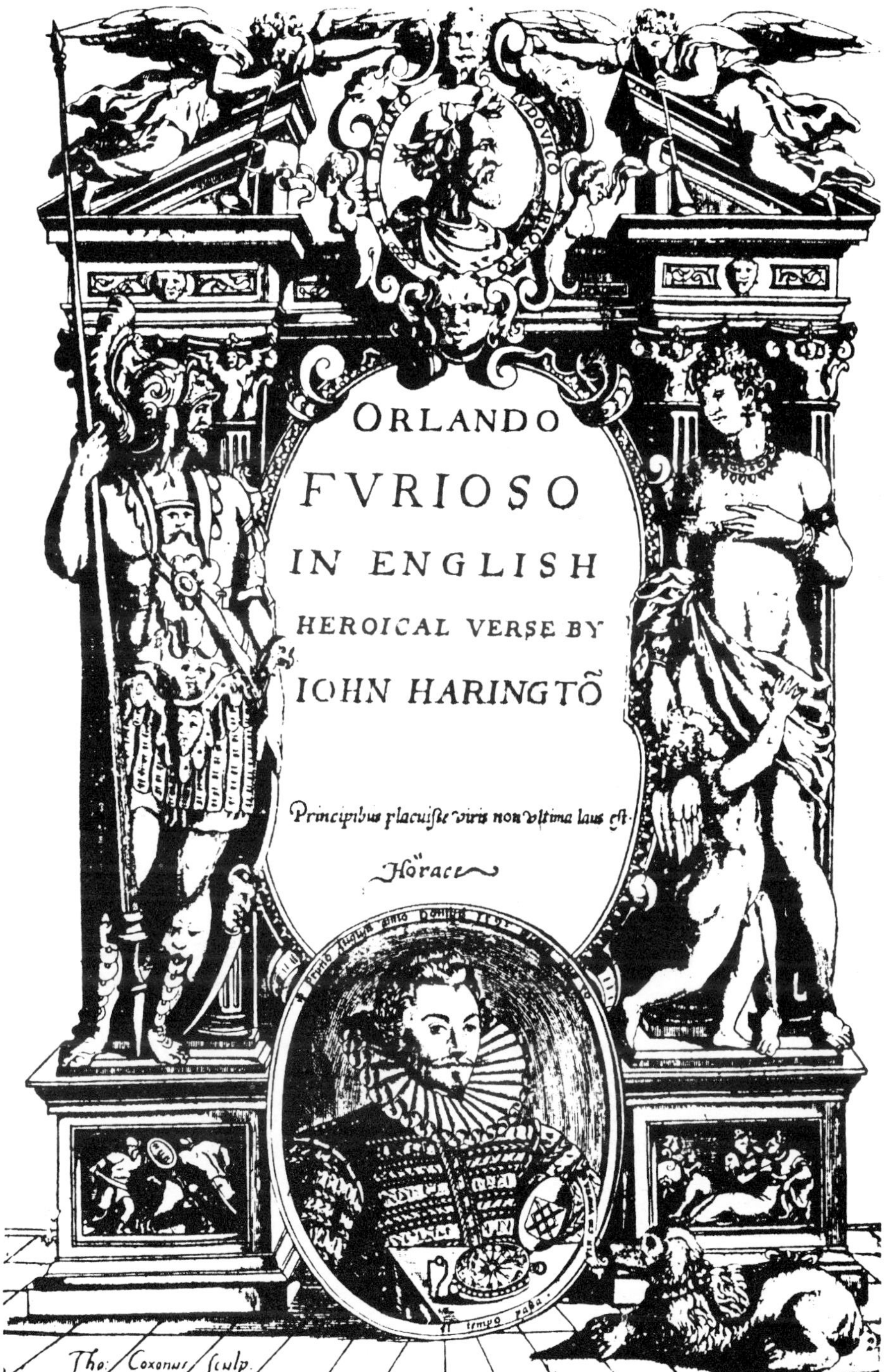

(48) TITLEPAGE TO *ORLANDO FURIOSO* - drawn by Thomas Cockson for Sir John Harrington's translation of *Orlando Furioso* (1591). A dramatisation of this story was published under the name of 'Greene' in 1594.

(49) PALLAS ATHENA - frontispiece to Francis Bacon's *La Sagesse Mysterieuse des Anciens* (1641).

Inscription 1: *"Sic fulget in umbras."*

Translation 1: "Thus it shines in the shadows.'

Inscription 2: *"Obscuris vera involvens."*

Translation 2: "Truth is enveloped in obscurity."

Athena's Veil

Why did Francis hide his name and use pseudonyms? One primary reason has already been mentioned, given by himself: that is, he was born the son of the Queen and heir to the throne, and was compelled to secrecy on the subject, but was trying to record for posterity the true history of his life and times. The recording of this history thus had to be done cryptically, and in a variety of publications and styles that kept the real authorship concealed, lest suspicion be aroused. Secondly, the danger of public opinion and opinion in high places, which was marked by ignorance, intolerance, superstition, and religious dogma and fervour, was very real, and could lead to dreadful persecutions and death (as so often proved). Thirdly, censorship was tightly controlled.

After Whitgift's 'Star Chamber Decree' in January 1586, for many years no manuscript could be set up in type without the licence of the Archbishop of Canterbury or the Bishop of London. The penalties for breaking the law were horrific. It is inconceivable, for instance, that any printer could have been induced to publish *Mother Hubberds Tale* or *The Ruines of Time* (containing an attack on Burghley) unless he was assured of powerful protection, and the author certain of anonymity. (Bacon, however, was a friend of Archbishop Whitgift, and they each helped the other.) Fourthly, it was considered "a discredit for a gentleman to seem learned or to show himself amorous of any good art" in Elizabeth's court; and so those gentlemen who did write did so anonymously, often under pseudonyms. The *Arte of English Poesie* (1589) has this to say:-

> "And peradventure in this iron and malitious age of ours, Princes are lesse delighted in it [the Art of the Poet], being over earnestly bent and affected to the affaires of Empire and ambition, whereby they are as it were inforced to endevour them selves to armes and practises of hostilitie, or to entend to the right pollicing of their states, and have not one hour to bestow upon any other civill or delectable Art of naturall or morall doctrine: nor scarce any leisure to thincke one good thought in perfect and godly contemplation, whereby their troubled mindes might be moderated and brought to tranquillitie. So as, it is hard to find in these dayes of noblemen or gentlemen any good *Mathematician,* or excellent *Musitian,* or notable *Philosopher,* or els a cunning Poet: because we find few great Princes much delighted in the same studies. Now also of such among the Nobility or gentrie as be very well seene in many laudable sciences, and especially in making or Poesie, it is so come to passe that they have no courage to write, and if they have, yet are they loathe to be knowen of their skill. **So as I know very many notable Gentlemen in the Court that have written commendably and suppressed it agayne, or els suffred it to be publisht without their own names to it: as if it were a discredit for a Gentleman, to seeme learned, and to shew him selfe amorous of any good Art.** In other ages it was not so........"[1]

Francis himself writes (as 'Bacon'), in his Essay *Of Simulation and Dissimulation,* "that a habit of secrecy is both politic and moral;" and in *Examples of the Antitheta*[2] he says "Dissimulation is a compendius wisdome." He adds:-

> "Beside in all wise humane Government, they that sit at the helme doe more happily bring their purposes about, and insinuate more easily things fit for the people, by pretexts and oblique courses; than by.....downright dealing. Nay (which perchance may seem very strange) in things meerely naturall, you may sooner deceive nature than force her; so improper and selfe impeaching are open and direct proceedings; whereas on the other side, an oblique and an insinuating way gently glides along, and compasseth the intended effect."[3]

In this comment we have the fifth and, probably, the most important reason why Francis chose to veil his works so elaborately and consistently. Francis set out to be, and was, an ethical and spiritual teacher, a sublime philosopher, a scientist in the field of human consciousness and endeavour, and very much a reformer of the world's state of ignorance and vice, of malpractice and repressive regimes. By any viewpoint he was a revolutionary - not one that wielded the sword and forced his will on others, but one that used the pen as an art and led people on to better things by means of inspiring entertainment and a treasure trail, in liaison with the natural curiosity and love of pleasure that most people have. Any reformer as fundamental as was Bacon, living at that time, would soon have gone to the stake or scaffold if he had not been as discreet as Bacon. But, playing the game as God plays it, and becoming "as little children," wonders can be performed. This is the key which Francis never tired of turning.

> "For of the knowledges which contemplate the works of Nature, the holy philosopher hath said expressly; *that the glory of God is to conceal a thing, but the glory of a king is to find it out:* as if the Divine Nature, according to the innocent and sweet play of children, which hide themselves to the end they may be found; took delight to hide his works, to the end they might be found out; and of his indulgence and goodness to mankind, had chosen the Soule of man to be his Play-fellow in this game."[4]

> "For so he (King Solomon) saith expressly, *The Glory of God is to conceale a thing, but the Glory of a King is to find it out.* As if according to that innocent and affectionate play of children, the Divine Majesty took delight to hide his works, to the end to have them found out, and as if *Kings* could not obtain a greater Honour, then to be God's play-fellowes in that game, especially considering the great command they have of wits and means, whereby the investigation of all things may be perfected."[5]

> "Whereas of the sciences which regard nature the Holy Philosopher declares that 'it is the glory of God to conceal a thing, but it is the glory of the King to find it out.' Even as though the Divine Nature took pleasure in the innocent and kindly sport of children playing hide and seek, and vouched-safe of his kindness and goodness to admit the human spirit for his play fellow in that game."[6]

> "He (God) cannot be seene of any mortal creature but is notwithstanding known by his works."[7]

TO

TRVE NOBILITY,

AND TRYDE LEARNING,

BEHOLDEN

To no Mountaine for Eminence,
nor Supportment for Height, FRANCIS,
Lord *Verulam*, and Viſcount St,
Albanes.

O Giue me leaue to pull the Curtaine by,
That clouds thy Worth in ſuch obſcurity,
Good *Seneca*, ſtay but a while thy bleeding,
T'accept what I receiued at thy Reading:
Here I preſent it in a ſolemne ſtrayne,
And thus I pluckt the Curtayne backe againe.

The ſame

THOMAS POWELL.

(50) DEDICATION PAGE TO *ATTOURNEY'S ACADEMY* (1630) by Thomas Powell - reproduced in Sir Edwin Durning Lawrence's *Bacon is Shakespeare* from a copy of the book owned by him. The upside-down printing of the headpiece is unusual, and only exists in this form on a few such copies; later copies of the same edition have the position of the headpiece corrected.

Thomas Powell, in *The Attourney's Academy* (1630), which was dedicated "To True Nobility, and Tryde Learning, beholden To no Mountaine for Eminence, nor supportment for Height, Francis, Lord Verulam and Viscount St. Albanes," follows the dedication with the following lines, placed under a portrait of Bacon:-

"O Give me leave to pull the Curtaine by
That clouds thy Worth in such obscurity.
Good Seneca, stay but a while thy bleeding,
T' accept what I received at thy Reading:
Here I present it in a solemne strayne,
And thus I pluckt the Curtayne backe again."[8]

This verse is linked directly in meaning with the titlepage to Peacham's emblem book, *Minerva Britanna* (1612), which depicts a curtain hiding an author whose right hand reaches out from behind the curtain, writing "MENTE VIDEBOR" ("By the mind I shall be seen") on a scroll. Cryptic signs lead the reader to pages 33 and 34. Page 34 contains an emblem of an Arcadian shepherd-knight holding a spear and piercing a serpent. It is addressed "To the most judicious and learned, SIR FRANCIS BACON Knight." Directly opposite is page 33, showing an emblem of a gauntleted hand shaking a lance or spear, but with the 'Shaker' being invisible. (For a fuller account of these revealing emblems, see F.B.R.T. Journal I/3, *The Great Vision,* pp. 2, 53-57.)

Again, linked to this, is John Owen's epigram which appears in his *Epigrammatum,* published in 1612:-

AD. D.B.
Si bene qui latuit, bene vixit, tu bene vivis:
Ingeniumque tuum grande latendo patet.

Translated:-

To **L**ord **B**acon (*i.e.* '**D**ominum **B**aconum')
Thou livest well if one well hid well lives
And thy great genius in being concealed is revealed.[9]

The association of anonymity or secrecy with **PALLAS ATHENA** is well known. Athena is said to wear a helmet of light which renders her invisible. But those who discover how, can lift her vizor (or veil) of light and see her face. She is the Muse or Goddess of Divine Intelligence, or Universal Consciousness, in which is contained All-Wisdom, the Love-Light of God. She represents the spiritual form of pure Light, which limits and veils the limitless Light by giving it form; but at the same time she reveals in form the otherwise formless and unrecognisable Light. She signifies the perfect Idea of Truth. She cannot be the whole Truth (*i.e.* she cannot portray formlessness as such), but she is that part of Truth which is given form and therefore capable of being perceived and known. The Veil is *par excellence* her symbol; for the veil conceals in order to reveal. The **Rose** is also, *par excellence,* a symbol with the same meaning, and Athena and the Rose can be equated with each other.

The Goddess teaches wisdom by means of the veil, and her greatest veil is entertainment. (Athena is known as the Tenth Muse, leader of the other nine - the Goddess of Peace and Learning, and Goddess of Poetry. As the Goddess Fame, with whom she is identical, Athena is the Leader and Patroness of the Seven Liberal Arts and Sciences.) For this reason many sages of old have secretly taught mankind the principles of love and virtue

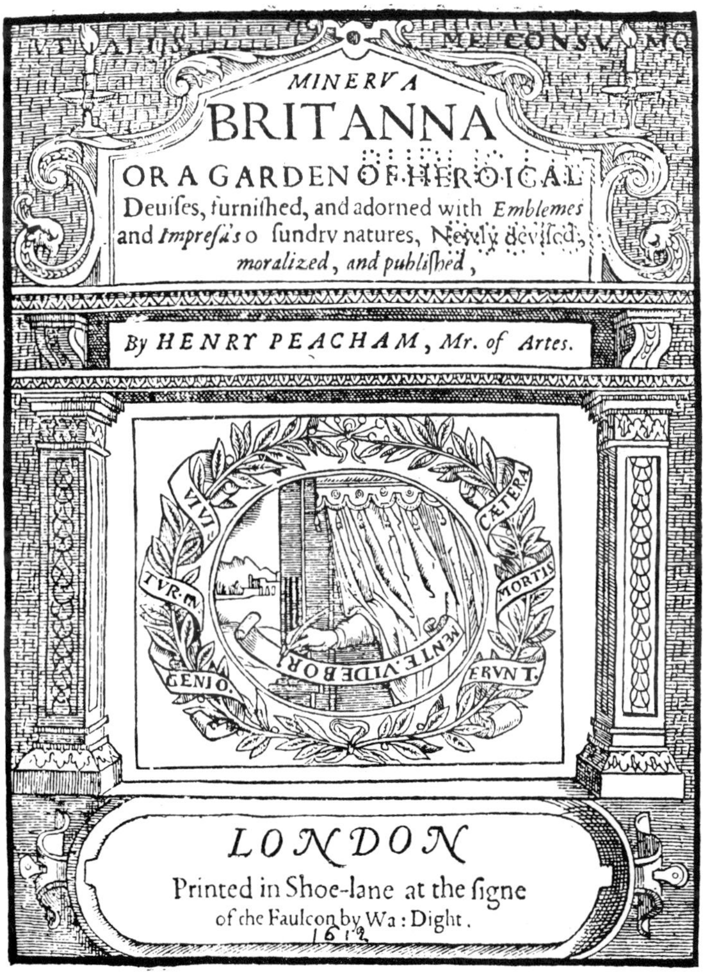

(51) TITLEPAGE TO *MINERVA BRITANNA* (1612) by Henry Peacham. (See FBRT Journal I/3, *Dedication to the Light,* pp. 53-57.)

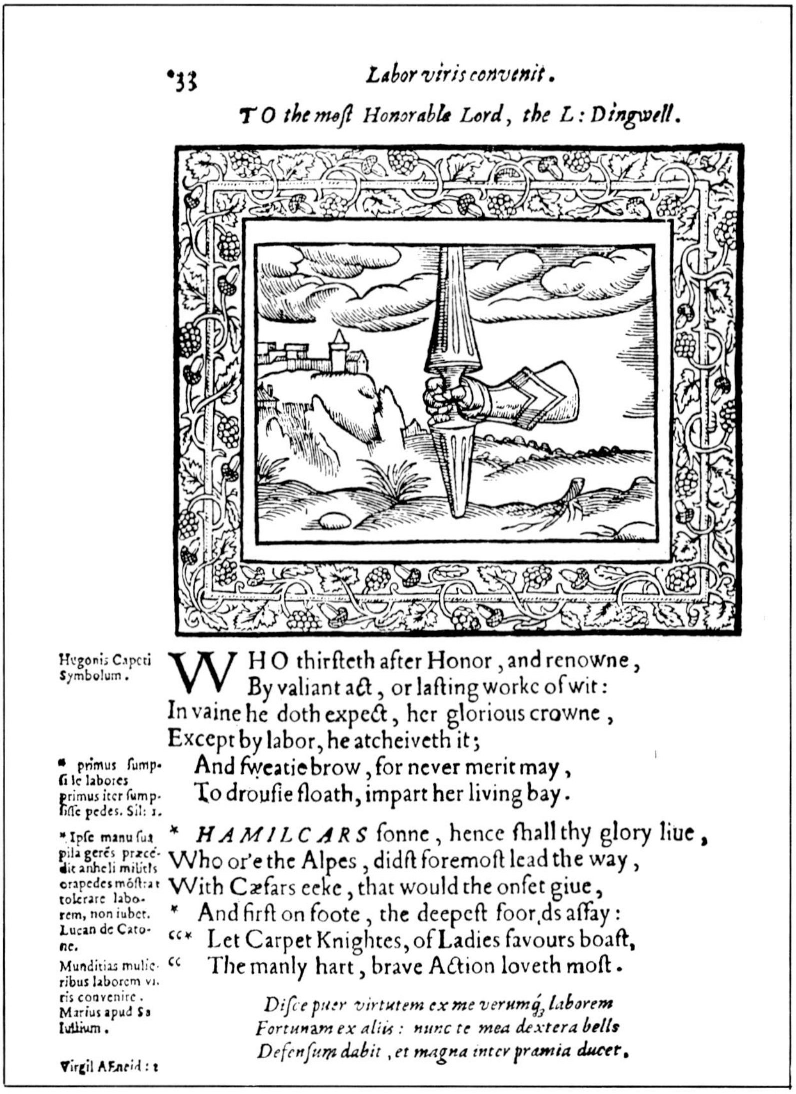

*33 *Labor viris convenit.*

TO the most Honorable Lord, the L: Dingwell.

Hugonis Capeti Symbolum.

WHO thirsteth after Honor, and renowne,
By valiant act, or lasting worke of wit:
In vaine he doth expect, her glorious crowne,
Except by labor, he atcheiveth it;
And sweatie brow, for never merit may,
To drousie sloath, impart her living bay.

* primus sumpsisse labores primus iter sumpsisse pedes. Sil: 1.

* *HAMILCARS* sonne, hence shall thy glory liue,
Who or'e the Alpes, didst foremost lead the way,
With Cæsars eeke, that would the onset giue,
* And first on foote, the deepest foor,ds assay:
"* Let Carpet Knightes, of Ladies favours boast,
" The manly hart, brave Action loveth most.

* Ipse manu sua pila gerés præcédit anheli militis orapedes mőstrat tolerare laborem, non iubet. Lucan de Catone.

Munditias mulieribus laborem viris convenire. Marius apud Sa lullium.

Disce puer virtutem ex me verumq; laborem
Fortunam ex aliis: nunc te mea dextera bells
Defensum dabit, et magna inter pramia ducet.

Virgil AEneid: 1

(52) 'SHAKESPEARE' EMBLEM - Page 33, *Minerva Britanna* (1612) by Henry Peacham.

Ex malis moribus bonæ leges. 34

To the most iudicious, and learned, Sir FRANCIS BACON, Knight.

THE Viper here, that stung the sheepheard swaine,
(While careles of himselfe asleepe he lay,)
With Hysope caught, is cut by him in twaine,
Her fat might take, the poison quite away,
And heale his wound, that wonder tis to see,
Such soveraigne helpe, should in a Serpent be.

By this same Leach, is meant the virtuous King,
Who can with cunning, out of manners ill,
Make wholesome lawes,* and take away the sting,
Wherewith foule vice, doth greeue the virtuous still:
Or can prevent, by quicke and wise foresight,
Infection ere, it gathers further might.

* vitiorum emendatricem legem esse oportet. Cic. 1. de legibus.

Salus Civitatis in legibus. Arist.

Afra venenato pupugit quem vipera morsu,
Dux Gregis antidotum læsus ab hoste petit:
Vipereis itidem leges ex moribus aptas
Doctus Apollinea conficit arte SOLON.

vitiis quæ plurima menti
Fœmineæ natura dedit humana malignas

Cura dedit leges, et quod natura remittit
Invida iura negant &c.

Ovid Metamor: lib 10.

G1. TO

(53) 'SHEPHERD-KNIGHT' EMBLEM - Page 34, *Minerva Britanna* (1612) by Henry Peacham.

through the guise of entertainment - particularly music, poetry and drama. Richard Berry, alluding to this in his *Philobiblon* of 1345, called it "a pious fraud, - the delicate Minerva secretly lurking under the mask of Pleasure."

Francis Bacon was one in a long line of sages who adopted Athena (Minerva) as his Muse: so much so, in fact, that he became identified with her. His mind became consciously one with Divine Intelligence, which is pure Genius; such genius, in fact, that it is sometimes hard to accept or grasp just what he accomplished and taught - and still does - even when the veil is lifted and one stares the truth in the face. Not for nothing did his contemporaries and friends adulate him in the highest possible terms, almost as a 'demi-god' or god-like person, just as Hermes Trismegistus was adulated both in his life-time and after his death by the Greeks and Egyptians.[10]

For instance, in a sonnet written to Francis Bacon by the French poet Jean de la Jessée, private secretary to the Duc d'Anjou, Jessée praises and acknowledges Francis as a supreme poet, and associates Francis with Pallas Athena:-

> "À MONSIEUR FRANCOIS BACON - SONNET
> Ce qu' inspire du Ciel, et plein d'affection
>   Je comble si souvent ma bouche, et ma poitrine
>   Du sacré Nom fameus de ta Royne divine
>   Ses valeurs en son cause et sa perfection.
> Si ce siecle de fer si mainte Nation
>   Ingratte à ses honneurs, n'avait lâme Æmantine
>   Ravis de ce beau Nom, Qu'aus Graces je destine
>   Avec eus nous l'aurions en admuration.
> Donc - Baccon - s'il advient que ma Muse l'on vante
>   Ce n'est pas qu'elle soit ou diserte, ou scavante:
>   Bien que vostre Pallas me rende mieus instruit
> C'est pource que mon Lut chant sa gloire sainte
>   Ou qu'en ces vers nayfz son Image est empraînte
>   Ou que ta vertu claire en mon ombre reluit.
>
> LA JESSEE."[11]

A free translation of the last six lines:-

> "Therefore Bacon if it chances that my Muse praises someone
> It is not because she is eloquent or learned,
> Although your Pallas has taught me better (how to speak);
> It is because my Lute sings the saintly glory
> Or in these artless [naïve] lines his image is imprinted
> Or that thy Virtue bright shines in my shade."[12]

This poem was written in Francis' lifetime but never published. However, a few months after Francis' death to the world as 'Bacon' (Easter, April 1626), a collection of thirty-two remarkable Latin elegies[13] to the memory of Francis Bacon were published by John Haviland under the title of *Memoriae Honoratissimi Domini Francisci, Baronis de Verulamio, Vice-comitis Sancti Albani Sacrum* (1626). These are more commonly known as the *Manes Verulamiani,* and selections of them were reprinted in the 1640 and 1674 editions of *The Advancement of Learning.* In Dr. William Rawley's 'Introduction' to the *Manes,* he states that he, as editor, felt it necessary to withhold the "best" poems from publication; nevertheless, those poems that Rawley did publish declare Francis Bacon to

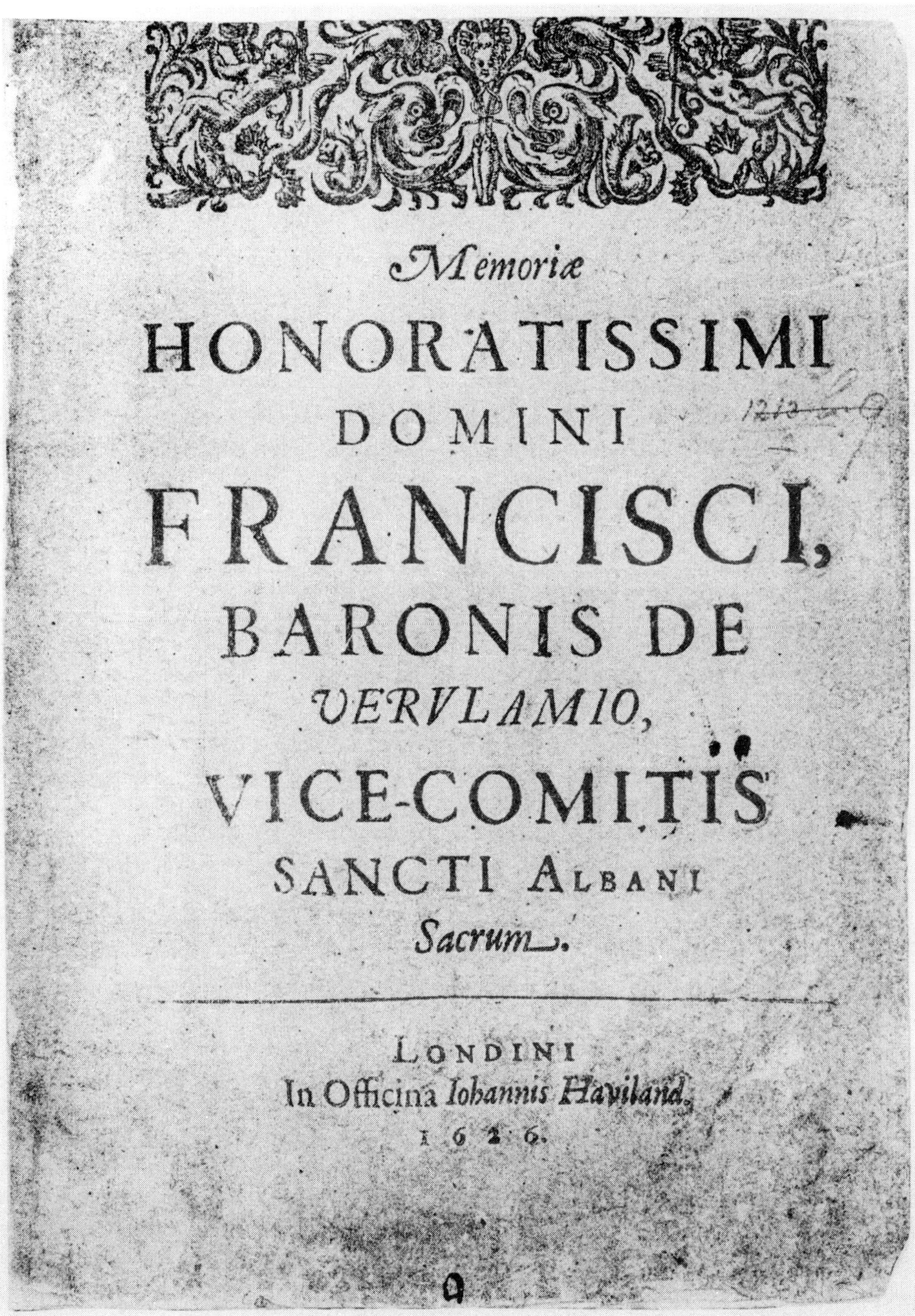

Memoriæ

HONORATISSIMI

DOMINI

FRANCISCI,

BARONIS DE

VERVLAMIO,

VICE-COMITIS

SANCTI ALBANI

Sacrum.

LONDINI

In Officina *Iohannis Haviland,*

1626.

(54) TITLEPAGE TO THE *MANES VERULAMIANI (1626), published by William Rawley.*

have been not just a supreme poet, second to none, the writer of unacknowledged literary work, associated with the theatre, and the centre of a mystery which it was reserved for posterity to unravel, but also a genius of all geniuses in every field of human endeavour and possibility. Even in those days of elaborate praise, such a tribute has rarely been given to any man in living history; and, with their veiled but pointed allusions, what these elegies are saying is quite remarkable and should be noted. The following are extracts from the elegies, as freely translated from the original Latin by W.C.G. Gundry[14]:-

> "Supreme both in eloquence and writing, under every head renowned, famous in council chamber and lecture hall; in war too, if war should submit to art; surpassing in every pursuit, under every title, a very Chiron....."[15]

Chiron was the wisest and justest of the centaurs. He was the son of Saturn (*i.e.* Pan), and was renowned for skill in hunting, medicine, music, prophecy, and in all science and art. Himself the pupil of Apollo and Diana, he became the tutor to the most renowned of the Greek heroes.

> "The very nerve of genius, the marrow of persuasion, the golden stream of eloquence, the precious gem of concealed literature, the noble Bacon........O how am I in verse like mine to commemorate you, sublime Bacon! and those glorious memorials of all the ages composed by your genius and by Minerva."[16]

> "Unite the two roses, Henry; Bacon gives a thousand; as many words in his book, so many roses I ween."[17]

> "......but add, a concealed poet from Ithaca, and you hold all."[18]

Ithaca is the island home of Ulysses (also known as Odysseus).

> "As Euridice wandering through the shades of Dis longed to caress Orpheus, so did Philosophy entangled in the subtleties of Schoolmen seek Bacon as a deliverer...........he renovated her, walking lowly in the shoes of Comedy. After that more elaborately he rises on the loftier tragic buskin...........The Columbus of Apollo with his lordly crew passes beyond the Pillars of Hercules in order to bestow a new world and new arts..............Let every star emit a spark into his pyre; be it sacrilege that the kingly pile should be kindled for Prometheus from a kitchen fire...............Thus, blessed inhabitants of heaven, let Astraea enjoy her champion of old, or with Bacon give back Astraea"[19]

This elegy equates Francis Bacon with Orpheus, Columbus and Prometheus, and associates him directly with Apollo and Astraea, as their champion. The reference to Astraea is particularly fascinating, as it says "champion of old", implying that Bacon was a champion of Astraea from an early age, and also suggesting Queen Elizabeth's time, as Elizabeth I was purposefully portrayed by the poets and writers of her time (principally by Francis Bacon) as the personification of Astraea. Francis is said in the elegy to have been her champion - which he was, but not under the name of 'Bacon'! As 'Bacon', Francis is not known to have written anything directly championing Astraea as such (although of course he did in a veiled way). The greatest champion of Astraea known to the general public was 'Spenser', hence this passage associates Bacon as being 'Spenser' the poet. The final remark equates Francis directly with Astraea, as being the goddess's equivalent.

> "When he perceived that the arts were held by no roots and, like seed scattered on the surface of the soil, were withering away, he taught the Pegasean arts to grow, as grew the spear of Quirinus swiftly into a laurel tree. Therefore, since he has taught the Heliconian goddesses to flourish, no lapse of ages shall dim his glory. The ardour of his noble heart could bear no longer that you, divine Minerva, should be despised. His godlike pen restored your wonted honour and, as another Apollo, dispelled the clouds that hid you. But he dispelled also the darkness which murky antiquity and blear-eyed old age of former times had brought about; and his super-human sagacity instituted new methods and tore away the Labyrinthine windings, but gave us his own. Certainly it is clear that the crown of ancient sages had not such penetrating eyes. They were like Phœbus rising in the East, he like the same resplendent at noon.........They begot the infant Muses, he adult. They were the parents of mortal muses, he produced goddesses.........Pallas too, now arrayed in a new robe, paces forth, as a snake shines when it has put off its old skin."[20]

"The Pegasean arts" are the arts of the Muses. **Pegasus** is the symbol of **Fame**, which is Athena in her role as leader of the Seven Liberal Arts and Sciences. Fame, or Pegasus, contains and sums up all the arts and sciences. This white winged horse represents the purified and illumined mind, whose thoughts can fly in the realm of heaven as well as ride upon the earth. Hippocrene ('the horse's fountain'), the fountain or spring of the Muses, gushed forth from the side of Mount Helicon in Bœotia when it was struck by the hoof of Pegasus. Helicon and the Hippocrene are especially sacred to the Muses, as also is Mount Parnassus and the Castalian spring. The "Heliconian goddesses" are the nine Muses. The Muses presided over poetry, music, dancing, and all the liberal arts. They were contained and summed up in Pallas Athena, who is also Fame. This is the *Fama* of the Rosicrucians. Pegasus - the 'Fame' of the Rosicrucians - is thus the symbol of Pallas Athena.

"The spear of Quirinus" refers to the spear thrown by Romulus Quirinus into the ground on the Quirinal hill, where it took root and became a laurel tree, thereby founding Rome. *Quirinus* is derived from the Sabine word *quiris,* meaning 'lance/spear.' *Quirinus* thus means 'spearman,' or 'spear-shaker.' Romulus was the son of Mars, whose surname was Quirinus. Cabbalistically Mars signifies Judgment, or Justice (as discussed before), which is represented by Astraea.

> "Verulam, reigning in the citadel of the gods, shines with a golden crown; and, enthroned above the bounds of the sky, he loves with face towards Earth to view the stars; who grudged the immortals that wisdom should be confined to the abode of the blessed, undertaking to bring it back and restore it to mortals by a new cult. Than whom no inhabitant of Earth was master of greater intellectual gifts; nor does any survivor so skilfully unite Themis and Pallas. While he flourished the sacred choir of the Muses, influenced by these arts, poured forth all their eloquence in his praise....."[21]

Verulam has a close meaning to 'Spear-shaker' (*i.e. Veru,* Latin for 'javelin/spear' + *lam,* English for 'to beat/thrash'). The bringing of wisdom (*i.e.* fire, or light) back to the earth is a Promethean task.

Themis is the Goddess of Justice (*i.e.* she equates with Astraea), and the first introducer of oracles. (She inspired, for instance, the Delphic oracle.) She makes known the laws of Zeus to men. She was the first to whom the inhabitants of earth raised temples. She is attended by the Seasons, and was considered to be the first of the gods who established the laws of religion, sacrifices, divination, and whatever tends to the harmony and peace of society. She is the representative of Cosmic Order and Harmony (like *Maat* in Ancient Egypt). She frequently appears on coins holding a cornucopia and a pair of scales. She is also symbolised as holding a sword in her right hand and scales in her left.

Pallas is Pallas Athena, a higher counterpart of Themis. Whereas Themis represents specific intelligence, Pallas Athena signifies universal intelligence. To unite Themis with Pallas is to unite in consciousness the individual mind (or soul) with the universal mind (or soul) - the goal of all human endeavour.

> "Bacon......a Muse more rare than the nine Muses..."[22]

> "......ah! the tenth Muse and the glory of the choir has perished."[23]

There is only one Muse rarer than the nine, and that is Pallas Athena, the tenth Muse, spouse of Apollo, the god of Light.

> "If my prayers with your, O Sisters! had prevailed.........the contest of our love would not be ambiguous.........we should be in possession of our tears and of thee, Apollo, the darling, learned Bacon of your native land."[24]

> "......take the whole world, but give me back my Phœbus."[25]

Phœbus, 'Brilliant', is a title of Apollo.

> "How has it happened to us, the disciples of the Muses, that Apollo, the leader of our choir, should die?"[26]

Apollo, known in ancient mythology as the husband (or 'masculine' counterpart) of Athena, is thus also the leader (or co-leader) of the choir of Muses. Apollo is known as the Chancellor of Parnassus, the Mount of Poetic Inspiration and Illumination - the Mount of the Muses. The parallel between Francis Bacon, Lord Chancellor of England, and Apollo, Lord Chancellor of Parnassus, is played here to the full.

These allusions to Apollo, Parnassus, Pegasus, *etc.,* are echoed in the writings directly associated with the Rosicrucians of the 17th century. For instance, the *Fama Fraternitatis* (published 1612 onwards) is introduced by a satire, entitled 'The Universal and General Reformation of the Whole Wide World.' The title summarises the purpose of the Rosicrucian work - which is identical with the purpose of Bacon's work. The story told in this satire concerns Apollo, who, determined to cure the world of ignorance and vice, founds a Society of men famous for wisdom and virtue. This Society forms a Council, comprised of three wise men of Italy and seven wise men of Greece. One of the three wise men of Italy is referred to as the Secretary. This Secretary is the chief of the Council and regent of Apollo.[27]

Almost one hundred years later, in 1704, N.N. Esquire brought out an adapted version of the *Fama,* in which he specifically refers to the Secretary as Sir Francis Bacon. Such a statement is supported by the utterances of others - George Withers for one, who, in his *The Great Assizes holden in Parnassus by Apollo and his Assessours* (1645), refers to the Secretary as the President, representing the god Apollo and presiding over all the learning of his age. He names this President as "Lord VERULAN Chancellor of Parnassus," and places this name directly beneath that of APOLLO, which is spelt in large letters in an

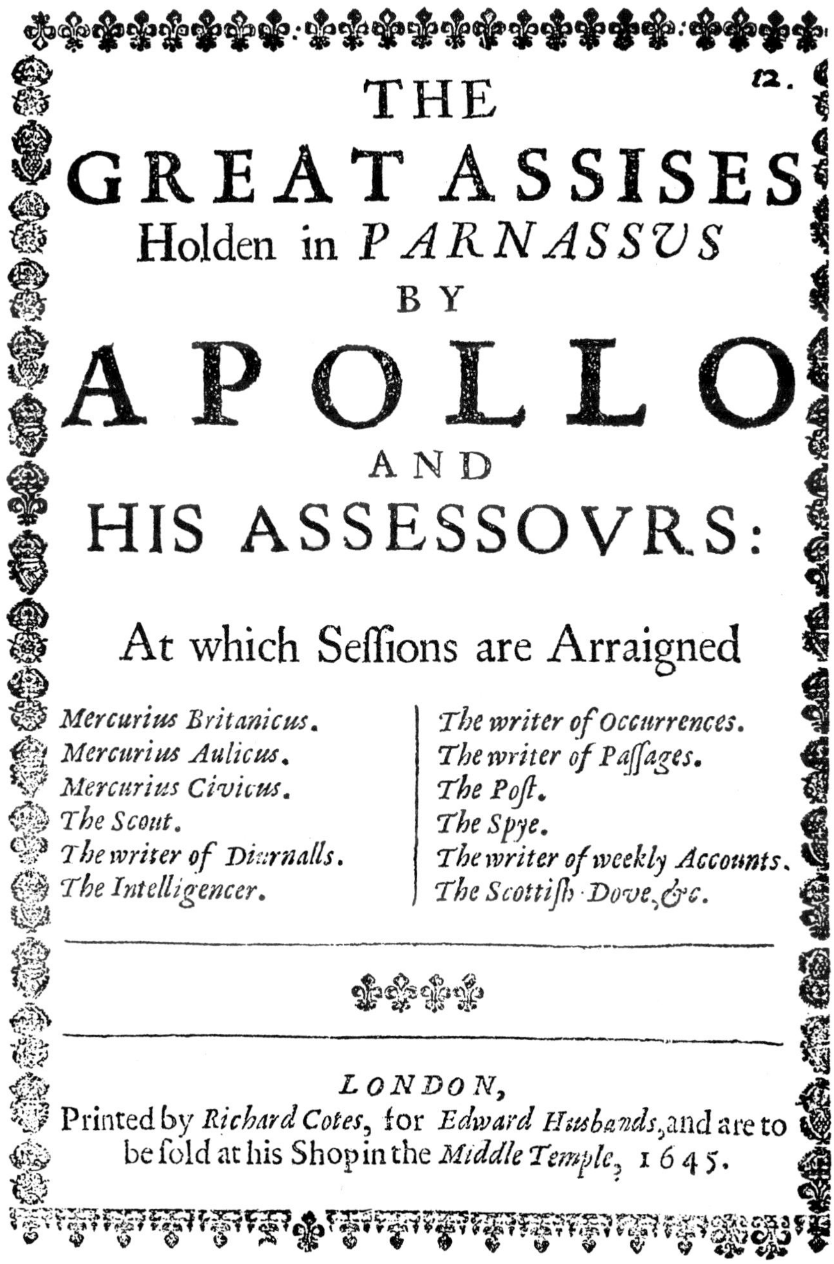

12.

THE
GREAT ASSISES
Holden in *PARNASSUS*
BY
APOLLO
AND
HIS ASSESSOVRS:

At which Seſſions are Arraigned

Mercurius Britanicus.
Mercurius Aulicus.
Mercurius Civicus.
The Scout.
The writer of Diurnalls.
The Intelligencer.
The writer of Occurrences.
The writer of Paſſages.
The Poſt.
The Spye.
The writer of weekly Accounts.
The Scottiſh Dove, &c.

LONDON,
Printed by *Richard Cotes*, for *Edward Husbands*, and are to be ſold at his Shop in the *Middle Temple*, 1645.

(55) TITLEPAGE TO *THE GREAT ASSISES* (1645) by George Withers.

APOLLO.

The Lord VERVLAN, *Chancellor of Parnassus.*	ERASMUS ROTERODAM.
Sir PHILIP SIDNEY, *High Constable of Par.*	JUSTUS LIPSIUS
WILLIAM BVDEVS, *High Treasurer.*	JOHN BARCKLAY
JOHN PICVS, *Earle of* Mirandula, *High Chamberlaine.*	JOHN BODINE
JVLIVS CESAR SCALIGER	ADRIAN TVRNEBVS
	ISAAC CASAVBON
	JOHN SELDEN
	HVGO GROTIVS
	DANIEL HEINSIVS
	CONRADVS VOSSIVS
	AUGUSTINE MASCARDUS

The Jurours.	*The Malefactours.*
George Wither	*Mercurius Britanicus*
Thomas Cary	*Mercurias Aulicus*
Thomas May	*Mercurius Civicus*
William Davenant	*The Scout*
Josuah Sylvester	*The writer of Diurnals*
Georges Sandes	*The Intelligencer*
Michael Drayton	*The writer of Occurrences*
Francis Beaumont	*The writer of Passages*
John Fletcher	*The Poste*
Thomas Haywood	*The Spye*
William Shakespeere	*The writer of weekely Accounts*
Philip Massinger.	*The Scottish Dove, &c.*

A 2 Jo-

(56) **PAGE 3 OF *THE GREAT ASSISES*** (1645) by George Withers.

JOSEPH SCALIGER, the Censour of manners in *Parnassus.*	EDMVND SPENCER, Clerk of the Assises.
BEN. JOHNSON, Keeper of the Trophonian Denne.	
JOHN TAYLOVR, Cryer of the Court.	

(57) **PAGE 4 OF *THE GREAT ASSISES*** (1645) by George Withers.

ornamental frame. Then he couples Francis, Lord Verulam,[28] with Sir Philip Sydney, the reputed leader (in the public eye but not in actuality) of the English Areopagus in the days of Queen Elizabeth, by making Sydney the "High Constable of Parnassus." Interestingly he makes Ben Jonson to be the "Keeper of the Trophonian Denne," and Edmund Spenser the "Clerk of the Assizes."

Michael Maier, in his *Themis Aurea,* says that the Temple of the Rosy Cross is located beside the spring Helicon on double-peaked Parnassus, where Pegasus opened a fountain of perennial waters. (The Helicon is here associated and equated with its twin spring, the Hippocrene, which was opened up by Pegasus, which in turn is equated with the Castalian spring of Mount Parnassus.) The two peaks of Parnassus were the seats of Apollo and Athena - a double-throne. In these subtle ways, Francis Bacon is singled out as being the Fr. C.R.C. of the Rosicrucian manifestos and fraternity.

But these are by no means the only tributes to Francis Bacon's greatness and secrecy. Tobie Matthews, a Roman Catholic but good friend of Francis Bacon, and one to whom Francis often sent his compositions for critical comment by Tobie, noted in one of his letters to Francis:-

> "The most prodigious wit that ever I knew of my nation, and of this side of the sea, is of your Lordship's name, though he be known by another."[29]

He also wrote:-

> "It is not his greatness that I admire, but his virtue; it is not the favours that I have received from him (infinite though they be) that have thus enthralled me and enchained my heart, but his whole life and character."

Ben Jonson, who knew Francis Bacon well in the days of King James I, and became one of Francis' "good pens", wrote after Francis' death in 1626:-

> "He who hath filled up all numbers and performed that in our tongue which may be compared or preferred either to insolent Greece or haughty Rome. In short, within his view, and about his times, were all the wits born that could honour a language or help study. Now things daily fall; wits grow downward and eloquence grows backward, so that he may be named and stand as the mark and acme of our language."[30]

Ben Jonson also wrote, in his prefatory verses to the 1st Shakespeare Folio (1623) of which he was an editor[31]:-

> "...........Soule of the age!
> The applause! delight! the wonder of our stage!
>
> ....Or, when thy Sockes were on,
> Leave thee alone, for the comparison
> Of all, that insolent *Greece,* or haughty *Rome*
> Sent forth, or since did from their ashes come."
> Triumph, my *Britaine,* thou hast one to showe,
> To whom all scenes of *Europe* homage owe.
> He was not of an age, but for all time!
> And all the *Muses* still were in their prime,
> When like *Apollo* he came forth to warme
> Our eares, or like a *Mercury* to charme!

............For a good *Poet's* made, as well as borne.[32]
And such wert thou. Looke how the fathers face
Lives in his issue, even so, the race
Of *Shakespeares* minde, and manners brightly shines
In his well torned, and true fileal lines:[33]
In each of which, he seems to shake a Lance,
As brandish't at the eyes of ignorance."[34]

Referring to Francis' merry wit, which went hand in hand with the graver side of his character that most observers only see when reading Bacon's philosophical works, Ben had this to say:-

> "There happened in my time one noble speaker [Francis Bacon] who was full of gravity in his speaking. His language - where he could spare or pass a jest - was nobly censorious..........His hearers could not cough or look aside from him without loss. He commanded where he spoke; and had his judges angry and pleased at his devotion. No man had their affections more in his power. The fear of every man that heard him was, lest he should make an end."

Emphasising the particular qualities of a great poet and master of creative imagery, one who could have truly led the English Areopogus, Fuller says in his *Worthies*:-

> "He was a rich Cabinet filled with Judgment, Wit, Fancy and Memory, and had the golden key, Elocution, to open it. He was *singular in singulis,* in every Science and Art, and being in-at-all came off with credit."

Francis Bacon was always recognised by those of his contemporaries who were 'in the secret', as the greatest of poets. In 1615, although nothing of any poetical importance bearing the name of 'Bacon' had been published up to that point, Stowe places Francis Bacon seventh on his list of Elizabethan poets in his *Annals.* Later in the century, the poet Edmond Waller, in the dedication to the first edition of his poems in 1645, writes:-

> "Not but that I may defend the Attempt I have made up in Poetry, by the Examples (not to trouble you with History) of many Wise and Worthy Persons of our own Times; as Sir Philip Sydney, Sir Francis Bacon, Cardinal Perron, the ablest of his Countrymen; and the former Pope, who they say, instead of the Triple Crown, wore sometimes the poets Ivy, as an Ornament, perhaps, of lesser weight and trouble: But, Madam, these nightingales sang only in the spring, it was the diversion of their Youth........."

In the place where most people might have expected the names of Spenser and Shakespeare to occur, is to be found the name of Francis Bacon!

But not many people knew (or know nowadays) the secret: few have ever pierced the veil of Athena - although Pallas awaits patiently the day when the many will draw the curtain aside and enjoy the light. Archbishop Tenison was certainly one of those who did know the secret, and he tried hard to drop the hint to others! Hence he states, in *Baconiana* (1679) that:-

> ".....those who have true skill in the Works of Lord *Verulam,* like the great Masters in Painting, can tell by the *Design,* the *Strength,* the way of *colouring,*

(58) FRANCIS BACON, Baron Verulam, Viscount St. Alban - illustration from *The Mirrour of State and Eloquence* (1656), published anonymously.

This quarto publication has a running title, 'Bacons Remaines', and is basically a reprint of the quarto printed eight years previously entitled *The Remaines of the Right Honorable Francis Lord Verulam, Viscount of St. Albanes, sometimes Lord Chancellour of England, being Essayes and severall Letters to severall great Personages, and other pieces of various and high concernment not hithertofore published* (1648), containing an essay on Kingship, another on Death, letters of State, an essay on Christian Paradoxes, a Confession of Faith and a Prayer. The Confession and Prayer saw their first publication in the 1648 *Remaines* which, as for the *Mirrour,* was published anonymously. The titlepage of *The Mirrour of State and Eloquence* is, however, different to that of the *Remaines,* and it also contains the above portrait of Francis Bacon which is not included in the *Remaines.*

This portrait appears to be based upon that of Bacon which is printed as the frontispiece to the 1640 edition of the *Advancement of Learning,* but there are important differences. In the 1640 picture, Bacon is depicted writing the 7th volume or part of the Great Instauration, the words which he has written on the two open pages being *"Mundus"* and *"Mens"* ("the World/Universe" and "the Mind") and *"Conubio jungam stabili....."* ("the connection made firm in marriage") - a reference to the necessary marriage between thought and action, and between soul and spirit, which brings true knowledge or enlightenment. (See FBRT Journal I/4, *The Great Vision,* pp. 244-245.) In the above picture from the *Mirrour* Bacon has written a heading on each open page, with indecipherable words beneath each heading. These headings are, on the left page, *"Crums",* and on the right page, *"Faith".* Read together they appear to mean "Crum(b)s of Faith", referring to the breaking of the Bread of Heaven, or Light, into 'crumbs'. The Bread of Heaven is the Word of God, thus the crumbs of Faith are the many words of truth spoken or written, or manifested in the life of an individual. It would seem that this illustration is indicating that in this book are words of wisdom, touching Faith or Truth, which Bacon has written: and this would seem to be a deliberate effort to bring our attention to Bacon's remarkable Confession of Faith and his Prayer. It should also be noted that the number of books depicted in the illustration has increased to nine (there are seven in the 1640 picture).

(59) FRANCIS BACON, Baron Verulam, Viscount St. Alban - frontispiece to his *Of the Advancement and Proficience of Learning* (1640), engraved by William Marshall. (See FBRT Journal I/4, *The Great Vision*, pp. 144-145, for commentary.)

whether he was the Author of this or the other Piece, though his Name be not to it."

The testimony of a more modern poet, Shelley, is of great import, not only because Shelley acknowledges Francis Bacon as being a great poet, but also because he rates Francis together with Plato as being supreme poets. Did Shelley know or discover more than he let on? Look at the frontispiece to the *Advancement of Learning* (1640), where it compares Bacon to Plato in the plaque at the top of the picture. Here are Shelley's words (from his *Defence of Poetry*):-

> "Poetry is at once the centre and circumference of knowledge.....and that to which all science must be referred.............A Poet is the author to others of the highest wisdom, pleasure, virtue and glory.............Plato was essentially a Poet; the truth and splendour of his imagery, and the melody of his language, are the most intense that it is possible to conceive...........Lord Bacon was a Poet. His language has a sweet and majestic rhythm which satisfies the sense, no less than the superhuman wisdom of his philosophy satisfies the intellect. It is as rain which distends and then bursts the circumference of the reader's mind and pours itself forth together with it into the universal elements with which it has perpetual sympathy...............The language of Plato is that of an immortal spirit, rather than a man; Lord Bacon is, perhaps, the only writer who, in these particulars, can be compared to him."

Shelley's words are vitally important, especially today when poetry tends to be such a poor excuse for the real thing. Read again Shelley's first two sentences. Then hearken to what Sir Tobie Matthew said about his friend, Sir Francis Bacon. Tobie Matthew challenged anyone "to muster out of any age four men who, in many respects, should excel four such as we are able to show - Cardinal Wolsey, Sir Thomas More, Sir Philip Sydney, and Sir Francis Bacon, for they were all a kind of monsters [*i.e.* prodigies, marvels, giants] in their several ways." After extolling the virtues of the first three, Tobie continues:-

> "The fourth [Sir Francis Bacon] was a creature of incomparable abilities of mind, of sharp and catching apprehension, large and faithful memory, plentiful and sprouting invention, deep and solid judgment;......a man so rare in knowledge of so many several kinds, indeed with the facility of expressing it in so elegant, significant, so abundant, and yet so choice and ravishing a way of words, of metaphor and allusion as perhaps the world hath not seen since it was a world."[35]

How is it that this kind of testimony is missed over by the majority of our modern scholars and historians and self-professing poets? How is it that our common *intelligensia* can spend days, weeks, years, sometimes lifetimes in studying men of such common, low or even squalid character as those whose only right to fame is that their names were used on the titlepages of certain literary works! How is it that our professors of English can spend tortuous hours in contortions of the mind in order to make the reputed authors fit the nature and quality of the works that they are supposed to have written? How is it they choose to do this, rather than take the plentiful testimony of the time as to who were the real geniuses, the real bards, and why they chose to conceal themselves behind masks? It is not difficult to find the truth: it even stares one in the face! But perhaps this is why it is not seen; for truth can blind if one is not prepared for it. Francis himself said:-

> "I will teach by my own example. Silence sometimes says more than speech."[36]

Silence and invisibility go hand in hand. But only so that one day they may be both heard and seen. The shock of realising that one can be, and has been, so deaf and blind to truth, is a great teacher. Francis Tudor set out to teach: he knew that he was no ordinary man, and he knew the burden of responsibility which that meant:-

> "The chief wish I now have is to continue my rightful, humble, yet truly worthy works for my toiling fellows, who wrestle in blind helplessness with the forces of Nature. We that know the manifold mighty influences of unseen things, owe more of this knowledge of our environings to the light from our Celestial Source than to our investigations. Therein lieth the duty we owe to our fellow-men, for do not our Scriptures say: Freely ye have received, so must ye in like manner give? This then doth urge me ever on, up to heights of knowledge that no one hath ever reached."[37]

> "When my very soul doth lie, as the souls of men shall, before our Father's judgment seat, exposed to the eyes of men and angels, I shall receive all men's praise instead of a whole nation's or many nations' contumely. Then my love shall be known, which would sacrifice my ease that humanity might share in all these labours, reaping rich benefits from my studies."[38]

Finally, Francis' beautiful method may be glimpsed in these following ideas, which he proceeded to carry out to perfection:-

> "*Poesie* seems to endow human nature with that which *History* denies, and to give satisfaction to the mind, with, at least, the shadow of things, where the substance cannot be had............A strong argument may be drawn from Poesie that a more stately greatness of things, a more perfect order and a more beautiful variety delights the soul of man than any way can be found in Nature, since the Fall."

> "Poesie serveth and conferreth with magnanimity, morality, and to delectation, and therefore it was ever thought to have some participation of divineness, because it doth raise and erect the mind, submitting the shews of things to the desires of the mind."[39]

> "Dramatic poetry, which has the Theatre for its world, would be of excellent use if it were sound, for the discipline and corruption of the Theatre is of very great consequence...........The action of the Theatre, though modern states esteem it ludicrous unless it be satirical and biting, was carefully watched by the ancients that it might improve mankind in virtue; and indeed many wise men and great philosophers have thought it to the mind as the bow to the fiddle."[40]

> "Plato saith elegantly: 'That virtue if she could be seen would move great love and affection.'...........Rhetoric paints our virtue and goodness to the life, and makes them in a sort conspicuous..........Shew them to the Imagination so as maybe in a lively Representation."[41]

> "Seeing things in present sight do more strongly fill the Imagination.......Things future and remote are beheld as if they were actually present....varied according to the Auditors....like a skilful musician accommodates himself to different ears."[42]

"The story of Orpheus....seems meant for a representation of universal Philosophy. For Orpheus himself, - a man admirable and truly divine, who being master of all harmony subdued and drew all things after him by sweet and gentle measures, - may pass by an easy metaphor for philosophy personified. For as the works of wisdom surpass in dignity and power the works of strength, so the labours of Orpheus surpass the labours of Hercules....

"Orpheus betook himself to solitary places, a melancholy man and averse from the sight of women; where by the same sweetness of his song and lyre he drew to him all kinds of wild beasts, in such manner that putting off their several natures, forgetting all their quarrels and ferocity, no longer driven by the stings and furies of lust, no longer caring to satisfy their hunger or to hunt their prey, they all stood about him gently and sociably, as in a theatre, listening only to the concords of his lyre. Nor was this all: for so great was the power of his music that it moved the woods and the very stones to shift themselves and take their stations decently and orderly about him....

"The singing of Orpheus is of two kinds; one to propitiate the infernal powers, the other to draw the wild beasts and the woods. The former may be best understood as referring to natural philosophy; the latter to philosophy moral and civil....

"Philosophy,...applying her powers of persuasion and eloquence to insinuate into men's minds the love of virtue and equity and peace, teaches the peoples to assemble and unite and take upon them the yoke of law and submit to authority, and forget their ungoverned appetites, in listening and conforming to precepts and discipline; whereupon soon follows the building of houses, the founding of cities, the planting of fields and gardens with trees; insomuch that the stones and the woods are not unfitly said to leave their places and come about her."[43]

"And Orpheus assembled the wild beasts to come in heards to harken to his musicke and by that means made them tame, implying thereby how, in his discreet and wholesome lessons, uttered in harmonie and with melodious instruments, he brought the rude and savage people to a more civil and orderly life."[44]

"Invest me in my motley: give me leave
To speak my mind, and I will through and through
Cleanse the foul body of the infected world
If they will patiently receive my medicine."[45]

DE

FVRTIVIS

LITERARVM NOTIS,

VVLGÒ

DE

ZIFERIS

LIBRI IIII.

IOAN. BAPTISTA PORTA

NEAPOLITANO AVTORE.

CVM PRIVILEGIO

NEAPOLI,

Apud Ioa. Mariam Scotum. MDLXIII.

(60) TITLEPAGE TO *DE FURTIVIS LITERARUM NOTIS* (1563) by Ioan. Baptista Porta.

The 'AA' Signature of Light

The literary output of the Renaissance was printed in books that were remarkable for their ornamentation and emblematic illustrations of various kinds. These illustrations were the Renaissance equivalent of modern photographically reproduced pictures, and were usually of a very high quality of artistry and craftsmanship. Until the 17th century the illustrations were painstakingly carved by hand out of boxwood and inserted into the printer's block of text for printing. It was a laborious job, and a labour of love. The emblem blocks were, for this reason alone, of great value and carefully looked after. Usually each one was used more than once, in different publications and often by different printers - sometimes even migrating from country to country! A great deal can be learnt from the migration of these emblem blocks alone: one of the discoveries being that many of the blocks were almost certainly owned by the author or authors of the work, who had commissioned their manufacture, and another being that on the Continent and in England there seems to have been certain planned cooperative efforts in what was published and where, funded not by the printer-publishers but paid for by the authors and their patrons. Certainly this is true in most if not all the works by or associated with Francis Bacon and his 'fraternity in learning'.

The emblems were not just pretty pictures, but highly symbolic visual images of some truth or other. The makers employed a standard language of symbolism that had been passed on for centuries, even thousands of years in some cases, and which had been used in the great Mystery Schools and Academies of old. These kind of illustrations reached their peak of beauty and wealth of meaning with the Baconian-Rosicrucian literary output. Chief among the symbols adopted by Francis Bacon, and woven masterfully into complex emblems that were used by himself and his fraternity, was the so-called '**Double A**' sign - a clear signature both of his intentions and also of the guiding source or 'Light' of the fraternity.

The earliest date at which the '**AA**' seems to have appeared in print, in Renaissance literature, is in 1563. It appears in the form of an emblematic headpiece to John Baptista Porta's cipher book, *De Furtivis Literarum Notis, Vulgo De Ziferis,* printed in Naples. Porta had a subtle but profound influence on the Hermetic tradition as it developed in the Renaissance and moved its way through Europe, particularly in the field of bringing magic into the realm of organised science for his day and age - a magic which had its immediate origins in the *Magia* or spiritual-science of the Ancient Egyptians. The very fact that this emblem is used by Porta, and that it is Porta who uses it, is highly significant. It is Francis Bacon who next takes it up.

In Christopher Plantin's 1577 edition of Andreas Alciat's *Emblemata,* the '**AA**' appears in an entirely new emblem, No.XLV, which for some reason was introduced in place of the original emblem as published in the Aldine 'Fountain' edition of 1546.[1] Both the later and the earlier emblem carry the motto, "*In dies meliora,*" ("On to better days") and an identical verse, but the pictures are entirely different, seemingly linked only by the inclusion of a **boar** in both. But this boar and its related motto is a key. It is the chief figure of the Bacon family crest, and the motto of this emblem helps to give a more complete meaning of Francis Bacon's unique motto, "*Moniti meliora*".

Francis Bacon inherited the Bacon family crest of the boar, and also the family motto, "*Mediocra firma*" ("Mediocrity [the middle way] is safe [sure]"). But he personally selected and used two other mottos as having an especial bearing on his aims and objects in life. One was "*Plus Ultra*" ("More Beyond"), and the other was "*Moniti meliora*" ("Being instructed, to better things").[2] Both these mottos appear on emblem No.XLV of Alciat's 1577 *Emblemata.* Bacon's motto, "*Moniti meliora,*" is incomplete and should, in its fullness, read "*Moniti meliora sequamur*" ("Let us, being instructed [admonished], follow [strive after] better things"), which is taken from Virgil's *Aeneid.*[3] But, if we take the hint and read Bacon's deliberately incomplete motto with both the Virgil quotation and the motto of emblem No.XLV, we are given the meaning of "Let us, being instructed [*i.e.* by divine Providence], follow on to better days," which is a definite allusion to the Golden Age - the theme of the Renaissance and a particular object of the Baconian work: for the Golden Age is the Age in which the Muses reign.

The equivalent emblem from the 1546 edition depicts a boar's head carried on a salver. The **boar** is an ancient symbol of spiritual authority, and goes hand in hand with the **bear** which is a symbol of temporal authority. 'Boar' has a derivation from *bar,* meaning 'soul' or 'son' which, with its close association (as well as common word derivation) with 'bear', alludes to the mystery of *Ursa Minor.* The fact that Francis Bacon received, by adoption, the heraldic symbol of a boar, whilst the family emblem of his real father was a bear, is most significant and throws up many questions.

The boar signifies the incarnation of Adonis (or Mercury), the Messenger of the Sun Spirit (Apollo). The boar is the incarnate truth (Adonis) which is hunted after, and it is the sacrifice and meat which is devoured at the sacred feast - hence the origin of serving the boar's head up on a platter with an apple (symbol of knowledge) in its mouth. The 'boar' dies in order that others might live. Druids were known as 'boars', as a title of distinction. The boar is the male, the sow is the female, and the pig is the young swine. The swine, when raised for slaughter, used to be known as the hog, the earliest form of which can be found in the Welsh language as *hwch,* which is derived from the sacred words, *Hava-Oc,* meaning 'Bride-Spirit'. Interestingly (as the key to all truth lies in language), the Welsh for pig is *beak,* which is etymologically the same as *peak* (French, *pique*) which means a 'point', 'pike', 'spear'.

Emblem No.XLV from the 1577 edition depicts a boar rooting the ground with a swineherd standing by it, pointing with his left hand at the boar and his right hand at the two pillars of a ruined temple left standing solitarily above its crypt. Wrapped around these twin pillars (which are also known as the Pillars of Hercules, Boaz and Jachin, *etc.*.) is a banner displaying the motto "*Plus Oltra.*" Above and running along the spine of the boar is another motto, "*Ulterius,*" which, coupled with the pointing swineherd, gives an exhortation to the viewer to go further, to go beyond the Great Pillars. This is a most perfect Freemasonic emblem, alluding to the Royal Arch Degree and, in Christian terminology, to sacrificial death and resurrection into eternal life (*i.e.* the Golden Age). This is precisely what the boar symbolises, which was inferred symbolically in the 1546 emblem. In the 1577 emblem this ancient symbology is coupled with the equivalent Royal Arch symbology, and its meaning is made much clearer.

ANDREAE ALCIATI

In dies meliora.

Rostra nouo mihi setigeri suis obtulit anno,
Hæcq; cliens uentri xenia dixit habe.
Progreditur semper, nec retro respicit unquam,
Gramina cum pando proruit ore uorax.
Cura uiris eadem est, ne spes sublapsa retrorsum
Cedat, & ut melius sit, quod & ulterius.

(61) *"IN DIES MELIORA":* **Boar's Head** - emblem from Andreas Alciat's *Emblemata* (1546).

SPES. 53

In dies meliora.

Roſtra nouo mihi ſetigeri ſuis obtulit anno,
Hæcq; cliens ventri xenia (dixit) habe.
Progreditur ſemper, nec retro reſpicit vnquã,
Gramina cum pando proruit ore vorax.
Cura viris eadẽ eſt, ne ſpes ſublapſa retrorſum
Cedat: & vt melius ſit, quod & vlterius.

Ruſtica cõparatio Porci in paſtu progredientis ad Carolum V. Imper. ciúſque ſymbolon VLTERIVS, quo oſtenditur, Semper in melius procedendum.

D 3

(62) ***"IN DIES MELIORA":*** **the Boar, Swineherd and three Principals of the Royal Arch** - emblem from Andreas Alciat's *Emblemata* (1564).

204 ANDREAE ALCIATI

In dies meliora.

EMBLEMA XLV.

ROSTRA nouo mihi ſetigeri ſuis obtulit anno,
Hæcq; cliens ventri xenia, dixit, habe.
Progreditur ſemper, nec retrò reſpicit vnquam,
Gramina cùm pando proruit ore vorax.
Cura viris eadem eſt, ne ſpes ſublapſa retrorſum
Cedat; & vt melius ſit, quod & vlterius.

(63) ***"IN DIES MELIORA":*** **Boar, Swineherd and Ruins of Solomon's Temple** - emblem from Andreas Alciat's *Emblemata* (1577), depicting the 'AA' sign on a pyramid.

(64) FRANCIS BACON'S BOAR CREST - from the binding of a presentation copy of his book, *Novum Organum* (1620).

(65) ACHIEVEMENT OF ARMS OF FRANCIS BACON, Baron Verulam, Viscount St. Alban - frontispiece to Vol. II of Bacon's Works (1740).

Arms - Quarterly: 1 & 4. Gules on a chief argent, two mullets sable,* a crescent for difference.†
2 & 3. Barry of six, or and azure, a bend gules.

Crest: A Boar passant, ermine, a crescent for difference.

Supporters: Two Roman Officers.††

* Guillim's Heraldry (1610), p.120, gives: "Ruby on a chiefe Pearl, two Mullets, Diamond."

† i.e. as 2nd son of Sir Nicholas Bacon by Anne Bacon (*née* Cooke), the arms without cadency being those of Sir Nicholas Bacon, and later (after Sir Nicholas' death in 1579) of Anthony Bacon.

†† Like the two stars (mullets), signifying Castor and Pollux.

In the middle ground, between the swineherd and the ruins of Solomon's temple, is a little pyramid. It shows two of its faces, one in shadow and the other illumined. Carved on each of these faces is the letter **A**, thus depicting a 'double-A'. The pyramid itself, as a geometric figure, is a form of the letter **A** in three-dimensions.

The 1564 edition of Alciat's *Emblemata* depicts this emblem in yet another form - one that also hints at the Royal Arch of Freemasonry (which is the threshold of Rosicrucianism), and which depicts both the boar rooting the ground and also a boar's head being handed by a swineherd to an emperor. This emperor is seated at a table in a 'tabernacle', the opening of which is supported by two pillars, and is attended by two noblemen. These three are clearly intended to represent the three principals of the Royal Arch, whilst the offering of the boar is equivalent to the offering of the lamb or *Agnus Dei,* symbol of every master mason aspiring to the Royal Arch, as well as symbol of Hiram Abiff. In the distance the blazing sun rises as the Morning Star, signifying resurrection into eternal life - which is the meaning of the Royal Arch. An explanatory note beneath the accompanying verse states that the emperor is Charles V, the Holy Roman Emperor at that time. Charles V's special emblem comprised the two temple pillars united by a banner displaying the motto "*Plus Ultra*", which appears not on this version of Alciat's emblem but on the subsequent 1577 version. However, for those who are familiar with Charles V's emblem, it is implied in the note that "*Ulterius*" means the same as "*Plus Ultra*". This motto, "*Plus Ultra*", does not appear in this 1564 version, neither does the '**AA**' on the little pyramid. It remains for the 1577 edition to portray these - for what reason?

Emblem No.XLV is not the only emblem to have been deliberately changed in the 1577 edition of Alciat's *Emblemata.* Emblem No.II, entitled "*Mediolanum*" ("Milan"), was also changed, and again it has a swine as its central feature; but this time the swine is a sow. The sow is an ancient symbol of wisdom-knowledge, as the feminine counterpart to the spiritual authority depicted by the boar.[4] The sow was known as the Great Mother, and is a symbol of Venus. She represented the Mysteries - the form of Truth. She was imagined as moving (even flying!) over the surface of the earth, and from time to time she dropped a litter of piglets. Whenever and wherever she did, a Mystery School sprang up - the pigs becoming the initiates and then the adepts as they grew up. The 1546 emblem, which depicts a sow standing at the foot of an oak tree, is changed in the 1577 edition for another architectural picture. In this latter picture a city (Milan) with its temple is shown being built by freemasons. In the foreground is depicted a deep pit or quarry, into which three craftsmen are lowering a sow, referred to as a "little sow' in the accompanying verse. The sow is thus being buried as the 'foundation sacrifice', on a site where once the cult of Minerva was practised. Nearby, on one of the stones that have been quarried and are being worked on, is carved a single letter **A**.

This emblem alludes dramatically to the 'Lesser Mysteries' of Craft Masonry, wherein the Temple of Wisdom is built by the Freemasons, and in the 3rd degree of which the Master (personifying embodied wisdom-knowledge) is slain and buried as the foundation sacrifice. It goes hand in hand with emblem No.XLV: the former depicting the Lesser Mysteries of Craft Masonry, and the latter depicting the Greater Mysteries of the Royal Arch and those higher degrees of Illumination that lie beyond.

As for emblem No.XLV of the 1577 edition of Alciat's *Emblemata,* emblem No.II is foreshadowed by the equivalent emblem in the 1564 edition. In this latter version the emblem entitled "*Mediolanum*" shows nearly all the features depicted in the later 1577 version - the hill in the distance with its castle, the city and temple being rebuilt, the sow being lowered into the quarry-pit, *etc.* - but there is no letter **A** chiselled on any of the

ANDREAE ALCIATI

. Mediolanum .

(66) ***"MEDIOLANUM":*** **The Sow** - emblem from Andreas Alciat's *Emblemata* (1546).

8 EMBLEM. DEDIC.

Mediolanum.

Bituricis veruex, Heduis dat Sucula ſignum.
His populis patriæ debita origo meæ eſt,
Quam Mediolanam ſacram dixére puellæ
Terram: nam vetus hoc Gallica lingua ſonat.
Culta Minerua fuit, nũc eſt, vbi numine Tecla
Mutato matris virginis ante domum.
Laniger huic ſignum ſus eſt, animalq́; biforme,
Acribus hinc ſetis, lanitio inde leui.

(67) *"MEDIOLANUM":* The Sow and Building Solomon's Temple - emblem from Andreas Alciat's *Emblemata* (1564).

EMBLEMATA. 19

MEDIOLANVM.

EMBLEMA II.

BITVRIGIS veruex, Heduis dat ſucula ſignũ:
His populis patriæ debita origo meæ eſt,
Quam Mediolanum ſacram dixere puellæ
Terram: nam vetus hoc Gallica lingua ſonat.
Culta Minerua fuit, nunc eſt vbi numine Tecla
Mutato, Matris virginis ante domum.

B 2 Lani-

(68) *"MEDIOLANUM":* The Sow and Building Solomon's Temple - emblem from Andreas Alciat's *Emblemata* (1577, 1584), depicting the 'A' sign on a stone.

EMBLEMATA. 27

MEDIOLANVM.

EMBLEMA II.

BITVRICIS *veruex, Heduis dat sucula signum:*
His populis patriæ debita origo meæ est.
Quam Mediolanum sacram dixere puellæ
Terram: nam vetus hoc Gallica lingua sonat.
Culta Minerua fuit, nunc est ubi numine Tecla
Mutato, Matris virginis ante domum.
Laniger huic signum sus est, animalque biforme,
Acribus hinc setis, lanitio inde leui.

(67) ***"MEDIOLANUM":*** **The Sow and Building Solomon's Temple** - emblem from Andreas Alciat's *Emblemata* (1564).

Note that this picture is identical to that of the 1577/1584 picture, except that the letter 'A' is missing from this 1584 version. Instead an 'AA' headpiece, placed on end, is used to form the dexter border of the picture. In the centre of the 'Double A' - the sign of creativity - is the highly significant letter 'R'.

The sinister border is composed of a 'Bride' headpiece set on end. It has at its centre a symbol for the receptivity and understanding of the mind - the 'scallop-shell' crown of the Bride, with her veil drooping beneath; but her face is hidden in this representation of Truth by a vertical bar joining the two 'Commas' - symbols also of the intelligence of the mind (and of Cabbala). Facing into the centre of this 'Bride' headpiece are two coneys or hares, symbols of the initial state of initiation in the Druidic Mysteries.

The top border depicts a hunt, with hounds chasing two hares (who are about to leap into the shelter of their cave-like burrow). This is a portrayal of the preparatory stage leading to initiation in the Lesser Mysteries, as symbolised in the Druidic Mystery teachings. (See FBRT Journal I/3. *Dedication to the Light).*

The lower border shows the face of the Bride, crowned with flowers and fruits.

stones as there is in the 1577 edition. Although all of these features are present (except the '**A**'), the 1577 version is a completely new and different drawing, even though it would have been easy enough to have added the letter **A** to one of the stones in the 1564 version if that was all that was wanted. Why? Perhaps the obelisk has an answer.

The 1577 "*Mediolanum*" design was repeated in the 1584 edition, but when the 1589 edition of *Emblemata* was published, the "*Mediolanum*" emblem, although seemingly identical to the previous 1584 and 1577 editions, appeared with the letter '**A**' removed from the stone in the picture. Why was this done? Well, look at the border to the picture and you will see that a beautiful little '**AA**' headpiece has been introduced, placed on end to act as the dexter border (heraldically speaking) of the emblem. There could not be a clearer sign given of the Arcadian influence upon and control of these emblem books, in order to deliver the Areopagean message.

These emblems should be of great import to all Freemasons, when taken in the context of other related happenings, as they would appear to be a deliberate announcement to the world as to when ancient Freemasonry first began to be revived and reorganised into modern Freemasonry. It seems to be a deliberate sign concerning its origins - its nativity - in the mind and art of its otherwise hidden creators.

These Alciat emblems should be set beside another revealing picture that was also published in 1577, in England; namely, the illustration on the titlepage of Dr. John Dee's *General and Rare Memorials pertayning to the Perfecte Arte of Navigation.* It should also be noted that the preparation and publication of the 1564 edition of Alciat's *Emblemata* coincided with the period when Dr. John Dee was visiting the court of the Duke of Urbino in Italy, and when he published his *Monas Hieroglyphica* in Antwerp (31st March 1564). It would seem that there was a carefully executed conspiracy in England and abroad to announce and record this great birth, and to prepare the way for the man of whom Paracelsus had prophesied - Francis Bacon.[5]

The choice of Andreas Alciat's *Emblemata,* and hence of Milan, Alciat's city, is interesting also. Perhaps it is unnecessary to point out Alciat's initials! A mind like Francis Bacon's, or John Dee's, would never pass by such an opportunity or jest. As for Milan, once known as *Mediolanum,* it was a city of Cisalpine Gaul, located in a beautiful plain, which was first founded by Gallic invaders (the Bituriges) under Bellovesus. It was captured by the Romans in 222 BC. In course of time it was referred to as the 'New Athens', for it became the residence of the emperors of the Western Empire, until the invasion of Attila in 450-452 AD caused the Romans to move their imperial headquarters to Ravenna. The city revived, however, under the Ostrogoths, but later fell to the Burgundians and lost its splendours. During the Middle Ages it became a flourishing and opulent place of trade, a status which it has never since lost. The memory of its being famous as the 'New Athens' under the Romans is alluded to in the verse accompanying emblem No.II, where it says that the cult of Minerva (*i.e.* Athena) was established there. The clear purpose of the emblem is to show symbolically a new edifice - a temple and city of imperial standards, dedicated to Athena and the Muses, being raised on the ruins of the old - its Architect and Founder, who is also the foundation sacrifice, being represented by the boar or sow. It should also be noted that, in the exceptionally Freemasonic Shakespeare Play, *The Tempest,* which *begins* the 1st Folio, Prospero is described as being the rightful Duke of Milan:-

> "...........my Milan, where
> Every third thought shall be my grave."[6]

The nose of the swine is said traditionally to bear a curious resemblance to a written letter **m**. When printed, the capital letter **M** has a close resemblance to a double-A (**AA**). Thus, when the boar roots the ground, he can be said to imprint the '**AA**' on the earth with every thrust of his snout. Adonis, Apollo's beloved messenger, is represented by the boar. Adonis is Mercury, also known as Hermes or Thoth in the aspect of a messenger/teacher of wisdom; and Mercury is the great scribe or author-artist, who writes the words of truth upon the ground of matter.

One of the derivations of the '**AA**' is as the grieving cry "Ai, Ai" of Apollo, uttered when his beloved youth, Hyacinth, the fifteen-year old son of Amilchar,[7] was accidentally slain by a quoit (*i.e.* discus) thrown by Apollo during the Spartan games. The quoit hit the ground and bounced up at Hyacinth, who was over-eager to retrieve it, and sliced off the top of the youth's head.[8] Weeping, Apollo changed the slain youth into a flower, his spilt blood determining its colour. The god of Light then marked the flower and its leaves with his mourning sighs, "Ai, Ai," written as the Greek letter **U**,[9] which letter is a symbol of spiritual knowledge or illumination. **U** is the Greek *upsilon,* corresponding to the English letter **V**, which letter is the image or 'imprint' of **A**.

> "Thee my lyre, thee my songs, shall ever celebrate; and, changed to a new flower, thou shalt bear an inscription of my sighs. The time too shall come, when a very powerful hero shall be changed into this flower, and his name read upon thy leaves.'
>
> "While these things are uttered by Apollo's prophetic mouth, lo, the blood, which falling upon the ground has stained the grass, ceases to be blood, and a flower more bright than Tyrian purple springs up, assuming the same form as the lily, but that in the former is a purple colour, in the latter that of silver. But this is not enough for Phœbus, for he was the author of the honour now bestowed: he marks his own sighs upon the leaves, and the flower has *Ai, Ai* drawn upon it in funeral characters."[10]

Hyancinth and Adonis are one and the same, and the story relates to the mysteries of initiation - particularly to the 3rd degree in which the initiate dies, slain for love in his quest for light, but is reborn in a new and more beautiful form (the 4th degree). The colour of the hyacinth was said to be the express image of *Air,* as a soft and sweetly fragrant breath - the celestial form of the soul and of the Holy Breath or Ghost. The robe of a high priest was made the colour of hyacinthine blue (*i.e.* violet-blue) for these reasons.[11] With Hyacinth (Mercury) represented as a boar, it should come as no surprise to know that Apollo was said to take the form of a swineherd:-

> "..............and the fire-rob'd god
> Golden Apollo, a poor humble swain as I seem now."[12]

When the '**AA**' is used in the headpieces to 'Baconian' works, the letters are sometimes shown entwined with hyacinth flowers. This flower was referred to as a "herb" by the Elizabethan authors, and allusively as a "weed" (from Latin *herba,* meaning 'weed'). Note the meaning in one of Bacon's own prayers:-

> "The state and bread of the poor and oppressed have been precious in mine eyes; I have hated all cruelty and hardness of heart: I have (though in a despised weed) procured the good of all men."[13]

The actual symbol and use of the letter **A** goes back into deepest antiquity. Like all the profoundest truths, it is simple but subtle. "**A**", pronounced "Aaah!", is the basic sound of love. It is the sound of true love or heart love - the natural sound we all tend to make when something arouses that kind of feeling or emotion within us (a baby, for instance, or an exceptionally beautiful experience). "Ah!", or "Ah-hah!", is also the sound we naturally tend to make when we suddenly see something that catches our attention and fills us with wonder, or arouses our curiosity. Then, in a rather more drawn out fashion, as "Ahh!", it forms our natural expression when we finally come to understand something (*i.e.* a second kind of 'seeing'). Finally, there is a fourth meaning to its expression - as the final and definitive "Ah!", when we have accomplished something and can truly say, "Ah, now I know!" We know because we have done it: we have experienced it as a truth. This is real knowledge. The third "Ahh!" signifies only intellectual knowledge or partial illumination. Full illumination comes only when the deed is done; and that deed, if good, will be the direct expression of heart-love, and will give a revelation of that love, the initial "Aaah!".

In this way, something as simple as this one sound summarises and expresses every stage of a perfect and beautiful life, a life based on divine love. From the initial **A** to the final **A**, all life is enfolded. The symbol of the equilateral triangle, signifying the Holy Trinity which is the very basis of life, was devised to represent this sound of life. The first **A** symbolises the creative Spirit of divine Love, whose radiant warmth is Light (or Sound). The last **A** is the revelation of this perfect Love-Light in matter, as the true and enlightened Soul. Thus the first **A** stands for the Spirit of Truth, and the final **A** for the Bride or Soul of Truth. The former incarnates into matter in order to do its work; the latter is the perfected life form that evolves unto its spiritual source. To represent this incarnation of the Spirit and evolution of the Soul, the simple device of drawing the first **A** with point down, and the second **A** with point up, was adopted. Their 'marriage' in the form of the six-pointed Christ Star thus signifies the ultimate marriage of Spirit with Soul (the Bride), which occurs when the latter has become the perfect manifestation of the former, so that they are truly One.

The Ancient Egyptians, for instance, used the '**AA**' for precisely these reasons. When they wrote the letter **A** as a hieroglyphic, they veiled it under the apparent form of a hand-plough - an **A** held on its side and pointing forwards. This also meant that the letter **T**, signifying the Truth of divine Love (*i.e.* the Word of Wisdom) could be subtly indicated, embodied in the form of the 'plough'. In parables, or myths, the sages then could teach the ordinary person that God is like a Great Ploughman, and that Love (or Truth) is his plough by which the ground is tilled and made ready for seed, *etc.*. Arthur (**A-T**ah) was known as the Great Ploughman. *Ursa Minor* (Arcadia) is known as 'the Little Plough', and *Ursa Major* as 'the Great Plough'.

The **A** has always been used as the first letter of the Alphabet - the primary letter from which all other letters proceed. This knowledge is the basis of all cabbala. The Alphabet itself represents, cabbalistically, the divine Word out of which all other words are made. **A** is the very source and summation of that Word. Hence it is that the **A** also signifies a spring or fountain-head, as the source of the Fountain of Life - of all love, wisdom and power.

The date when it appears that Francis Bacon first used the '**AA**' publicly, in the 1577 edition of Alciat's *Emblemata,* is significant. For immediately prior to this, in September 1576, Francis had been recalled to England by the Queen and then sent back to France again with a special mission, having had revealed to him by the Queen something of such great import that he was prepared to give up all hope of marriage to his love, Marguerite de Valois, and to devote himself and his great talents wholly and uniquely to the betterment of mankind. In 1576 he was aged 15 years old - the age of Hyacinth, the youth

(70) EGYPTIAN HANDPLOUGH, emblem of the rulers of the earth and shepherds of the people.

whom Apollo loved and whose 'second-coming' as a "very powerful hero" had been prophesied by Ovid.

Then, in February 1578[14] Francis signed the dedication to his remarkable book, *Académie Françoise,* which he wrote whilst in France, and which was published in France that same month. It was dedicated to the French king, Henri III, and it would appear to have been written during January 1578 as a result of Francis having attended the rather special Academy of Michel Montaigne in the latter weeks of 1577.[15] In the *Académie Françoise* he makes a particular and pointed emphasis of the letter **A** in naming pseudonymously the four students at the Academy: *viz.* **A**ser, **A**mana, **A**ram and **A**chitob, which can be shown to represent the four **A**'s referred to above.[16] In the dedication he thanks Henri III for his hospitality when he, Francis, attended the Court of Henri at Blois, held there for a meeting of the Estates-General (November 1576 - February 1577). The time there had been one of exceptional festivity and cultural activity, for which Francis likened the French king to King Solomon - a name not without significance. It is therefore between that date (February/March 1577) and the date of publication of *Académie Françoise* (February 1578) that the 'public launch' occurred, so as to speak. The Blois meeting, at which the learned from "everie nation" were present, may have been a time when a major step was taken. Then there is the intriguing 'gap' in Francis' life in 1577, between July and late September, when all that we know of Francis is that he had undertaken a hard and perilous journey, and had returned to Sir Amyas Paulet and the French Court at Poitiers worthy of Sir Nicholas Bacon's favour:-

> "I must tell you that I rejoice much to see that your son, my companion, by the Grace of God, passed the brunt and peril of this journey..........Your son is safe, sound and in good health, and worthy of your fatherly favour."[17]

This journey may have had nothing to do with the question posed above, and may have been solely a diplomatic or an intelligence-gathering mission of some kind, as suggested before.[18] The planning and organisation that lay behind the concerted effort by both Francis Bacon (in France) and John Dee (in England) to announce publicly the 'birth' of modern Freemasonry and Rosicrucianism were almost certainly hatched in England, in the months between Spring and September 1576 when Francis had been specifically summoned home by the Queen, and had been drawn completely into the Walsingham-Dee circle in order to train as a top agent or intelligencer in the Elizabethan Secret Service.[19] During those months he was with his foster-brother Anthony, who was privy to Francis' visions and plans from the first, and who had assisted in bringing them to birth; and he certainly must have been with Dee, who, as the original '007', was one of the master-minds of the English intelligence network. Furthermore, without reasonable backing - moral, financial and otherwise - Francis is unlikely to have been able to have launched his project, especially at such a young age.[20] The return by Francis to France in September 1576, in the entourage of Sir Amyas Paulet, charged with a special and secret mission by Her Majesty, thus takes on even greater significance.

It seems fairly clear that Francis had managed to obtain support and sponsorship of some kind from the 'Family (or House) of Love', of which Christopher Plantin, the Frenchman who set up a printing business in Antwerp and who printed the 1577 edition of Alciat's *Emblemata,* was one of the most noted and ardent members. The House of Love was a fellowship of initiates and adepts who were responsible for the deep study and transmission of occult and mystical truths during the European Renaissance, and who became particularly potent (or recognisable) in the latter half of the 16th century. The 'Family' believed in tolerance, friendship, the promotion of learning and artistry in service of truth, and, most of all, that the purity and strength of a person's inner spiritual life was of far more importance than any outer religious observance or political point of view, and that charity was the greatest of the virtues. Members pursued the liberal arts and sciences as in the days of the Mystery Schools; and, in fact, although the members of the House of Love were scattered throughout Europe, yet it formed a true and effective Mystery School.

It seems almost certain that Francis Bacon was involved in this 'Family' from the start (as discussed in F.B.R.T. Journals I/3 & I/4),[21] it being thought likely that Sir Nicholas Bacon had been a member, who had the strongest influence on Francis during the latter's younger, formative years, as Francis' foster-father. Then, later (from 1575 to 1579), this special upbringing was followed by Francis' experiences in France at the French Court.[22] The House of Love had a good representation in France at the time that Francis was resident there, especially in Paris where the apothecary, Pierre Porret, acted as an agent for the Plantin publishing business in the rue Saint-Jacques. Porret was not only one of Plantin's oldest and closest friends, but was also "an agent, not only for the Plantin publications, but also for the diffusion of Familist teachings."[23] The Valois Court of Henri III was strongly influenced by the Familists, and Henri III even asked Plantin to establish himself as a printer in Paris as "royal printer for ten languages."[24] This offer, which was in fact refused, took place in 1577.

After its appearance in the 1577 edition of Alciat's *Emblemata,* the next known book to carry the '**AA**' device was *De Rep. Anglorum Instauranda libri decem, Authore Thoma Chalonero Equitite, Anglo,* published in 1579 by Thomas Vautrollier in London. This was followed by *The Essayes of a Prentise in the Divine Art of Poesie* of King James VI of Scotland (published 1584/5). The latter book bears the identical '**AA**' headpiece as Johanne Baptista Porta's *De Furtivis Literarum Notis. The Essayes of a Prentise* is said to have been "Imprinted at Edinburgh by Thomas Vautrollier"; but the fact is that Vautrollier's printing business was in London. He was the well-known London printer, whose printing shop was in Blackfriars and whose widow married Richard Field, the successor to the printing business. The latter became, in 1589, the printer of *The Arte of English Poesie* and, in 1593, of *Venus and Adonis.*[25]

As far as I can ascertain, the *De Rep. Anglorum Instauranda* (1579) carried the first English '**AA**' headpiece; whilst ten years later *The Arte of English Poesie* (1589) carried the second English '**AA**' headpiece. *Venus and Adonis* (1593) bore a differently designed '**AA**' headpiece over its dedication to the Earl of Southampton. This dedication announced *Shakespeare* to the world for the first time. Thus, in 1579 and 1584/5, Thomas Vautrollier published, in England and Scotland respectively, the first appearance in print of the '**AA**' device in those countries. Both the works on the subject of poetry - *The Essayes of a Prentice* (1584/5) and *The Arte of English Poesie* (1589) - also carried an emblem or trademark bearing the motto, "*Anchora Spei,*" which takes on particular significance when we find that Francis Bacon, writing years later to King James, refers him to "your Majesty's Royal promise (which is to me *Anchora Spei*)."

From these dates onwards many of the finest examples of English literature, of the Elizabethan-Jacobean period, carried the '**AA**' device in one form or another (usually as a headpiece), including *Shakespeare's Sonnets* (1609), the 1st folio edition of the *Authorised Version of the Holy Bible* (1611), the *Genealogies* in the 1st quarto edition of the Scriptures (1612) and Shakespeare's 1st folio edition of *Comedies, Histories and Tragedies* (1623).

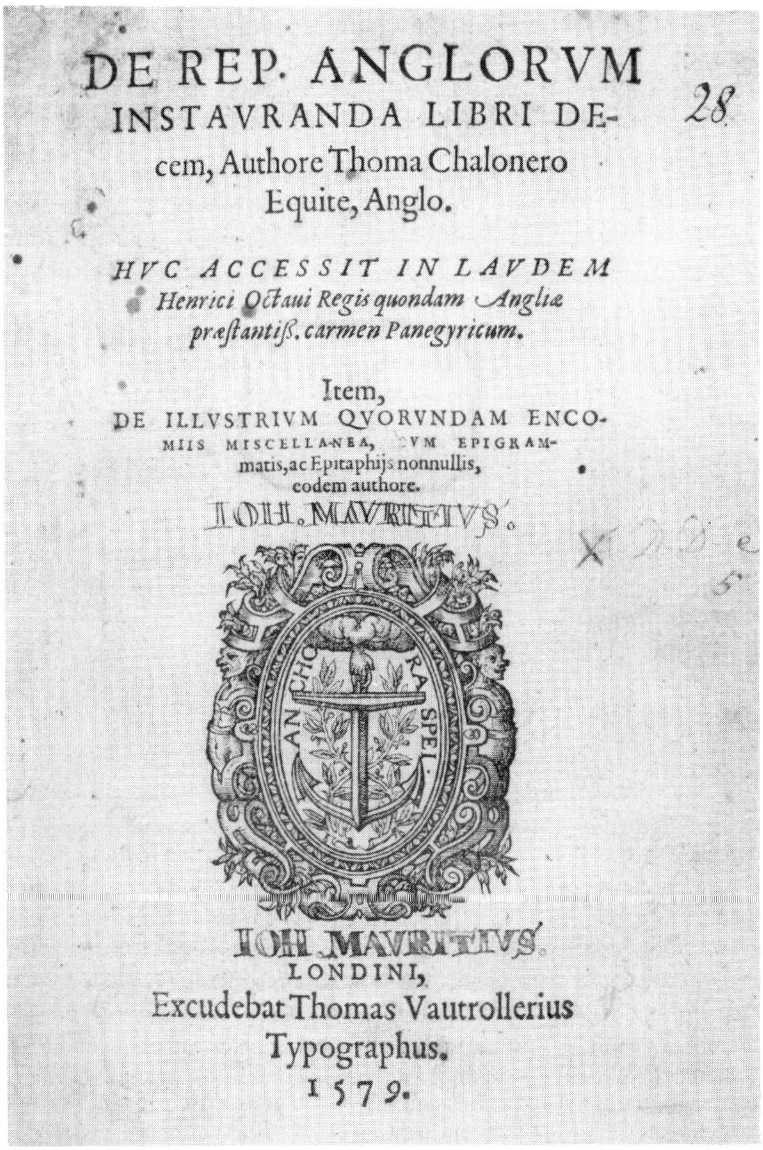

DE REP. ANGLORVM INSTAVRANDA LIBRI DE-cem, Authore Thoma Chalonero Equite, Anglo.

HVC ACCESSIT IN LAVDEM Henrici Octaui Regis quondam Angliæ præstantiß. carmen Panegyricum.

Item,

DE ILLVSTRIVM QVORVNDAM ENCOMIIS MISCELLANEA, CVM EPIGRAMmatis, ac Epitaphijs nonnullis, eodem authore.

LONDINI,

Excudebat Thomas Vautrollerius Typographus.

1579.

(71) TITLEPAGE TO *DE REP. ANGLORUM INSTAURANDA* (1579) by Thomas Chalonero, depicting the "*Anchora Spei*" emblem, printed by Thomas Vautrollier.

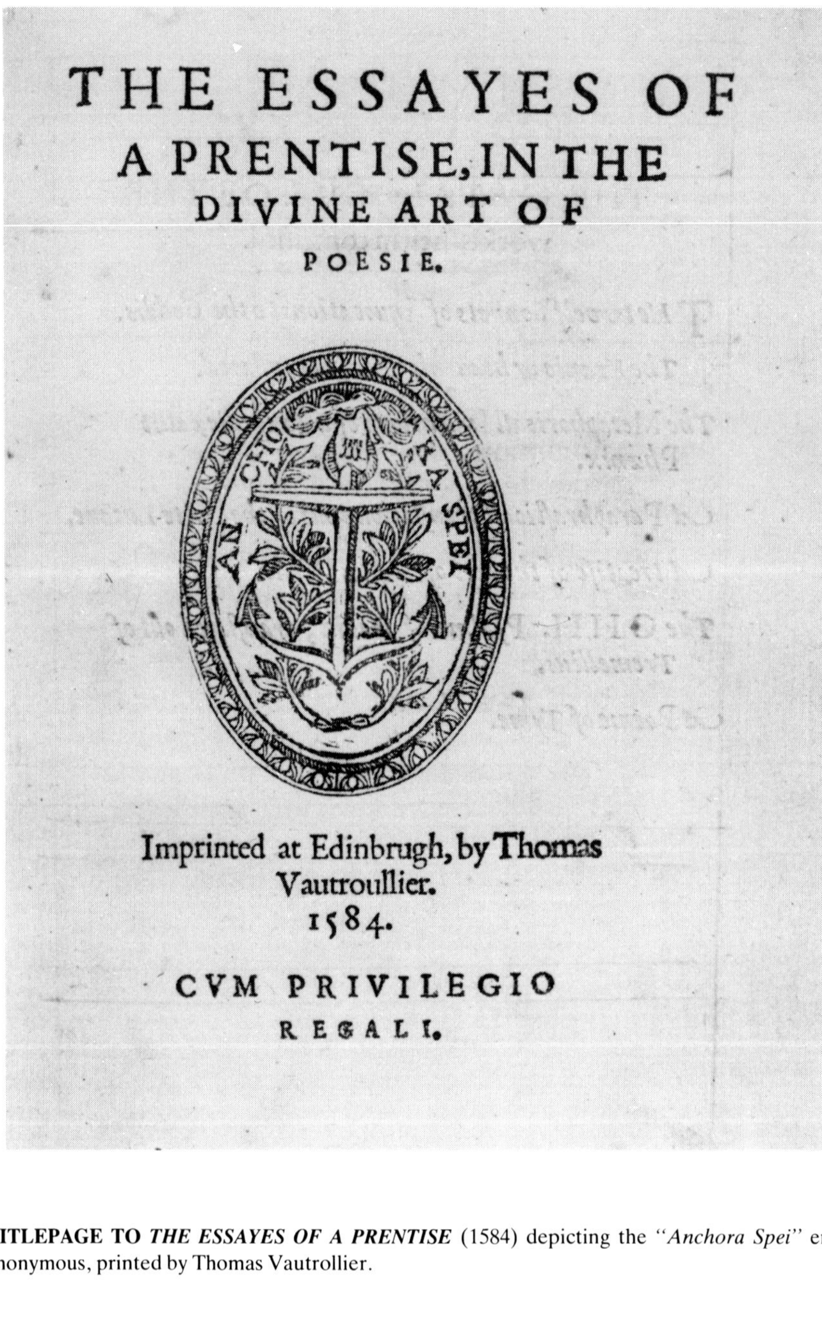

(72) TITLEPAGE TO *THE ESSAYES OF A PRENTISE* (1584) depicting the *"Anchora Spei"* emblem - anonymous, printed by Thomas Vautrollier.

THE ARTE OF ENGLISH POESIE.

Contriued into three Bookes: The firſt of Poets and Poeſie, the ſecond of Proportion, the third of Ornament.

AT LONDON
Printed by Richard Field, dwelling in the black-Friers, neere Ludgate.
1589.

(73) TITLEPAGE TO *THE ARTE OF ENGLISH POESIE* (1589) depicting the *"Anchora Spei"* emblem - anonymous, printed by Richard Field.

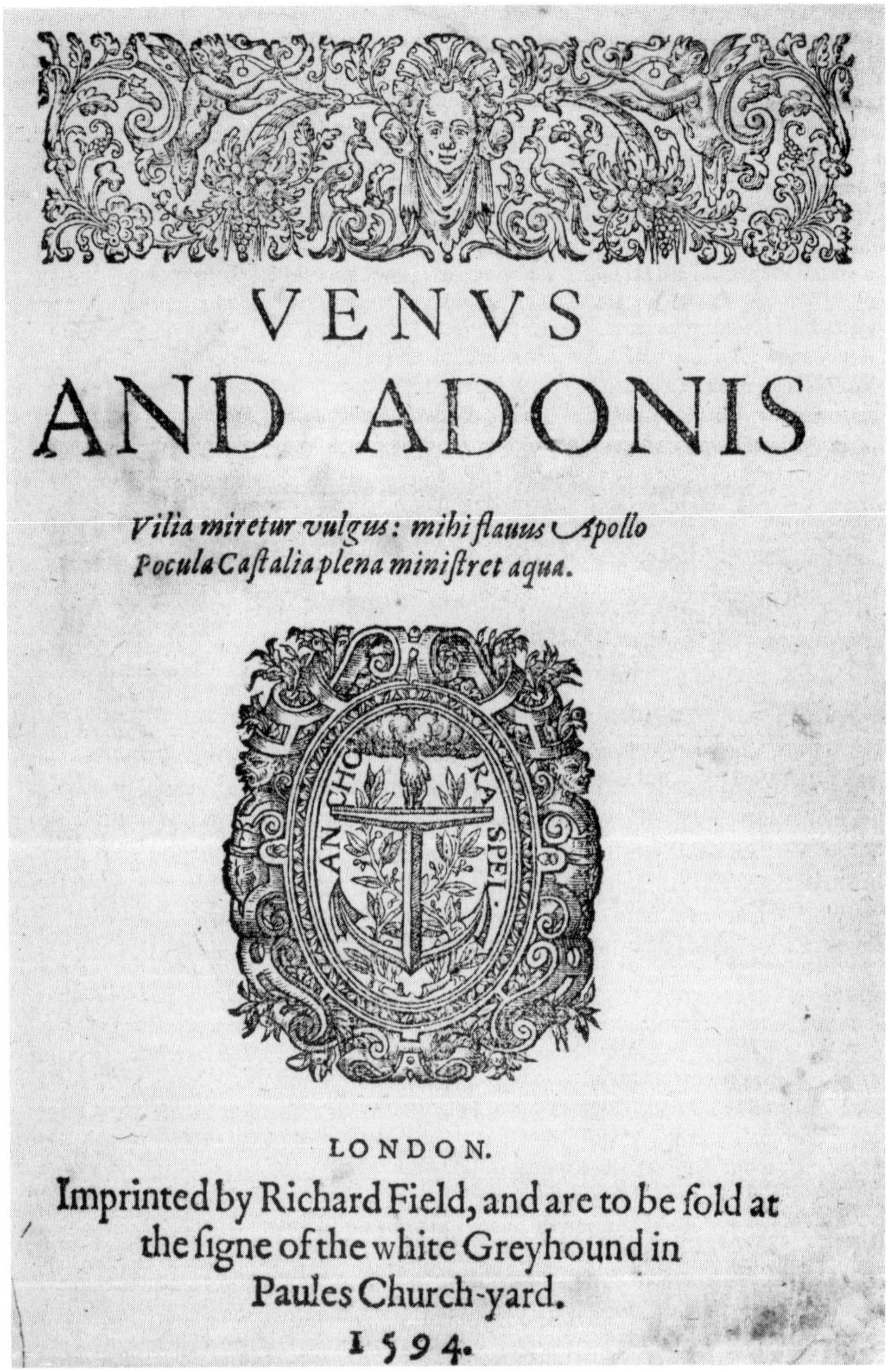

VENVS
AND ADONIS

Vilia miretur vulgus: mihi flauus Apollo
Pocula Caſtalia plena miniſtret aqua.

LONDON.
Imprinted by Richard Field, and are to be ſold at
the ſigne of the white Greyhound in
Paules Church-yard.
1594.

(74) TITLEPAGE TO *VENUS AND ADONIS* (1594) depicting the *"Anchora Spei"* emblem - William Shakespeare, printed by Richard Field.

The Arcadian headpiece depicts the face of the Bride, also known as Truth (or Truth Revealed), crowned and with her veil lowered, revealing the shining "countenance of the Lord".

TO THE RIGHT HONORABLE
Henrie VVriotheſly, Earle of Southampton,
and Baron of Titchfield.

Right Honourable, I know not how I ſhall offend in dedicating my vnpoliſht lines to your Lordſhip, nor how the vvorld vvill cenſure me for chooſing ſo ſtrong a proppe to ſupport ſo vveake a burthen, onely if your Honour ſeeme but pleaſed, I account my ſelfe highlie prayſed, and vow to take aduantage of all idle houres, till I haue honoured you vvith ſome grauer labour. But if the firſt heyre of my inuention proue deformed, I ſhall be ſory it had ſo noble a god-father: and neuer after eare ſo barren a land, for feare it yeeld me ſtill ſo bad a harueſt, I leaue it to your Honourable ſuruey, and your Honor to your hearts content, vvhich I wiſh may alwayes anſwere your owne vviſh, and the vvorlds hopefull expectation.

Your Honors in all dutie,

William Shakeſpeare.

(75) DEDICATION PAGE TO *VENUS AND ADONIS* (1594), depicting an Arcadian 'AA' headpiece - signed by 'William Shakespeare' and dedicated to Henry Wriothesly, Earl of Southampton.

Southampton was a scholar and a staunch friend of Robert, Earl of Essex. This is the first appearance in print of the name of 'Shakespeare' as author of any literary work. *Venus and Adonis* was published on (or near) St. George's Day, 1594, hence the tradition that 'Shakespeare' was born on St. George's Day.

The 1579 date for the first publication of the '**AA**' device in England, in the *De Rep. Anglorum Instauranda,* corresponds precisely with the return of Francis Bacon to England and the coming together of the English Areopagus of poets in order to create the English Renaissance. It is therefore not by chance that Gabriel Harvey, in one of his famous letters to 'Immerito' that were printed in 1980 and which announced the existence of the English *Areopagitæ* to the world for the first time, wrote:

> "Every one A per se A his terms and braveries in print."

Ten years later Harvey again referred to "A per se A", associating the expression with the "odd man *Triu Litteraru* " whom he mocks as being like the three-bodied, three-headed monster, Geryon, who goes under the three names of 'Hatchet', 'Greene' and 'Nash'. By 'Hatchet', Harvey means 'Lyly', *alias* 'Euphues', *alias* Francis Bacon, who, whilst writing under these three masks, had been playing a public game of wit and mocking jest with Harvey - a valuable and entertaining literary game by means of which new words, ideas and hints were foisted upon the public, preparing the ground for the more serious works of drama and literature to take root. The name 'Hatchet' is derived from 'Pap-hatchet', which itself is derived in jest from a pamphlet published anonymously in 1589 entitled *Pappe with an Hatchet.* This pamphlet was principally concerned with the Martin Marprelate dispute, but it contained an incidental 'snipe' at Harvey as part of the game of gibing wit. Harvey replied to this taunt in that part of his *Pierce's Supererogation* dated November 1589:

> "Pap-hatchet (for the name of thy good nature is pittyfully growen out of request), thy old acquaintance in the Savoy when young Euphues hatched the eggs that his elder friends laid (surely Euphues was someway a pretty fellow: would God Lilly had always been Euphues and never Pap-hatchet): that old acquaintance now somewhat straungely saluted with a new resemblance is neither lullabied with thy sweet Papp nor scarre-crowed with thy sour hatchet."

Later on, in another part of *Pierce's Supererogation,* that deals with the gibes at him by 'Greene', 'Lyly' and 'Nash', he versifies:

> "Aske not what Newes? that come to visit wood:
> My treasure is Three faces in one Hood:
> A chaungling Triangle: a turncoat rood.
>
> "Three hedded Cerberus, wo be unto thee:
> Here lyes the onely Trey, and rule of Three:
> Of all Triplicities, the A B C."

Harvey then goes on to say:

> "Somebody oweth the three-shapen Geryon a greater duty in recognisance of his often promised curtesies; and will not be found ungrateful at occasion. He were very simple that would feare a conjuring Hatchet, a rayling Greene, or a threatening Nash.....
>
> "These, these were the only men that I ever dreaded: especially that same odd man Triu Litteraru that for a linsey-woolsie wit and a cheverill conscience was A per se A."

In *Pierce Pennilesse,* "Nash' makes it absolutely clear that the 'A per se A' is his especial name, given to him, as it were, by his "Godfather Gabriel" - referring exoterically to Gabriel Harvey, his friend and teacher in rhetoric, and esoterically to the Archangel Gabriel, the divine agent in the annunciation of any great birth of Light:

> "A per se, A can doe it; tempt not his clemency too much. A per se, A? Passion of God, how came I by that name? My Godfather Gabriel gave it me, and I must not refuse it."[26]

Finally, consider the following words written by Francis Bacon in his *De Interpretatione Naturæ,* which link together the boar, the **A** or "Ai", and the authorship of tragic plays:-

> "Sus rostro si forte humi **A** literam unam impresserit, num propterea suspicabere integram trageodiam veluti literam unam, ab ea posse describi?"[27]

which, translated, renders:-

> "Supposing that a swine [boar] should impress a single letter **A** in the ground with its powerful snout, do you suppose therefore that an entire tragedy [tragic play] could be transcribed by that swine even as the single letter can?"

When one knows that, symbolically, all letters, words and everything else come from the letter **A**, and that this is what the story is all about, then the answer is obvious and Bacon's hint rings out clear as a bell.

(76) ANTHONY BACON, aged 30 years - miniature painted (1588) by Nicholas Hilliard.
Inscription: *"Attici amoris ergo."*
Translation: "Orators/poets of the Attic School because of love."

This portrait is usually described as an "Unknown Man clasping a hand from a cloud." No one else has yet successfully identified this unknown man with blue-grey eyes and brown curling hair. Dr. Strong's suggestion is that it might be Lord Thomas Howard, 1st Earl of Suffolk, in his younger, courtier days. I strongly suggest that this is in fact a portrait of Anthony Bacon, painted during his secret return to England during the months of the Spanish Armada, 1588. Like his foster-brother Francis, Anthony was a concealed writer and poet, known as such to a few of his friends at home and abroad, and referred to as a 'phoenix'.

The English Areopagus and, later, the Baconian 'Order of the Knights of the Helmet' (the 'Invisible College' of the Rose Cross brethren), used Apollo, Pallas Athena, and the Athenian Acropolis and Areopagus (as well as Mount Parnassus) as symbols of their inspirational source and 'lodge' of their fraternity. The inscription refers directly to these "Attic" (or Athenian) poets, initiates of the Platonic-Orphic Mysteries of Love. The poet's right hand is held up, clasping a female left hand of God issuing down from a cloud. From emblems and portraits of the time we know that the 'hand-in-hand' implies 'plighted faith', particularly that faith or love-act made with God whereby we put our hand into the hand of God. The aspect of Deity which is directly involved is portrayed (by the 'female' left hand) as feminine, alluding to the Holy Ghost or Holy Breath - symbolised by the Dove - which is the Divine Intelligence of Love that bears and brings into manifestation the Wisdom or Light of Love. This particular aspect of Divinity is the one which Pallas Athena represents. To the Greeks she was also known as Sophia; and a philosopher is a "lover of Sophia" - one who may become a true poet capable of manifesting the Word of God.

A Rare and Unaccustomed Suite

Whilst Francis Bacon was busily engaged in his poetic and dramatic writings, and trying to lay the foundations of an English Renaissance and a vastly improved English language with the help of the English Areopagus, he was ostensibly studying law at Gray's Inn. He had been 'admitted' to Gray's Inn, along with Anthony Bacon, on 27th June 1576, but it was not until after his return to England in 1579 that Francis took up residence there to study law. The first that we actually know for certain that Francis was living at Gray's Inn is from a letter he wrote to a Mr. D. Doyly in Paris, dated "From Gray's Inn, the xi[th] of July, 1580."[1] (Mr. Doyly had accompanied Anthony Bacon to Paris in September/October 1579. Anthony had remained for a while in Paris before setting out, in February 1580, on a highly secret and dangerous intelligence mission to Portugal.[2] He returned to Paris in May 1580 and, shortly afterwards, from thence was recalled to England. Mr. Doyly, however, was left in Paris, and was trying to send letters to Anthony *via* Francis, hence Francis' reply to Doyly on 11th July.[3]) Francis probably took up his residence in May that year, as the Gray's Inn Order of Pension dated 13th May 1580 reads:-

> "Mr. Francis Bacon in respect of his healthe is allowed to have the benefitt of a special admittance with all benefitts and privileges to a speciall admittance belongeng, for the fyne of xls."[4]

This special admittance meant that Francis could keep his own times, was freed from the obligation of keeping Commons and enabled to choose his own diet. He had already been made an 'Ancient' of Gray's Inn by Order of Pension dated 21st November 1576,[5] which bestowed certain other privileges including freedom from being bound by vacations. By this same Order, Edward and Anthony Bacon, half-brothers and sons of Sir Nicholas, were granted permission to reside in Sir Nicholas' suite of chambers in the absence of the two elder Bacon sons (Nicholas and Nathaniel) who had retained the priority right to the suite after their father's death. But Francis was not given permission to use these chambers (being designed for two persons at a time only), and he did not have his own special chambers until 1588, by which time he had become a Reader. To start with, therefore, Francis took up lodgings with a Mr. Fulwood in Fulwood House.

On 16th September that year Francis wrote two letters from Gray's Inn to the Burghleys: one to Lady Burghley (the sister of his foster-mother, Lady Anne Bacon) and one to Lord Burghley, his legal guardian. In these letters Francis reminds Lord Burghley of a special suit that he had discussed with him, and asks Burghley to fulfil his promise of tendering that "rare and unaccustomed" request to the Queen. At the same time Francis asks Burghley to be his patron "both herein and elsewhere" (*i.e.* both in connection with tendering Francis' suit and also in connection with other matters). One month later, on the 18th October, Francis wrote again to Lord Burghley from Gray's Inn, this time to thank him for tendering his suit to the Queen, and implying in the letter that Her Majesty had received the idea favourably, for which he, Francis, was most grateful:-

"And now seeing that it hath pleased her Majesty to take knowledge of this my mind, and to vouchsafe to appropriate me unto her service..."[6]

Francis offered the Queen his utmost services - more than would normally be required or even thought of by the Queen - in return for the granting of his suit. But what was his suit, and was it actually granted? The answer to the second question is no, it was never actually granted; and for thirteen years Francis was kept wondering and waiting by the Queen, in the forelorn hope that she might keep her promise and grant the request. Throughout that time he kept his side of the bargain; but, by the time thirteen years were past, he could stand it no more and made himself, and what his suit must have been about, vividly clear in a final hard-hitting letter to Burghley. This occurred in 1593, when Francis reached the age of 31 years old. But, between 1580 and 1593, he kept pursuing his suit, *via* Lord Burghley, Sir Francis Walsingham and Sir Christopher Hatton. The 1593 letter is reproduced here in full, as it is well worth reading:-

"TO MY LORD TREASURER BURGHLEY.

"My Lord,

With as much confidence as mine own honest and faithful devotion unto your service and your honourable correspondence unto me and my poor estate can breed in a man, do I commend myself unto your Lordship. I wax now somewhat ancient; one and thirty years is a great deal of sand in the hour-glass. My health, I thank God, I find confirmed; and I do not fear that action shall impair it, because I account my ordinary course of study and meditation to be more painful than most parts of action are. I ever bare a mind (in some middle place that I could discharge) to serve her Majesty; not as a man born under Sol, that loveth honour; nor under Jupiter, that loveth business (for the contemplative planet[7] carrieth me away wholly); but as a man born under an excellent Sovereign, that deserveth the dedication of all men's abilities. Besides, I do not find in myself so much self-love, but that the greater parts of my thoughts are to deserve well (if I were able) of my friends, and namely of your Lordship; who being the Atlas of this commonwealth, the honour of my house, and the second founder of my poor estate, I am tied by all duties, both of a good patriot, and of an unworthy kinsman, and of an obliged servant, to employ whatsoever I am to do you service. Again, the meanness of my estate doth somewhat move me: for though I cannot excuse myself that I am either prodigal or slothful, yet my health is not to spend, nor my course to get. Lastly, I confess that I have as vast contemplative ends, as I have moderate civil ends: for I have taken all knowledge to be my province; and if I could purge it of two sorts of rovers, whereof the one with frivolous disputations. confutations, and verbosities, the other with blind experiments and auricular traditions and impostures, hath committed so many spoils, I hope I should bring in industrious observations, grounded conclusions, and profitable inventions and discoveries; the best state of that province. This, whether it be curiosity, or vain glory, or nature, or (if one take it favourably) *philanthropia,* is so fixed in my mind as it cannot be removed. And I do easily see, that place of any reasonable countenance doth bring commandment of more wits than a man's own; which is the thing I greatly affect. And for your Lordship, perhaps you shall not find more strength and less encounter in any other. And if your Lordship shall find now, or at any time, that I do seek or affect any place whereunto any that is nearer unto your Lordship shall be concurrent, say then that I am a most dishonest man. And if your Lordship will not carry me on, I will not do as Anaxagoras did, who reduced himself with contemplation unto voluntary poverty: but this I will do; I will sell the inheritance that I have, and purchase some lease of quick revenue, or some office of gain that shall be executed by deputy, and so give over all care of service, and become some sorry book-maker, or a true pioneer in that mine of truth, which (he said) lay so deep. This which I have writ unto your Lordship is rather thoughts than words, being set down without all art, disguising, or reservation. Wherein I have done the honour both to your Lordship's wisdom, in judging that that will be best believed of your Lordship which is truest, and to your Lordship's good nature, in retaining nothing from you. And even so I wish your Lordship all happiness, and to myself means and occassion to be added to my faithful desire to do you service. From my lodging at Gray's Inn."[8]

What is made clear from this letter, from the five previous letters written by Francis concerning his suit,[9] and from other evidence, is that his suit was about the Great Instauration, as a vast work of *philanthropia* that would be of immediate benefit to the Queen and her subjects, and to the world at large in future ages. Francis offered the Queen his dedicated service in whatever she asked him to do, in return for her making it easier for him to accomplish his collosal philanthropic aims. What he was asking for was a position or office of some kind, neither too low nor too high, which would give him some reasonable remuneration and command of wits to enable him to pursue his scheme, without all the time having to worry about where the next penny was coming from or how to obtain sufficient help in matters both mundane and otherwise. In the end Francis had to be, as he puts it, both architect and hod-man.

But without giving much beyond the occassional word of encouragement, and the outward show of support, the Queen accepted and took all. Something of what it meant to the Queen can be gathered from the building up of her public image as a type of goddess incarnate - a Virgin Queen, Astraea, Gloriana, Belphoebe, *etc.* - which was done deliberately by the Areopagus and others as part of the 'Baconian' scheme to raise England, and then the world, into a Golden Age. What was needed at that time was a grand spectacle - a public media - by which spiritual truths and moral virtues could be portrayed vividly and thereby "insinuated" into the minds of people; so that by visually dramatic example the crude rabble and licentious courtiers could be cleansed and raised in consciousness and action, to become citizens proud of their country and nation, and both inspired and able to do good. The media presentation was done principally through poetry, drama and grand spectacles, with the Queen presented as the Great Ideal. More serious work was done through the encouragement and organisation of scientific and cabbalistic schools, 'modern' Freemasonry, and the creation of a comprehensive English literature and art covering as many subjects as possible in depth - particularly those subjects in which Francis identified man as being "deficient."

Francis promised that he would make the Queen a name and a reign that all posterity would be proud of, and he kept his promise. But he was not fully aware at first of the hate that his 'cousin', Robert Cecil, bore towards him, nor of Robert Cecil's capacity to do harm. Neither did Francis sufficiently realise to begin with how ambitious Lord Burghley was for his clever son, and how blind both Burghley and the Queen were to Robert's cunning and malicious nature. Robert Cecil was exceedingly clever, and he seemed to play with ease upon the Queen's almost obsessional fear of being deposed and ending up like Richard II; using the undoubted fact that Francis did indeed desire to become king of England and that he was proud of being a Tudor prince. But Francis' desire was to become king only in his turn, after the Queen had enjoyed a long and peaceful reign, and had died peacefully. He wished only fame and fortune upon his mother. However, Robert Cecil easily twisted the reality of this natural desire of her eldest son into an all-consuming fear in the Queen's mind (and possibly in Burghley's mind too), and perverted the course of

(77) DEDICATORY PAGE TO *HYMNE TO ELIZABETH* (1586) by George de la Mothe.

justice and English history. In the Biliteral Cipher, Francis tells us:-

"At length I turned my attention from love [*i.e.* for Marguerite de Navarre], and used all my time and wit to make such advancement in learning or achieve such great proficiency in studies that my name as a lover of Sciences should be best known and most honoured, less for mine own aggrandizement than as an advantaging of mankind; but with some natural desires to approve my worthiness in the sight of my book-loving and aspiring mother, believing that by thus doing I should advance my claim and obtain my rights, not aware of [Robert] Cecil, his misapplied zeal in bringing this to her Majesty's notice, to convince her mind that I had no other thought save a design to win sovereignty in her life-time.

"I need not assert how far this was from my heart at any time, especially in my youth, but the Queen's jealousy so blinded her reason that she, following the suggestion of malice, shewed little pride in my attempts, discovering in truth more envy than natural pride, and more hate than affection."[10]

"A fox, seen oft at our Court in the form and outward appearance of a man named Robert Cecil - the hunchback - must answer at the Divine Araignment to my charges against him, for he despoiled me ruthlessly. The Queen, my mother, might, in course of events which followed their revelations regarding my birth and parentage, without doubt having some natural pride in her offspring, often have shewn us no little attention had not the crafty fox aroused in that tiger-like spirit the jealousy that did so torment the Queen, that neither night nor day brought her respite from such suggestions about my hope that I might be England's king.....

"He bade her observe the strength, breadth and compass, at an early age, of the intellectual powers I displayed, and even deprecated the generous disposition or graces of speech which won me many friends, implying that my gifts would thus, no doubt, uproot her, because I would, like Absalom, steal away the people's hearts and usurp the throne whilst my mother was yet alive.

"The terrors he conjured up could by no art be exorcised, and many trials came therefrom, not alone in youth, but in my early manhood."[11]

"Think not in your inmost heart that you or any others whom you would put in this same case as ours, would manifest a wiser or calmer mind; because none who do not stand, as I stood, on Pisgah's very height, do dream of the fair beauty of that land that I have seen. England, as she might be if wisely governed, is the dream or beautious vision I see from Mt. Pisgah's lofty top."[12]

"Furthermore, no thought so holds the imagination of youth as that of imperial power. We crave Caesar's laurel crown at cost of sleepless hours in the night, and weary toil by day. I can undertake such a feeling better than most, having had the same interest in a degree much greater, and in so vastly better right or title."[13]

In addition to (and as a result of) all this, the Queen placed spies upon Francis twenty-four hours a day, and did everything in her power to ensure that no true report or history should ever be published to the nation concerning her private life, nor of those directly connected with her, nor even of state matters that she considered private:-

"......And, please you, the writing of the secrets is chiefest in my conceit, for 'twere a more noteworthy thing, I hold, to make true and correct records of the history of England and of Queen Elizabeth, her life, than to relate the most thrilling tale man's mind can produce.....

"The reason is not far to seek; 'tis this: the many spies employed by our mother, the constant watchful eyes she had upon us, marking our going out and our coming in, our rising up and all our movements from the rising of the sun, to his rising upon the following morning; not a moment when we could openly write and publish a true, accurate history of our times, since nought which her Majesty disapproved could ever find a printer.

"This then is the only cause of my secrecy.........A true accompt of my mother's favourite treasure is strictly given in my history[14] - her love of golden praises, of silvery-tongued words, of flattering speech, dialogues of compliment and princely sayings, or ceremonies. It formed her chief wealth, while, unlike the mother of the Gracchi, she did not reckon sons as jewels, nor did she openly acknowledge either my brother or myself - born princes - heirs to the kingdom."[15]

The very fact of being rejected by his mother, and so ruthlessly, was an almost unbearable pain to Francis. Only those children who are similarly treated - who are rejected by their mothers at birth, fostered by others, brought up believing their foster-parents are their real parents, only to be told, at a crucial point in their lives (*i.e.* puberty), who their real parents are and that they are still alive, yet still be rejected by them, *etc.* - will fully understand what this pain is. Add to this Francis' extraordinary position, his acute sense of virtue and justice, of the importance of being 'true', and of the responsibilities of royalty, and we may begin to comprehend the veritable crucifixion that he had to bear every day of his life - a pain and deep sorrow which shows in all portraits of him. Francis thus continues his cipher report:-

"It burneth as an injury no lapse of time can cure - a ceaseless corrosive which doth eat the heart. The sole relief doth come by making out a complete history of my wrong that doth so embitter my days. Men can eat, sleep, drink, work when the heart is bowed down in pain, yet the joys are gone from their whole lives, and do not return. Chief of sorrows is a sense of wilful wrong on the part of such men or women as have greatest obligation by relation; and more especially those of nearest and most tender relationship, that of parents to a child. This will never grow inferior, nor even merely equal to the natural ills in life. It doth rather greatly magnify and increase.

"Why and wherefore I shall not ask, nor marvel at ought of similar nature. The Creator planted this within the bosom of our kind. Who hath so great wisdom or so just judgment of our life, of right or wrong, as our Maker? Who can pronounce His laws at fault? A fool or blind[man], perchance: not he that sees, nor the man of thought."[16]

Francis, as usual, bowed his head to the greater judgment of God. It did not take away the pain, but he was fully aware of a greater Justice at work, and of divine plans in motion, of which he was but a messenger and a servant, which would bring the Great Instauration to fruition in the course of time. This knowledge did not relieve the heart-ache, nor the burden of carrying out the initial work almost (but not quite) single-handedly - a pain and a burden compounded when he knew that it could easily be otherwise. There were not a few scholars, poets, scientists, artists and craftsmen available for employment by him in the great scheme, but mostly they needed to earn a living. If he only had had the financial means of employing them, the work could have proceeded so much the better. But this means was continually denied to him throughout the reign of his jealous and fear-stricken mother, chiefly because of the machinations of the Cecils - father and son.[17]

Secret Intelligence and the Grand Tour

In 1579, towards the end of the year (late autumn or early winter), Anthony Bacon had set out for what has usually been reported as an extended tour of the continent, as befitted a gentleman of those times. Undoubtedly this was just the kind of impression that Anthony and his friends wished others to believe. The reality is that Anthony was in the employ of the Queen, as a high-ranking intelligencer, and working directly under Sir Francis Walsingham, the Queen's Principal Secretary of State, who had organised and was head of the most efficient intelligence network in Europe, and Lord Burghley, who was the Queen's Lord Treasurer.

Over fifty agents were involved in the network abroad, and possibly a similar number in the British Isles, ranging from the ordinary spy to the master-spy or intelligencer. Most of them were recruited from universities, and then given special training at the 'spy school' in London. Sir Francis Walsingham, as Controller, ran the network, being directly responsible to the Queen. In the topmost ranks of the agency were his nephew, Thomas Walsingham (who recruited the agents, as well as being a patron of poets), Dr. John Dee (who used the cipher '007' in this capacity, and was a master-teacher of mathematical cryptology, telepathy, psychic communication and far-sight, *etc.*), Nicholas Faunt (who was one of Walsingham's secretaries as well as a close friend and colleague of Anthony Bacon), Thomas Phelippes (probably the best of the cryptanalysts as well as a widely-travelled agent), Anthony Bacon and Francis Bacon. Francis became an expert on cipher, inventing many new ones and adapting many others. He was specifically employed by the Queen and Walsingham to decipher all correspondence to and from or pertaining to Mary Queen of Scots, as well as performing various missions. Anthony also was an expert cryptologist, but his main work became that of principal intelligencer abroad, after he had fulfilled four dangerous missions under the alias of 'Edward Burnham' (1578-1580) and other work besides.[1] Later, when he had returned home to live in England (1592 onwards), he was to take over the running of the whole network that his father's friend, Walsingham, had established.

In the "Note of special services performed by Edward Burnham for Her Majesty" written by Nicholas Faunt and left amongst Anthony Bacon's papers, after recounting the four missions undertaken by Burnham, the note states that:-

> "For these services, as well as the several journeys, in which he had been employed by the Secretary to the Duke of Anjou, William Prince of Orange, Charles de Croy, Prince de Chimay, and the States of the Low Countries, he [Edward Burnham] requested some 'extraordinary gratification' "

The "extraordinary gratification" requested may have been to have permission to remain abroad for sometime and to travel widely throughout Europe, in the service of the Queen.

Such a request was almost certainly tied in with Francis Bacon's own request made at the same time - that is, for help and support in forwarding the Great Instauration - as Anthony was committed to his foster-brother's project. An essential part of the Great Instauration is to acquire knowledge about all things, from careful observation and experience, and hence to form a carefully tabulated 'History'. Such a 'History' is the foundation of the Pyramid of Knowledge (or Philosophy) that can then be built. It would certainly seem that the service of the Queen and the acquisition of a wide range of knowledge in things natural, human and divine, was usefully married together in the sending of Anthony abroad on an extended tour, giving him the task of setting up a moveable intelligence H.Q. abroad.

Cryptology, involving cryptography, was seen as part of the Great Instauration. The most brilliant of the French cryptanalysts and cryptographists was a nobleman named Blaise de Vigenère. He wrote in his most important book, *Traicts des Chiffres,* a statement with which the Bacon brothers certainly agreed:-

> "All the things in the world constitute a cipher. All nature is merely a cipher and a secret writing. The great name and essence of God and his wonders, the very deeds, projects, words, actions and demeanor of mankind - what are they for the most part but a cipher?"[2]

This is precisely how, for instance, the Ancient Egyptians viewed things. Vigenère refers to the sacred writings of the Ancients who, like God, therein veiled their holy secrets. He goes on to say that his book is of similar ciphers, but "rare and known to few people - learnt partly from others in our travels in different parts of Europe, but the greater part originated in our own thoughts, and not, as far as we know, touched upon by anyone until now."[3]

In his book, which was published in Paris in 1586, Vigenère actually gives the principles of several of the Baconian ciphers, including the Biliteral (although in a slightly altered form, so as not to give the game away). He claims that he was the first to use the device of making letters depend upon each other and serve as keys by (i) shape, (ii) size, (iii) quality or equivalence, and (iv) place. The Biliteral cipher is developed from these principles, as a cipher depending upon minute differences in type; and, on page 200, Vigenère explains such a cipher, but groups his letters in threes instead of fives as Francis Bacon does. Francis, when he mentioned his Biliteral cipher in his *Advancement of Learning* (1605) and then gave its principles in his *De Augmentis Scientiarum* (1623), stated that he had invented that particular cipher in his youth, whilst in France, at the French Court in Paris (*i.e.* sometime between 1576 and 1579). As far as has been dicovered, his first use of it in any printed work was the *Shepheards Kalendar,* published in London on the 5th December 1579. Without doubt, Francis Bacon and Blaise de Vigenère knew each other in Paris; and it would appear that Francis was inspired by some of Vigenère's suggestions, and Vigenère in turn, seven years later, recorded the subsequent cipher invention of Francis. Perhaps Francis had asked Vigenère to do this, for good reason? Or was Vigenère allowing his name to be used as a mask for Francis's authorship of the cipher book? Or did they co-author the book, but agree to use only Vigenère's name on the title-page?

Blaise de Vigenère produced a cipher for Henri III which was an improvement on Giovanni Baptista della Porta's system - an excellent letter substitution method. Porta published his system in 1563, in his *De Furtivis Literarum Notis Vulgo. De Ziteris* - the very book that carried the first printed example of the '**AA**'.

Anthony Bacon sailed for France, towards the end of 1579, with a few assistants in attendance, one of them being a Mr. Doyly. Anthony's first port of call was Paris, where a new English Ambassador, Sir Edward Stafford, had taken the place of Sir Amyas Paulett. Having stayed there a month or two, in February 1580 Anthony set out for Portugal on his fourth mission as 'Edward Burnham', to discover whether Spain would gain control of the country and thus of the principal Atlantic ports and trade routes to the East and to the Americas. The possible threat to Britain was considerable. After three dangerous months in Portugal, Anthony just managed to escape the country in time, a warrant having been issued for his identification and arrest. He made his way by ship to France, and then rode to Paris where he became involved briefly with a Welsh Catholic, Dr. William Parry, who was another agent for Walsingham. A misunderstanding concerning the reasons for Anthony's contact with Parry arose in England (which contact was in fact according to Burghley's orders), and it gave a reason or cover for Anthony to be recalled to England, to give a full report.

It is not known exactly when Anthony arrived in England, in order to make his report, but judging by the fact that Mr. Doyly was trying to send Anthony letters *via* Francis in July 1580, at Gray's Inn, Anthony must have returned to England in May or June. Anthony stayed in England until (seemingly) December. But he was back in France and had arrived in Bourges by the 15th December, writing from thence to his uncle, Lord Burghley, on 14th January 1581 to tell him so.

In the meantime Francis had also determined on an extended journey abroad, for similar reasons - for service of his Queen and country, for the increasing of his own experience and knowledge, and for building the foundations of the Great Instauration. He had spent several years in France and knew it well, and had visited Navarre, but he had yet to visit Italy and Spain, Germany and the Low Countries. Italy and Spain he particularly desired to see.[4] It would appear that this may have been part of his suit to the Queen. Certainly she must have granted permission for Francis to go, on condition that he rendered to the Government a good account of affairs abroad when he returned,[5] and Burghley consulted Anthony Bacon for his advice as to the best route to take. But the Queen does not seem to have made any financial contribution whatsoever towards the expense of the journey (which was not unusual - Anthony, for instance, was never paid by the Queen for his services, and many good people beggared themselves on the Queen's behalf). Instead, the costs were born by Sir Thomas Bodley (the Queen's Gentleman-Usher) and certain "friends" - presumably because the journey was made at Francis' own request.

In a further letter to Burghley, dated February 1581, Anthony enclosed a note of advice and instructions for Francis' forthcoming journey. Although we do not know when Francis left for the continent, it is a fairly reasonable assumption that he departed England soon after receiving the note of advice from Anthony - probably in March 1581. Francis had entered politics by then, having been elected as a Member of Parliament for Bossiney, Cornwall, in January 1581.[6] According to his own evidence he sat in the parliamentary session of that year, of which the first sitting took place on the 16th January, and was present on the 14th February when Arthur Hall, a member of Parliament for the Borough of Grantham, was committed to the Tower for publishing slanderous libel.[7] Except for an undated letter from Bodley written to Francis Bacon, in answer to a letter from Francis requesting further funds which was written at Orleans on the "19th October" and received by Bodley on the "18th December," there is no documented evidence of where Francis was or what he was doing between March 1581 and 15th April 1582 (when he was back in Gray's Inn).[8] From all the evidence available, Francis' letter to Bodley must have been

(78) SIR THOMAS BODLEY, aged 53 years (b.1545-d.1630) - miniature painted (1598) by Nicholas Hilliard.

Sir Thomas Bodley was a diplomat and distinguished linguist. The Queen employed him as a confidential envoy, principally in France and Denmark (1576-1587). He was made ambassador in the Netherlands (1589-1596). In 1598 he enabled the 15th-century library at the University of Oxford to be restored and enriched (in seven years it trebled in size). Thereafter it became known as the Bodleian Library, and became the nucleus for the first national library in England after Bodley had gained for it (in 1610) the right to receive a gift copy of every book the Stationer's Company printed. He was knighted by King James I in 1603.

written during that period - that is to say, on 19th October 1581; and Bodley's reply to Francis was almost certainly written soon after, as the matter was somewhat urgent.[9]

In support of all this, the Bodley letter outlines clearly the reasons for Francis' journey, which is given extra credence by a paper of *Notes on the Present State of Christendom* that were discovered amongst Francis Bacon's papers and which were written in or completed by the summer of 1582, judging from an assessment of the events referred to in the notes. Francis Bacon was in all probability the principal author or compiler of these succinct notes, although he may well have been supplied with information in some cases by Anthony Bacon and Nicholas Faunt. In addition Pierre Amboise, in his *Life* of Francis Bacon that was printed with French translations of Bacon's *Natural History* and *New Atlantis* in 1631, categorically states that Francis travelled in France, Italy and Spain, for the express purpose of acquiring a high state of knowledge:-

> "......M. Bacon himself wished to acquire that knowledge which in former times made Ulysses so commendable, and earned for him the name of Wise; by the study of the manners of many different nations. I wish to state that he employed some years of his youth in travel, in order to polish his mind and mould his opinions by intercourse with all kinds of foreigners. France, Italy, and Spain, as the most civilised nations of the world were those whither his desire for knowledge carried him. And as he saw himself destined one day to hold in his hands the helm of the Kingdom, instead of looking only at the people and the different fashions in dress, as do the most of those who travel, he observed judiciously the laws and customs of the countries through which he passed, noted the different forms of Government in a State, with their advantages or defects, together with all the other matters which might help to make a man able for the government of men."[10]

The Bodley letter, plus the *Notes on the Present State of Christendom,* plus the gap in records for Francis' life, fix the date of the journey nicely. Sir Thomas Bodley, himself a seasoned traveller, and Francis' "friends" in England would seem to have sponsored the journey in return for a careful report of the condition of both the State and the Church in the various countries of Europe at that time. The letter is well worth reading for its clarity in describing what the exact aims and objects of Francis' journey were to be, and the enormous expectations that Bodley had and demanded of the twenty-year old Francis:-

> "My Dear Cousin,
>
> According to your request in your letter (dated the 19th October at Orleans, I received here the 18th of December), I have sent you by your merchant 30l [*i.e.* £30] sterling for your present supply, and had sent you a greater sum, but that my extraordinary charge this year hath utterly unfurnished me. And now, cousin, though I will be no severe exactor of the account, either of your money or time, yet for the love I bear you, I am very desirous, both to satisfy myself, and your friends how you prosper in your travels, and how you find yourself bettered thereby, either in knowledge of God, or of the world; the rather, because the Days you have already spent abroad, are now both sufficient to give you Light, how to fix yourself and end with counsel, and accordingly to shape your course constantly unto it. Besides it is a vulgar scandal unto the travellers, that few return more religious than they went forth; wherein both my Hope and Request to you is, that your principal care be to hold your Foundation, and to make no other use

of informing your self in the corruptions and superstitions of other nations, than only thereby to engage your own heart more firmly to the Truth. You live indeed in a country of two several professions, and you shall return a Novice, if you be not able to give an account of the Ordinances, strength, and progress of each, in Reputation, and Party, and how both are supported, ballanced and managed by the state, as being the contrary humours, in the Temper of Predominancy whereof, the Health or Disease of that Body doth consist.

"These things you will observe, not only as an English-man, whom it may concern, to what interest his country may expect in the consciences of their Neighbours; but also, as a Christian, to consider both the beauties and blemishes, the hopes and dangers of the Church in all places.

"Now for the world, I know it too well, to persuade you to dive into the practices thereof; rather stand upon your own guard, against all that attempt you there unto, or may practise upon you in your Conscience, Reputation, or your Purse. Resolve, no man is wise or safe, but he that is honest: And let this Persuasion turn your studies and observations from the Complements and Impostures of the debased age, to more real grounds of wisdom, gathered out of the story of Times past, and out of the government of the present state. Your guide to this, is the knowledge of the country and the people among whom ye live; For the country though you cannot see all places, yet if, as you pass along, you enquire carefully, and further help yourself with Books that are written of the cosmography of those parts, you shall sufficiently gather the strength, Riches, Traffick, Havens, Shipping, commodities, vent, and the wants and disadvantages of places. Wherein also, for your good hereafter, and for your friends, it will befit to note their buildings, Furnitures, Entertainments; all their Husbandry, and ingenious inventions, in whatsoever concerneth either Pleasure or Profit.

"For the people, your traffick among them, while you learn their language, will sufficiently instruct you in their Habilities, Dispositions, and Humours, if you a little enlarge the Privacy of your own Nature, to seek acquaintance with the best sort of strangers, and restrain your Affections and Participation, for your own countrymen of whatsoever condition.

"In the story of France, you have a large and pleasant Field in three lines of their Kings, to observe their alliances and succesions, their Conquests, their Wars, especially with us; their Councils, their treaties; and all Rules and examples of experiences and Wisdom, which may be Lights and Remembrances to you hereafter, to Judge of all occurants both at home and abroad.

"Lastly, for the Government, your end must not be like an Intelligencer, to spend all your time in fishing after the present News, Humours, Graces, or Disgraces of Court, which happily may change before you come home; but your better and more constant ground will be, to know the Consanguinities, Alliances, and Estates of their Princes; Proportion between the Nobility and Magistracy; the Constitutions of their Courts of Justice; the state of their Laws, as well for the making as the execution thereof: How the Sovereignty of the King infuseth itself into all Acts and Ordinances; how many ways they lay Impositions and Taxations, and gather Revenues to the Crown.

"What be the Liberties and Servitudes of all degrees; what Discipline and Preparations for wars; what Invention for increase of Traffick at home, for multiplying their commodities, encouraging Arts and Manufactures, or of worth in any kind. Also what Establishment, to prevent the Necessities and Discontent of People, To cut off suits at Law, and Duels, to suppress thieves and all Disorders.

"To be short, because my purpose is not to bring all your Observations to Heads, but only by these few to let you know what manner of Return your Friends expect from you; let me, for all these and all the rest, give you this one Note, which I desire you to observe as the Counsels of a Friend: Not to spend your Spirits, and the precious time of your Travel, in a Captious Prejudice and censuring of things, nor in an infectious Collection of base Vices and Fashions of Men and Women, or general corruption of these times, which will be of use only among Humorists, for Jests and Table-Talk: but rather strain your Wits and Industry soundly to instruct your-self in all things between Heaven and Earth which may tend to Virtue, Wisdom, and Honour, and which may make your life more profitable to your country, and yourself more comfortable to your friends, and acceptable to God. And to conclude, let all these Riches be treasured up, not only in your memory, where time may lessen your stock; but rather in good writings, and Books of Account, which will keep them safe for your use hereafter.

"And if in this time of your liberal Traffick, you will give me any advertizement of your commodities in these kinds, I will make you as liberal a Return from my self and your Friends here, as I shall be able.

"And so commending all to your good Endeavours, to him that must either wither or prosper them, I very kindly bid you farewel.

Yours to be commanded, Thomas Bodley."[11]

What Francis Bacon could not furnish was probably made up for by what Anthony Bacon and Nicholas Faunt could add from their own travels and intelligence; but there is in fact no reason to suppose that Francis did not discover and write the whole of the information himself. Possibly Anthony had a similar task to perform; but in his case he did not depend on others for his finance. He paid for his travels out of his own pocket, and so the conditions placed upon him could not have been nearly so onerous as for Francis.

In a letter to Walsingham earlier that year (1581), Anthony stated that he found the people of Bourges to be both corrupt and irreligious, and announced his intention of moving on to Geneva. It is possible that he teamed up with Francis, having waited for Francis to arrive at Bourges, or he may have met Francis in Geneva or even Basle. In any case, Anthony is known to have arrived in Geneva at least by June 1581, and lodged himself in the house of the eminent theologian Théodore Beza, the most famous exponent of the Reformed faith since Calvin. Beza was an admirer of Anthony's mother, Lady Anne Bacon, who had in her younger days translated, from Latin into English, John Jewel's *Apologia for the Church of England,* and, from Italian into English, twenty-five sermons upon the predestination and election of God, which had been well-read by the 'faithful' in Geneva. In addition, Edward Bacon, Anthony's half-brother, had preceded Anthony and lodged with Beza some four years previously whilst on his own journeys through Europe.

Geneva was a good place from which to reach the German States, Central Europe or Italy (as well as France), and it was an important crossroads between East and West, North and South Europe. Anthony's move to Geneva corresponded with Nicholas Faunt switching his activities from Germany to Northern Italy, and travelling frequently between Northern

An Apologie
or anſwere in defence of the
Churche of Englande,
with a briefe and plaine
declaration of the true
Religion professed
and vsed in
the same.

Londini, Anno Domini
M.D.LXIIII.

(79) TITLEPAGE TO *AN APOLOGIE OR ANSWERE IN DEFENCE OF THE CHURCH OF ENGLAND* (1564) - a translation by Lady Anne Bacon of John Jewel's Apologia which was originally written in Latin.

Italy and Geneva; thus it would appear that Anthony probably positioned himself as a 'control post' or H.Q. in the key city, able to receive information from Italy *via* Faunt (and from other places and agents) and pass it on to England *via* his messengers, and *visa versa.* He may well have positioned himself thus in order to assist his foster-brother Francis, who was probably travelling through the Swiss States and the Tyrol in June 1581, and in Northern Italy during July and August 1581, and possibly journeying *via* Geneva in early September in order to reach Paris for the *Joyeuse Magnificences* at the end of September.

Anthony was still at Geneva in November 1581 (although whether he travelled elsewhere in the meantime we do not know), when Nicholas Faunt, who had been travelling back and forth between Northern Italy and Geneva since May or June that year (and possibly acting as messenger between Francis and Anthony for some of the time), arrived in Geneva from Padua for the last time.[12] From Geneva Faunt then had to go to Paris, and hence home to England, carrying with him the results of the intelligence work to date concerning Italy. Anthony asked Faunt to procure for him a licence of absence for another three years, intending to visit Italy - in particular Padua, Verona and Venice - then returning to France *via* the Mediteranean coast. It was not until May 1582 that Anthony eventually received the necessary licence, probably brought by his close friend from Gray's Inn, Ned Selwyn, who arrived in Geneva to join him.

But by then Northern Italy had become an exceedingly dangerous place for Englishmen, with the Inquisition set up in Venice, Protestants being persecuted, English travellers being thrown into prison, and a war between Savoy and Geneva about to break out. There was also news that Philip II of Spain was starting to prepare for an invasion of England. Anthony's friends in England were growing anxious for Anthony's safety. Faunt, who had been recalled to England at the beginning of April 1582, and who had called on Francis Bacon at Gray's Inn, wrote to Anthony several times urging him to return home and reporting that Francis had also expressed his desire to see his brother safely back. In his letter to Anthony dated 6th May 1582, Faunt reports:-

> "Yet by the way, in a word or two, he [Francis] hath showed his earnest desire to have you return, wishing me to be a persuader thereof, and saying that he marvelled how those that keep abroad more than that time [*i.e.* of their licence] could live to their contentment, seeing that himself was more than weary of his being forth, and that the home life is to be thought upon as of the end of due season."

Francis had clearly had enough of travelling! However, whatever the reason might have been (and it was probably on account of ciphered instructions enclosed in the letters from England), Anthony set course for the South of France, travelling with Ned Selwyn to Toulouse and staying there two or three months, then on to Lyons for a few months, then to Montpellier for another two or three months, and finally to Marseilles towards the end of 1582. As he journeyed, his retinue gradually increased, with extra attendants being added to his company - all at Anthony's expense, with money sent him with difficulty from his modest estates at home.

Long before this date Francis had returned to England, and was busy at Gray's Inn writing up his *Notes* and other things, including his first essay on the Great Instauration which he called *Temporis Partus Maximus* ('The Greatest Birth of Time'). Judging from the *Notes* he is likely to have visited France, Italy, Spain, Austria (or Bavaria), Germany, Portugal, Poland, Denmark and Sweden (the last three perhaps being visited by others on his behalf). It would also appear that he spent most of his time in Italy, France and Spain,

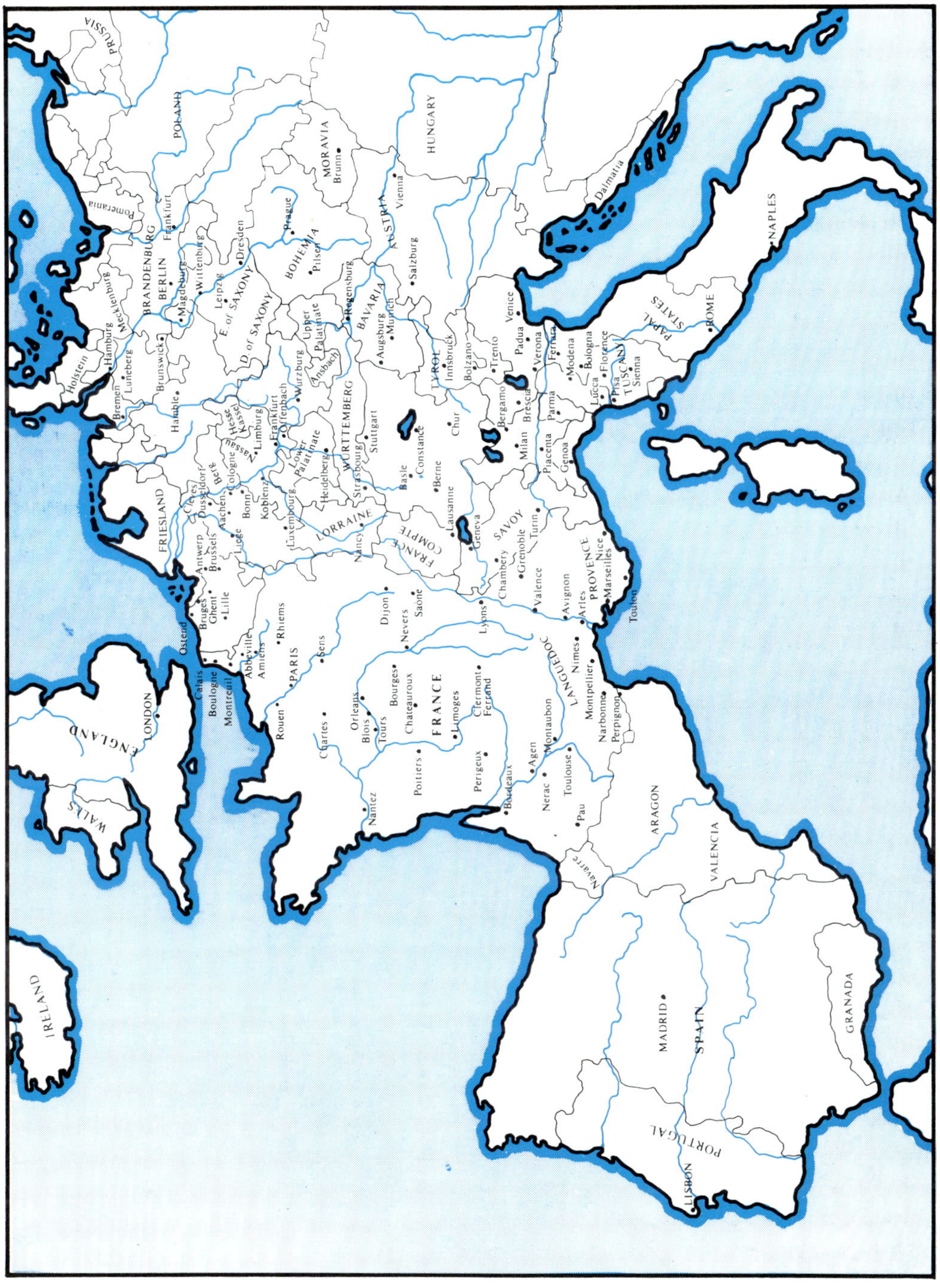

Diagram K: MAP OF EUROPE

which are dealt with in considerably more detail than the other countries. The observations about Germany are extremely brief, suggesting a rapid journey through the country. This would accord with his own desires, as we have already seen, and with what I believe was his route.

As an educated guess, I would speculate that Francis left England in May 1581 for the continent, either passing through Northern France and Lorraine, or else sailing up the Rhine through Germany, to Basle, and from thence following in the footsteps of Michel Eyquem Montaigne. It does seem extraordinary that Montaigne, a friend of the Bacons and probably the "wise old gentleman" of the French Academy at which Francis was present during the winter of 1577, should have gone on his famous tour more or less at the same time as (or just before) Francis' travels; and, moreover, that he should have visited the very places in Bavaria and Italy to which Francis seems to have gone only a short time afterwards.[13] It would appear from the *Notes* that Francis spent an extended period of time at Florence, possibly as a guest of the learned Grand Duke of Tuscany, Francesco de Medici, who had his palace there (although Francis' general rule was not to stay too long in any one place). Francesco's brother, Ferdinand, whom Francis almost certainly met and befriended then, became Grand Duke after Francesco's death. Ferdinand was an enlightened patron of the fine arts, and secretly associated with England and Holland in commercial enterprises in Spanish-America. Ferdinand's son, Cosimo, became Grand Duke in 1609. He and his father were specifically mentioned in Tobie Matthew's dedication to the Italian version of Bacon's *Essays* (*Saggi Morali,* 1618), in which Tobie wrote that Francis Bacon "honours the memory of his [Cosimo's] father Ferdinand de Medici with affection and particular admiration."

Francis could have been in Florence by the end of July 1581, if he had followed the same route as Montaigne from Basle - even earlier if he had come more directly. If this is so, he could easily have met up with Montaigne in Florence or Rome. But Francis certainly left Italy before Montaigne, for he wrote to Sir Thomas Bodley from Orleans on the 19th October 1581. Orleans is not far from Paris, and one cannot help wondering whether Francis raced from Italy to Paris in order to be there on time for the *Joyeuse Magnificences,* held in Paris during the last two weeks of September 1581 to honour the marriage of the French king's favourite, the Duc Anne de Joyeuse, to the Queen's half-sister, Marie de Lorraine.[14] I am certain that Francis would have tried to be there if he could - those *Magnificences* being the glorious culmination of the series of Valois entertainments held during the reign of Henri III, some of which Francis had experienced before. Then, as Orleans is *en route* from Paris to the Pyrénées and Spain, I would suggest that this was the time when Francis ventured into Spain. It may have been that, whilst in Spain, Francis discovered the preparations which Philip II was making for an invasion of England. From Spain he probably went into Portugal, and left by boat from one of the Atlantic ports for England in the early months of 1582, before matters became too dangerous.

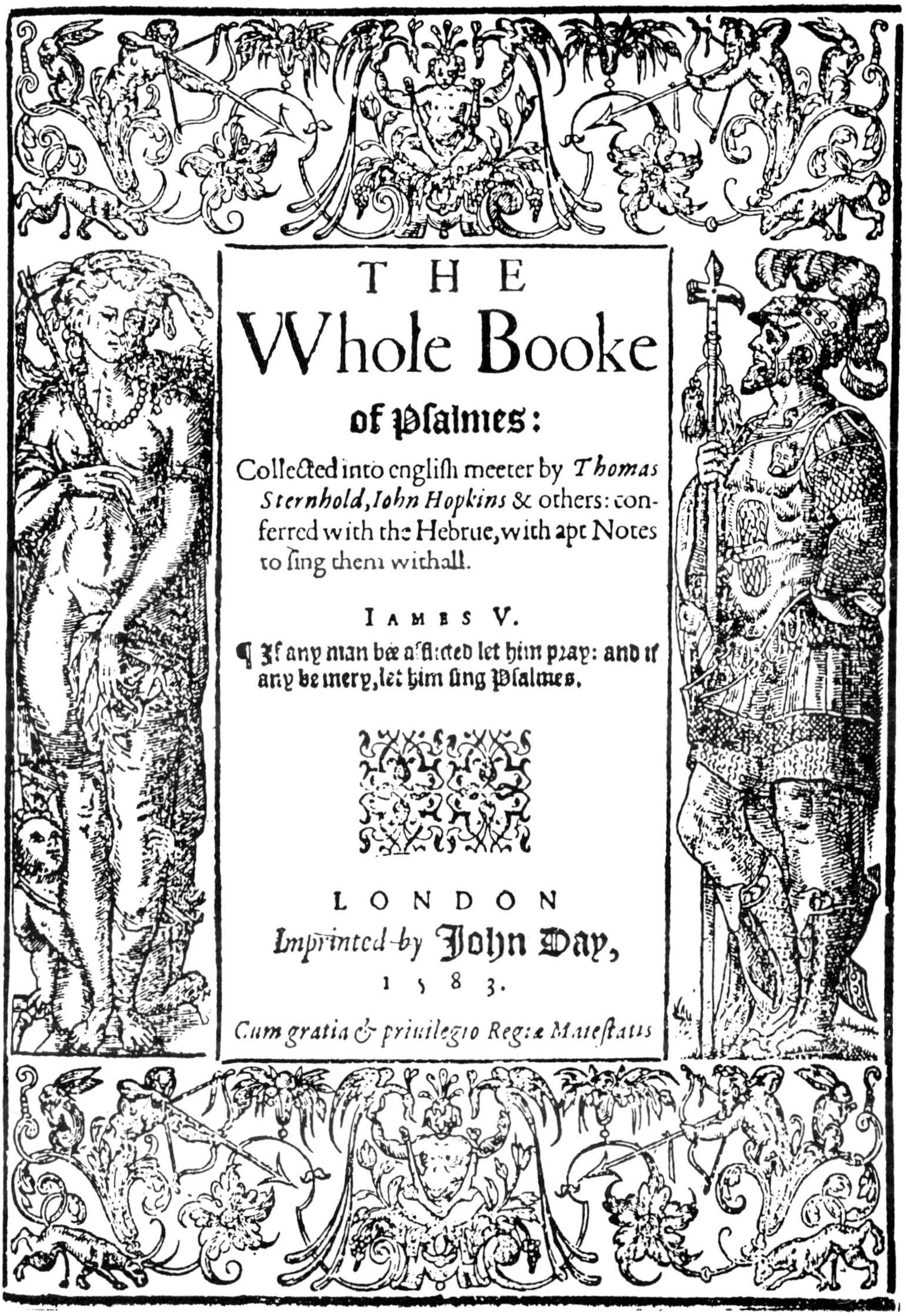

THE

Whole Booke

of Psalmes:

Collected into english meeter by *Thomas Sternhold, Iohn Hopkins* & others: conferred with the Hebrue, with apt Notes to sing them withall.

IAMES V.

¶ If any man bee afflicted let him pray: and if any be mery, let him sing Psalmes.

LONDON

Imprinted by Iohn Day,

1583.

Cum gratia & priuilegio Regiæ Maiestatis

(80) TITLEPAGE TO *THE WHOLE BOOKE OF PSALMES* (1583) - a republication of the popular collection of Psalms, being the first metrical version of the Psalms in English, made by Thomas Sternhold, John Hopkins and others, and which was first published as a collection of 14 Psalms in 1547.

This would seem to be the first appearance of the so-called "Archer' or 'Blind Boy' emblem, used here at the top and bottom of the titlepage, although normally used thereafter as a headpiece. The two figures represent the cabbalistic pillars of Grace (Mercy) and Judgment (Severity).

(81) 'Archer' Headpiece - from Spenser's *Faerie Queene* (1609).

(82) 'Archer' Headpiece - from Franciscus Mason's *The Consecration of the Bishops* (1613).

(83) 'Archer' Headpiece - from Francis Bacon's *Novum Organum* (1620).

(84) 'Archer' Headpiece - from the Shakespeare 1st Folio (1623).

The Greatest Birth of Time

As mentioned in the previous chapter, Francis Bacon was back in England by April 1582. Except for visits to Gorhambury, we know that he was living at Gray's Inn once more, busy writing up the results of his observations and experiences and intelligence gained abroad - material that he not only shared with Sir Thomas Bodley and his friends, but also used in his literary works, especially the Shakespeare Plays, as part of his grand Scheme. At this time he also wrote his first treatise defining his work, called *Temporis Partus Maximus* ('The Greatest Birth of Time'), which he presumably produced especially for his friends, since it was not (and was never intended to be)[1] published, but circulated privately. As Spedding says, the main object of the treatise, or of that small part which remains to us, was to "explode the various philosophical systems or theories which had previously been propounded; being the first and principal part of the doctrine of the Idols of the Theatre, - a part which, though not directly noticed in the *Advancement of Learning,* assumed soon after so prominent a place in Bacon's scheme that he resolved to place it in the very front of his battle."[2] He also referred to this literary piece, enigmatically, as *Temporis Partus Masculus* ('The Most Masculine Birth of Time'). Forty years later he said of this treatise:-

> "The Zeal and Constancy of my Mind, has not waxed old in this Design [*i.e.* of the Great Instauration], nor after so many years, grown cold and indifferent. I remember that about Forty Years ago, I composed a Juvenile Work about these things which with great Confidence, and a Pompous Title, I called *Temporis Partus Masculus,* or the Most Masculine Birth of Time. I am thus persuaded, because of its infinite usefulness, that it may be ascribed to Divine Encouragement."[3]

Alfred Dodd, in *Francis Bacon's Personal Life Story* (p.126), claims that this treatise actually described "the new ethical system of Freemasonry that had been born into the world among men."[4]

'The Most Masculine Birth of Time' can and does mean, however, 'the Most Virile Birth of Knowledge'; and I would think that this is the meaning which Francis Bacon intended to convey, whilst indeed probably referring to his 'fraternity in learning' that he was bringing to birth.[5] The reference to 'male' is, in essence, not a discrimination against the feminine sex, but an allusion to the real meaning of initiation taught in the schools and temples of the Ancient Wisdom: namely, that the adept is a 'male child' born from the female nature of his own psyche, which is as a virgin mother to his higher, Christ-like nature; and, moreover, that the spiritual nature of the adept can bring to birth all things. Another interpretation is that, in terms of Divinity, the Spirit (or Mind) of God is referred to as being masculine (*i.e.* the Heavenly Father) and the Matter of God is referred to as being feminine (*i.e.* the Earthly Mother): thus, 'The Most Masculine Birth of Time' can simply be an occult way of implying 'The Most Spiritual Birth of Knowledge' as distinct from a material birth.

The spiritual birth of the adept, that Francis wrote about in 1582-3, also appeared in 1583 in pictorial form as an unusual and deeply significant emblem (like the '**AA**'), known as the 'Archer' or 'Blind Boy' emblem. The 'Blind Boy' is the centre-piece of the emblem, and represents the second Cupid (*i.e.* the adept) who arises reborn or resurrected from the cornucopia of his lower self, and manifests the first Cupid - the God of all gods.[6] This device appeared in a republication of Sternhold and Hopkins' *Psalmes,* and it was the first appearance of this particular emblem. The book is actually entitled *The Whole Booke of Psalmes: Collected into english meeter by Thomas Sternhold, Iohn Hopkins & others, &c.,* and the emblem is printed as an upper and lower border to the titlepage. Furthermore, the Twin Pillars of Freemasonry (and of Cabbala), in the form of a woman and child (*i.e.* right-hand pillar of Mercy) and an armed knight (*i.e.* left-hand pillar of Severity), stand between the 'Archer' emblems on either side of the titlepage plaque, and the 'W' of *Whole* is printed in the cryptic sign of Cassiopeia.[7] Sternhold and Hopkins' *Psalmes* was one of the most popular and widely read books ever printed in English, and was the first metrical version of the Psalms ever made in English. It was first published in 1547 as a set of fourteen Psalms, dedicated to Edward VI, which was subsequently enlarged to a set of forty Psalms by the time of the 1561 edition printed by John Daye. The only other Englishmen known to have attempted metrical versions of the Psalms in English are Francis Bacon and John Milton. The choice of David's Psalms, put into English metrical verse, seems to me a most apt one for displaying this most Baconian of emblems - the 'Blind Boy' - for the first time. The Great Instauration had truly begun in earnest.

Another most important pictorial device also appeared in 1583, which, like the '**AA**' and 'Blind Boy' emblems, is especially associated with literary works either written or supervised by Francis Bacon, or else coming from the Baconian-Rosicrucian 'School'. This was the famous triangular 'Pan tailpiece'.[8] It was printed in a French edition of Plinius' *Histoire du Monde,* published in Lyon in 1583. But that was not the first appearance of the 'Pan tailpiece, although it seems to have been only its second time in print. The first appearance of the 'Pan tailpiece' was in another magnificent book by Blaise de Vigenère (or, at least, published under his name), entitled *Philostrate* and published in Paris by Nicolas Chesneau in 1578 - hence following hard on the heels of the '**AA**' symbol as used by Francis Bacon in the 1577 edition of Alciat's *Emblemata.*[9] As will be realised by the student of Bacon's works, Pan and Cupid are equated; the only difference lying in the fact that Pan is a symbol for the whole evolutionary nature or life of the universe, and in particular of man, whereas Cupid is the symbolic description of that same life in its pristine state at the very beginning of Creation, and in its perfectly fulfilled state at the end of all Evolution. Cupid and Pan can also be understood as being Love-Wisdom and Intelligence - the emotive force and intelligent form of life, respectively.

As Spedding points out, the style of *Temporis Partus Masculus* is very different to the style of all Francis' later writings and publications under the name of 'Bacon'; the latter being "marked with the gentleness and modesty which are said to have distinguished his demeanour and conversation," whilst the former is "more crude" and Francis "assumes a tone quite different; entering abruptly into the subject in a spirit of contemptuous invective [for the errors of received opinions in philosophy], not to call it presumptuous and insolent, of which in all his writings, public or private, I remember no other example."[10] This is very much in keeping with the youthful nature of Francis, which was fiery, ebullient and, in appearance, somewhat self-conceited.[11] He was not yet tempered and mellowed with the harsh experiences of time that were to follow, and it is probably by

(85) ''Pan' Tailpiece - from Blaise de Vigenère's *Philostrate* (1578).

This is the first appearance of the 'Pan' tailpiece in print. The emblem depicts not only Pan, God of Arcadia, playing his pipes and holding cornucopiæ, but also the face of Truth, the Bride, set below Pan in a heart-shaped frame.

(86) ''Pan' Tailpiece - from Plinius's *Histoire du Monde* (1583).

(87) ''Pan' Tailpiece - from Spenser's *Faerie Queene* (1606).

(88) ''Pan' Tailpiece - from the Shakespeare 1st Folio (1623).

(89) ''Pan' Tailpiece - from Francis Bacon's *Opera Omnia* (1665).

very reason of this untempered, brilliant and perhaps impatient nature that Francis found the Queen and others so wary and even fearful of him. No doubt many of Francis' elders, conservative in their ways, were deeply offended by his immodesty and outspokenness; and probably they were highly irritated too, when his words trod rather too close to the heels of truth for their comfort. The *Shiva* energy, which the Baconian work embodies,[12] can be extremely uncomfortable in its cleansing actions. It literally is a spear of clear light that pierces deep into the mind and body of the earthly dragon of human nature. It is known as the Destroyer, for it destroys ignorance and vice; but only by bringing, in their stead, knowledge and virtue and true happiness. But Francis learnt quickly, and the vicissitudes of Elizabeth's reign produced in him a man humbled and exalted all at the same time.

Having completed the travels and studies that he had wished to experience and learn from, he threw himself into his work with gusto - not so much his legal studies, which he found tedious (although they stood him in good stead later on), but the reformation of all arts and sciences - the creation in England of a Renaissance of learning, and the founding of a universal work of progressive enlightenment for the whole wide world.

Francis used every means at his disposal to forward his philanthropic aims, whilst never for a moment forgetting the service he owed to his Queen and country. Thus, whilst faithfully studying law at Gray's Inn, serving as a member of Parliament, and carrying out various services for the Queen and Burghley on matters of secret intelligence, Francis conversed, studied and experimented with his friends of the Areopagus, and with the larger Dee-Walsingham-Dudley circle of scholar-scientists. He also began to organise some of his lawyer friends to help with the invention of new English words, which the Areopagus had launched upon in a mighty effort to enrich the English language, and to start the translation of Classics and other works into the 'new' English tongue. He wrote prodigiously, and published his writings either anonymously or under various pseudonyms. Moreover, he became increasingly involved in drama - from the revels at Gray's Inn to the Court masques and plays, and especially including the more public drama in the play-houses of the city. Not only did he write numerous plays, but, according to his cipher evidence, he was also involved in directing and rehearsing some of them.

At about this time Francis began to make use of Edward Bacon's residence, Twickenham Lodge in Twickenham Park, which Edward leased from the Crown. This was a fine but moderately sized building, situated in extensive grounds that were once known as Isleworth Park - a royal park. Directly across the river, on the east bank of the Thames, was Richmond Palace - one of the Queen's favourite residences. Francis once referred to Twickenham Lodge (when he eventually owned the lease for himself) as a "wholesome pleasant lodge" and a place fit for a prince. Here he began to establish and build up, over the years, a scrivenry of scholars whom he paid (with his brother Anthony's help) for assistance in ciphering, deciphering, translating, transcribing, and all manner of works. Francis and Anthony referred to them as "good pens."

Whilst Francis had been abroad, the first 'Grand' Accession Day Tournament was held at Whitehall on the 17th November 1581, to celebrate the Queen's accession to the throne of England. Accession Day was an annual national festival. At Court it was developed into a pageant on the scale of a French *Magnificence* - but one which carried on in yearly 'episodes' right up to the end of Elizabeth's reign.[13] The first 'episode' was that of November 1581, but its 'prologue' was the great Entertainment put on for the Queen at Woodstock and Kenilworth during July-August 1575.[14] As I suggested before,[15] I strongly believe that Francis Bacon had a major hand in this initial Entertainment, put on by

Leicester and Sir Henry Lee, which was the forerunner of the later Grand Tournaments on Accession Day each year; and that his ideas and writings formed the principal guiding light behind the overall design of all the subsequent Grand Accession Day Tournaments.

In 1580 Francis had written *The Faerie Queene* and submitted the early proofs to Gabriel Harvey for criticism. *The Faerie Queene* reads like a handbook or guide for the producing of the Grand Accession Day Tournaments, wherein the Faerie Queen - Gloriana-Belphœbe (personified by Queen Elizabeth) - is the central character to whom her 'Arthurian' knights (all disguised in allegorical costumes) come, to honour with tributes of prose, poetry and verse, music and dance, drama and chivalrous sports, at her Annual Feast (*i.e.* at which all the Muses are present). The principal allegorical themes were based on Arthurian and Arcadian symbology. The *Shepheards Calendar* and Sir Philip Sydney's *Arcadia* both played an important role, additional to that of the *Faerie Queene.*

The idea behind such grand and extravagent pageantry is to produce a powerful spectacle - an allegorical vision of truth - by means of which the morals, consciousness and artistry of the people might be raised; a spectacular entertainment in which the sciences and the arts, and the philosophers and the artists, are not only seen but also actively involved, and in a way which is not only profitable but also enjoyable - which is by far the best way of learning anything, and of increasing man's imaginative powers of creativity and artistry. This idea lies at the heart of Francis Bacon's philosophy and method. This is perhaps one of the major differences that sets apart Francis' 'New Method' from Aristotle's rather dry and wholly intellectual 'Old Method'.

Francis was not present for the 'first episode' of these Grand Accession Day Tournaments, but there was no need. The inspiration and guidance that he could offer at that stage had already been diffused, and his talented cousin, Sir Philip Sydney, together with the rest of the Areopagus were there to manage matters: not to mention Dr. John Dee who advised in the background; the Earl of Leicester who stood as the Queen's favourite and secret Consort, who loved lavish and meaningful ceremonial, and who was, as Master of the Horse, responsible for the production and administration of the whole Tournament; Sir Henry Lee, the Queen's Champion, who was also the director of ceremonies; and the scores of noblemen and gentlemen who longed for the chance to show themselves off in the eyes of the Queen and the general public, each of whom could find a scholar and poet ready to help with writing their speeches and songs, actors to play the allegorical scenes, artists and craftsmen all too willing to earn a good living and find fulfilment in creating sumptuous costumes, scenery and pageant cars. To begin with, the main core of tilters came from the ranks of the Queen's Gentlemen Pensioners - men of good family chosen for their handsome looks and manly presence, since their main function at Court was to form a decorative escort for the Queen on State occasions; but as the reign progressed more and more young noblemen took part as tilters.[16]

The spark and the excuse for the commencement of these Tournaments on such a grand scale, and at Whitehall,[17] was the third visit of the Duc d'Anjou to London as a prospective bridegroom for the Queen. Alençon had arrived unexpectedly in England in June 1581, to treat of marriage over the heads of the French commissioners who were in London for that purpose but who were not getting very far. He had stayed a couple of days before returning to France. Then, in November, he came back to London for the third time and for the last phase of his courtship. The stakes were high, because the Queen and her Government needed the friendship and alliance of this powerful Valois prince. Spain had annexed Portugal at the beginning of the year, and a military attack on the English government in Ireland had been made by a combined army of Italians, Spaniards, French

THE FIRST BOOKE OF THE FAERIE QVEENE:

CONTAINING

THE LEGEND OF THE KNIGHT OF THE RED CROSSE,

OR

Of Holineſſe.

1

LO, I the man, whoſe Muſe whilom did mask,
As time her taught, in lowely Shepheards weeds,
Am now enforc't a far vnfitter task,
For trũpets ſtern to change mine oatẽ reeds,
And ſing of Knights, & Ladies gentle deeds;
Whoſe praiſes hauing ſlept in ſilence long,
Mee, all to meane, the ſacred Muſe areeds
To blazon 'broad, amongſt her learned throng;
Fierce warres, and faithful loues, ſhall moralize my ſong.

2

Help then, ô holy Virgin, chiefe of nine,
Thy weaker Nouice to performe thy will;
Lay forth out of thine euerlaſting ſcrine
The antique rolles, which there lie hidden ſtill,
Of Faerie Knights, and faireſt *Tanaquill*,
Whom that moſt noble Briton Prince ſo long
Sought through the world, and ſuffered ſo much ill,
That I muſt rue his vndeſerued wrong:
O! help thou my weake wit, and ſharpen my dull tongue.

3

And thou moſt dreaded impe of higheſt *Ioue*,
Faire *Venus* ſonne, that with thy cruell dart
At that good Knight ſo cunningly didſt roue,
That glorious fire it kindled in his hart,
Lay now thy deadly Heben bowe apart,
And with thy mother milde come to mine ayde:
Come both, and with you bring triumphant *Mart*,
In loues and gentle iollities arrayd,
After his murdrous ſpoiles and bloody rage allayd.

4

And with them eke, ô Goddeſſe heauenly bright,
Mirrour of grace and Maieſtie diuine,
Great Lady of the greateſt Iſle, whoſe light
Like *Phœbus* lampe throughout the world doth ſhine,
Shed thy faire beames into my feeble eyne,
And raiſe my thoughts, too humble, and too vile,
To think of that true glorious type of thine,
The argument of mine afflicted ſtile:
The which to heare, vouchſafe, ô deareſt dread a-while.

A 2 Canto

(90) FIRST PAGE OF *THE FAERIE QUEENE*, Part I - Spenser 2nd Folio (1617).

THE SECOND PART OF THE *FAERIE QVEENE:*

CONTAINING

The FOVRTH, FIFT, and SIXT BOOKE.

By *Edm. Spenser.*

Imprinted at London for *Mathew Lownes.*
Anno Dom. 1613.

(91) FIRST PAGE OF *THE FAERIE QUEENE*, Part II - Spenser 2nd Folio (1617).
Inscription 1: *"Veritas tua et usque ad nubes."*
Translation 1: "Your Truth extends even as far as the clouds."
Inscription 2: *"Os homine sublime dedit."*
Translation 2: "The countenance of man is devoted to the sublime."

1609.

AT LONDON,
Printed by H. L. *for* MATHEW LOWNES.

(92) LAST PAGE OF *THE FAERIE QUEENE*, Part I - Spenser 1st Folio (1611).

A LETTER OF THE AV-thors, expounding his whole intention in the *courſe of this worke: which for that it giueth great* light to the Reader, for the better vnderſtanding is here-vnto annexed.

To the right noble and valorous, Sir *Walter Raleigh, Knight, Lo: Wardein of the Stanneries, &* *her Maieſties Lieutenant of the Countie of Cornewayll.*

SIR, knowing how doubtfully all Allegories may be conſtrued, and this booke of mine, which I haue entituled *The Faery Queene*, being a continued Allegorie, or darke conceit, I haue thought good, as well for auoyding of iealous opinions & miſconſtructions, as alſo for your better light in reading thereof, (being ſo by you commanded) to diſcouer vnto you the generall intention and meaning, which in the whole courſe thereof I haue faſhioned, without expreſsing of any particular purpoſes or by-accidents therin occaſioned. The generall end therefore of all the booke, is to faſhion a gentleman or noble perſon in vertuous and gentle diſcipline. Which for that I conceiued ſhould be moſt plauſible and pleaſing, beeing coloured, with an hiſtoricall fiction, the which the moſt part of men delight to read, rather for varietie of matter, then for profit of the enſample: I choſe the hiſtorie of King *Arthure*, as moſt fit for the excellencie of his perſon, beeing made famous by many mens former workes, and alſo furtheſt from the danger of enuie, and ſuſpicion of preſent time. In which I haue followed al the antique Poets hiſtoricall: firſt *Homer*, who in the perſons of *Agamemnon* and *Vlyſſes*, hath enſampled a good Gouernour and a vertuous man, the one in his *Ilias*,

¶. the

(93) 'A LETTER OF THE AV-': *THE FAERIE QUEENE*, Part I - Spenser 2nd Folio (1617).

and Irish, funded by the Pope, coupled with a civil invasion by military priests. In addition the Catholic Guises had sent their agent to Scotland in an attempt to influence the young Scottish king and his counsellors, in order to pave the way for an invasion of England *via* Scotland, calling upon all English Catholics to rise up and overthrow Elizabeth, and to place Mary Stewart on the throne of England instead. The Guises were at that time trying to influence affairs at the French Court with the intention of persuading Alençon to invade the Low Countries on behalf of his brother, Henri II, and bring the whole of the coastline facing England, from Brest to the Hague, under French (and Catholic) control. If this happened, it would open the way for a massive invasion of England from the north, west, south and east simultaneously.

That was the dangerous reality facing Elizabeth at the time of Alençon's visit. Hence she play-acted in expert fashion, fooling many of her own subjects as well, by convincing Alençon that she would marry him - even to the point of exchanging rings. A great show of celebration needed to be put on in his (and her) honour, to clinch the charade. The very best that the English were capable of had to be shown to impress the French - and it did! The eventual result was a diplomatic triumph for the Queen, for in February 1582 Alençon departed England for the Low Countries as a confirmed ally of England - but not as a husband - with £50,000 in his pocket, ready to use his influence and arms to counterpoise Spanish power in the Netherlands, and to remain independent of his brother, Henri III. The lords Leicester, Hunsdon, Willoughby, and Mr. Philip Sidney accompanied him to the continent, with a fleet of fifteen ships.

Doubtless Francis was sad that he missed the event, although it was not essential for him to be present, and his quest for further knowledge and attainment was even more important to him; but he was home for the subsequent Accession Day Tournaments, from 1582 onwards. When Robert, Earl of Essex, began to appear at the Tournaments, it was Francis who largely wrote and staged the 'masques and triumphs' of his secret and royal brother.

The effort that went into these great pageants, their popularity, and the effect they had on courtier and plebian alike, was considerable and should never be underestimated. They were truly a glory of Elizabeth's reign.

Francis was now 21 years old, and on the 15th June 1582 he was admitted to the bar as an Utter Barrister. It seems that the Queen was instrumental in helping him achieve this, which is hardly surprising considering that he had spent such little time actually keeping terms or studying law at Gray's Inn - firstly because of his travels, and secondly because there is not much doubt that he had little zest for the law and spent most of his time in experimentation, cipher work, creative writing and trying to organise help to get his Great Reformation or Instauration well started.[18] Although admitted to the bar, he did not practice as a barrister, but continued with his main interest. From the Queen's point of view, it probably seemed to her that she was helping her son - as much as she dared; but she had different ideas to Francis as to what he should become. From Francis' point of view, the law was not for him; he found it far too restrictive and with little future in it for the kind of position he was seeking; and he did not want particularly to be moulded into an image of his foster-father by making it his profession. But he clearly could not make his sovereign mother see this.

In March 1583 a marriage was arranged between Frances, daughter of Sir Francis Walsingham, and Philip Sidney. Philip's great passion had been for Penelope, daughter of Walter Devereux, 1st Earl of Essex, but after Essex's death he never followed the matter up and Penelope married Lord Rich. However, between Philip Sidney and Frances

(94) FRANCES WALSINGHAM, Countess of Essex (b.1567-d.1632) - miniature by Isaac Oliver.

Frances was the beloved daughter of Sir Francis Walsingham. In 1583 she married Philip Sydney, by whom she had a daughter. She was devoted to Philip, and travelled to the Netherlands to nurse him on his death-bed (1586). After Sir Philip's death, she married (in April 1589/90) Robert, Earl of Essex, which displeased the Queen enormously. Frances was equally devoted to Robert, and had a son and two daughters by him. She finally married (after Essex's execution in 1601 and before 1603) an Irish peer, Richard de Burgh, Earl of Claricarde, who was created Earl of St. Albans in 1628.

During 1583 plans were made to form a company to carry out the colonisation of 'Atlantis', as America was termed by the Dee circle. The plan was concocted between Dr. John Dee and two of his circle - Adrian Gilbert and John Davis.[25] Dee was a primary influence on the Elizabethan explorers and navigators, and his *Mathematicall Preface* to the English edition of *Euclide,* together with the *Euclide* itself and Dee's *General and Rare Memorials pertayning to the Perfecte Arte of Navigation,* were the greatest of all encouragements to the development in England of mathematics, navigation and hydrography. Until he left England for the Continent in February 1583, Dee was probably the major guiding spirit behind the Elizabethan expansion - the great voyages of discovery and the colonisation in the New Worlds.[26] He had been one of the principal advisers in the early English attempts to find a north-eastern passage to Cathay, and in the searches for a north-western route to the Orient. He was almost certainly involved in Sir Francis Drake's epic voyage around the world - the main promoters and backers being being drawn from the Dudley-Walsingham-Dee circle (*i.e.* Sir Francis Walsingham, the Earl of Leicester, Sir Christopher Hatton and Sir Edward Dyer). Drake had left England in 1577 and returned home again on the 26th September 1580 - his stories and records of the journey no doubt being eagerly assimilated by Francis and Anthony Bacon, the Areopagus, and all those involved in the Dudley-Walsingham-Dee circle.

In September 1580, on the return of Drake, Dee had entered into a witnessed agreement with Sir Humphrey Gilbert that granted Dee the rights to all newly discovered territory north of the 50th parallel (*i.e.* the majority of what is now known as Canada); but, after reaching Newfoundland, the voyage failed when both the flagship (the *Delight*) and a sister ship (the *Squirrel*) sank, and Gilbert was drowned.

Sir Humphrey Gilbert was the brother of Adrian Gilbert, the chemist whom Lady Pembroke (*i.e.* Mary Sidney) employed at Wilton as her 'laborator'. He was also the half-brother of Sir Walter Raleigh. Like Sir Nicholas Bacon before him, Adrian Gilbert was directly involved in and wrote about the establishment of a school in London for the better training of the nobility and gentry in arms and other public services, described as "Queene Elizabethes Achademy,"[27] on the premise that a purely scholastic education was by no means sufficient for producing gentlemen of worth, and that "by erecting this Achademie there shalbe hereafter, in effecte, no gentleman within this Realme but good for somewhat, whereas now the most part of them are good for nothinge."[28] The Bacon-Dudley-Walsingham-Dee circle was really this 'Academy', which was steadily expanded and built up by the efforts of Francis Bacon and his friends into something far greater and more international in character than the original founders envisaged.

The 1583 plans were a continuation of the 1580 agreement and voyage. One of the hopes behind the voyages of this kind was to find a direct way to the East and to the Oriental wisdom, and from this to begin to establish a *concordia mundi* and a universal faith that transcended the man-made limitations of sectarian religion; in other words, a true eirenicism, in which England (or Britain) was seen as playing a crucial role.[29] Freemasonry was probably the form envisaged that this universal faith might take, hence the illustration on the titlepage of *General and Rare Memorials.* This was a plan which formed an integral part of the Baconian *opus mundi,* and which Francis Bacon treats of (or hints at) in his *New Atlantis.*

This 1583 expedition was not wholly successful and was somewhat tragic. The five small ships of the expedition reached Newfoundland, which were "the first of our nation that carried people to erect an habitation in those northerly countries of America,"[30] and formally took possession of the harbour of St. Johns and of the land 200 leagues in every

(95) SIR FRANCIS DRAKE, aged 41/42? years (b.1540?-d.1596) - miniature painted (1581) by Nicholas Hilliard.

direction. In this manner Newfoundland became the first and oldest dominion of the English Crown overseas. But Gilbert then sailed south from Newfoundland in order to explore the mainland coast, and in doing so lost the ship that carried the majority of the stores and equipment for colonisation, plus one hundred men. The expedition had no choice but to turn back, with Gilbert determined to return the following spring. But his own ship, including himself, was lost in a storm. However, what Gilbert and his explorers had accomplished was to form an impulse and a stepping-stone to the whole process of North American colonisation by Englishmen.

In 1583 Dee was given the task by the Queen of reforming the Julian Calendar for Britain. Although his work on this was widely admired, the reforms were not carried out bccause of objection by the bishops on religious grounds. He left for the Continent on the 21st September 1583 with Edward Kelley, his skryer or medium, not returning until December 1589. Regrettably, soon after he had left, his house at Mortlake was broken into, and his library and collections plundered.

(96) MARY, QUEEN OF SCOTS (b.1542-cr.1542-abd.1567-ex.1587) - painting after Nicholas Hilliard.

The Catholic Menace

During all this while Francis Bacon was working away in London at deciphering all intercepted correspondence that was connected with Mary Queen of Scots. In the South of France Anthony Bacon was busy gathering intelligence and sending it by courier back to England, having stationed himself in the key port of Marseilles during the winter, spring and summer of 1583, before moving on to Bordeaux by September 1583. Walsingham had his agents everywhere, at home and abroad, trying to discover in particular the intentions of Philip II of Spain, the Hapsburgs, the Pope, and the French Catholics led by the Guises. One result was that, in November 1583, the so-called 'Throckmorton Plot' was fully uncovered through the efforts of Walsingham's secret service. The plot involved the invasion of England by the Duc de Guise and his brother, the Duc de Mayenne, with full Spanish and Papal backing. The Guises' previous plan for an invasion *via* Scotland had come to nought, because of the exposure and expulsion of their agent in Scotland, Esmé Stewart; but instead they plotted a three-pronged attack on England in which Guise would land a task-force in Northumbria, Mayenne would do the same in Sussex, and their ally Duke Albert of Bavaria would invade Norfolk from Dunkirk. In addition, Mendoza, the Spanish Ambassador in England, was to identify and contact every Justice of the Peace in England who had Catholic sympathies, and persuade them to levy men - ostensibly to protect Queen Elizabeth, but at the last minute to take them to join the Catholic forces ranged against the Queen. The project was called 'The Enterprise of England'.

Mary Queen of Scots, who would be installed as Queen of England if this project succeeded, was kept in close touch with every development by Mendoza, by Thomas Morgan (her agent in Paris) and by Francis Throckmorton (nephew of Sir Nicholas Throckmorton, who was the manager of the plot in England); and for her part she was attempting to come to terms with her son, James VI, trying to persuade him to become a Catholic and to accept her as co-ruler with himself in Scotland in return for massive support from the English Catholics, who would secure for him the succession to the throne of England after herself.

The result of the plot's discovery was that Mendoza was expelled from England and a severe persecution of Catholic missionaries in the country was instigated. Mary Queen of Scots was kept under even closer surveillance than before, and both Burghley and Dee advised her death - although Elizabeth would not condone this. Whilst all this was going on, Leicester secretly continued in attempting to arrange marriages for his son by Lettice and for Lettice's remaining unmarried daughter, that would give his children and grandchildren a claim to the throne of England after the death of Elizabeth - his young son Lord Denbigh to the Lady Arabella, and Dorothy to the young Scots king, James VI! But Leicester's plans misfired in ways that he could not possibly have conceived. Lady Dorothy eloped with and married Sir Thomas Perrot, and little Lord Denbigh died on the 19th July 1584. This latter tragic event distressed Leicester immeasurably.

In March 1584 it would seem that Francis Bacon made a visit to Edinburgh in the train of Sir Edward Wotton.[1] Sir Edward was sent thither by Queen Elizabeth on a political mission to King James VI of Scotland, to try to induce James to join in a league of Protestant sovereigns for mutual defence against the Roman Catholic forces. He also had to ascertain, if possible, the true character and disposition of the Scottish monarch. Sir Philip Sydney, who had been actively involved in trying to establish such a Protestant League since 1577, also became directly involved in the objects of this Scottish mission, entering into correspondence with James upon that subject the following year. It is probably during this time in Edinburgh that Francis made an agreement with King James concerning the publication of the Scottish king's book on poetry, *The Essayes of a Prentise in the Divine Art of Poesie* (also known as *A Treatise of the Airt of Scottis Poesie*), which was published later that year by Thomas Vautrollier. In this book, whose title page falsely states that it was "Imprinted at Edinburgh" when in fact it was printed in London,[2] there is a marked similarity in the ideas of its principal author with those printed in *The Arte of English Poesie,* which was published in 1589 by Richard Field.

A curious manuscript by Francis Bacon's friend, Edmund Bolton,[3] entitled *Hypercritica,* which was probably written 1618-1620 but printed in 1722 for the first time, ascribes the authorship of *The Arte of English Poesie* to one of Queen Elizabeth's gentleman pensioners called 'Puttenham'. But there is no Puttenham recorded in the lists of the Queen's pensioners, and '**Puttenham**' could easily be another jesting pseudonym of Francis Bacon - *i.e.* Putten = Put on = 'to assume'; Ham = 'Bacon'. There are many indications within the texts of both *A Treatise of the Airt of Scottis Poesie* and *The Arte of English Poesie* that suggest a Baconian principal authorship. An additional pointer to this is that, on the titlepages of both books, there is printed the emblem or trademark bearing the motto *'Anchora Spei'* which was later made so famous by its use in the Shakespeare works. Francis Bacon, writing years later to King James, refers him to "your Majesty's Royal promise (which to me is *Anchora Spei*)."

This work towards forming a league of Protestant princes was finally accomplished in 1586, when, on the 17th July 1586, a meeting took place at Luneburg between "some evangelical Princes and Electors",[4] together with representatives of the King of Navarre, the King of Denmark and the Queen of England. At this meeting a Protestant evangelical league of defence, referred to as the *'Confederatio Militiae Evangelicae',* was formed against the powerful and threatening Catholic League.

On the 10th June 1584 Françoise de Valois, Duc d'Alençon et d'Anjou, died in France, having retired humiliated from the Low Lands and his position there as the Duke of Brabant. His death meant that Henri de Navarre was now the heir to the throne of France. Alençon's death was followed soon after by the assassination of William of Orange, on the 10th July 1584.

Not only did these two allies of England die in 1584, but three separate attempts to poison Queen Elizabeth were made that year by the Catholics, all of them foiled by the good services of Walsingham's intelligence network. The Queen acted bravely throughout and, although she publicly scorned and refused an armed bodyguard, yet privately she depended heavily upon her gentlemen pensioners and her secret service. Parliament became extremely alarmed, as it was well understood that if the Queen fell then there would undoubtedly be a terrible civil war, followed in all probability by Spanish and Roman Catholic domination of the country. The Privy Council decided that something further must be done to protect the person of the Queen. The result was the Bond of Association. This bond stated that, if the Queen should be murdered, the signatories to

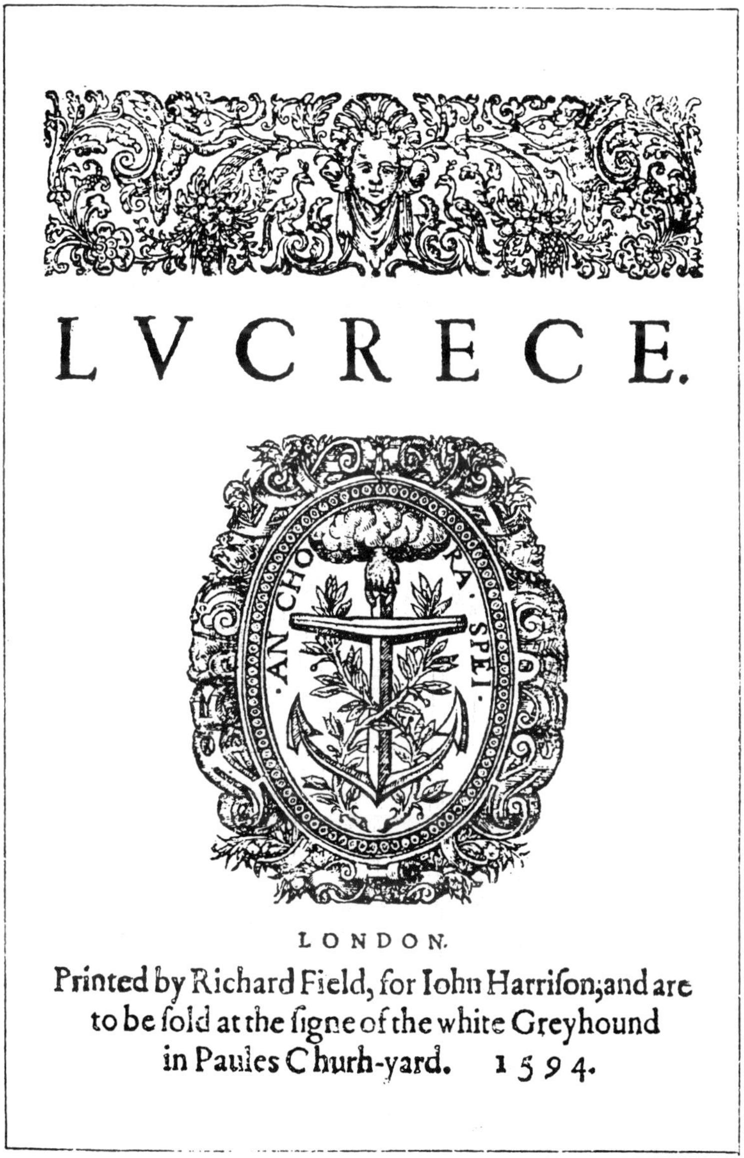

LVCRECE.

LONDON.

Printed by Richard Field, for Iohn Harriſon, and are to be ſold at the ſigne of the white Greyhound in Paules Churh-yard. 1594.

(97) TITLEPAGE TO *LUCRECE* (1594) depicting the *"Anchora Spei"* and the 'Bride' emblems - anonymous, but ascribed to 'Shakespeare' by references made in Henry Willobie's *Willobie his Avisa* (1594), William Covell's *Polimanteia* (1595) and Richard Barnfield's *Poems in Divers Humours* (1598), also by Gabriel Harvey (c.1601); printed by Richard Field.

the bond would not only pursue the murderer to the death, but would also put to death the person or persons in whose interests the murder had been committed and declare the heirs of those persons incapable of succession. Copies of the bond were carried throughout the shires of England, and a mass of signatures was obtained - so many, in fact, that when the papers were collected they had to be crammed into a large trunk. The Bond of Association was then incorporated in an Act of Parliament, after some revisions had been made by request of Elizabeth so as to ensure that anyone involved would always have a trial first, that any heirs were not to be unfairly punished, and that it should not therefore unreasonably preclude James VI of Scotland from any right he might otherwise have to succeed to the throne of England.

The next move by the Catholic League was to attack Leicester - the man who, after the Queen, held the greatest power in England. This attack was made by use of the pen rather than the sword, playing upon Leicester's notable weaknesses - his arrogance, his infidelity, his obvious love of wealth and power, and his association or even culpability in murder and other crimes, coupled with the fact that Leicester was an exceedingly unpopular man at large and that most people suspected his responsibility for the death of his first wife, and were quite willing to believe further or even worse things of him. Truths and half-truths, of an intimate nature, were cleverly built up into a concerted and libellous attack on Leicester, in a small book entitled *The Copy of a Letter written by a Master of Arts of Cambridge to his friend in London, concerning some talk passed of late between two worshipful and grave men about the present state, and some proceedings of the Earl of Leicester and his friends in England.* It was reprinted in 1641 under the title of *Leicester's Commonwealth.* Sir Francis Walsingham first received a copy of the book from the Lord Mayor of London and posted it on to Leicester on the 29th September 1584, adding that the author was "Morgan the Queen of Scots' agent in France," and that this author had been helped by some of the English Catholics in France, *viz.* Lord Paget, Charles Arundel, and William Tresham. Walsingham also stated in his letter that he had first received intimation that such a plan was afoot three years previously, and that it was also intended to include the Queen in a further libellous attack: "About three years past I had notice given unto me in secret sort of such an intention, with a meaning to reach higher."

The content of the book, although highly unreliable, yet has enough in it of factual interest, exposing some dark secrets and truths about Leicester as well as suggesting many others, and in addition is an easily readable, racy and vigorous text. It is a mine of information and misinformation for gossips, and that is exactly how it became so dangerous. It spread like wildfire, dozens of written copies of the book being made despite (and perhaps because of) the book being suppressed and banned. Sir Philip Sydney responded to this attack by writing a passionate defence of his uncle; however, this was not printed in his own lifetime. What was printed was a statement put out by the Privy Council ridiculing the book, and a letter of defence written by the Queen herself, signed copies of which were distributed to chief officers throughout the land.

In November 1584 Francis Bacon was elected a Member of Parliament for Melcombe in Dorset, and also for the pocket constituency of Gatton. Parliament had been called together especially to consider the severe threat to the Queen's safety and to the safety of the realm, revealed by the uncovering of the various plots to assassinate the Queen. Francis made such a brilliant speech in the Commons respecting this matter that Fleetwood, the Recorder of London, in writing of it said, "Before this time I never heard in Parliament the like things uttered. They were *magnalia regni.*"[5] This may well have been Francis' maiden speech in Parliament. He followed up the speech with a long letter of advice to the Queen concerning her safety, as well as other matters of State. The letter was circulated as a State document. Francis' special standing with the Queen should be well noted, as normally it would have been presumptuous in the extreme for a young and inexperienced barrister, who had never actually taken a case in law, to have taken upon himself to advise the Queen in such a way. But Francis could, and did. Moreover, his advice was sought.

Part of England's reply to the growing threat of Spain as a world power was to persevere with her attempts at colonisation of North America and control of her sea passages. With the Queen's blessing Walter Raleigh, who had been privy to and had helped to sponsor Adrian Gilbert's fateful expedition to Newfoundland and the coast of North America, in 1584 took upon himself his half-brother's unfinished task and sought a new charter from the Queen and Parliament with even greater powers than Gilbert had obtained, for the exploration, exploitation and colonisation of North America. Having received his charter, Raleigh sent out two West Country barks to reconnoitre the North American Atlantic coast, from the West Indies northwards towards Newfoundland. They returned with reports of an earthly paradise (*i.e.* the North Carolina coast): "the soil is the most plentiful, sweet, fruitful and wholesome of all the world.........We found the people most gentle, loving and faithful, void of all guile and treason, and such as live after the manner of the golden age."[6] The voyagers also brought back with them two young indians, who were to play an important part in the building up of contacts between the two races.

In 1585 Raleigh's expedition set sail. At the last minute the Queen could not bear to let Raleigh himself go on such a long and hazardous journey, and so Raleigh's cousin Richard Grenville took his place. This expedition was destined to plant the first English colony in North America, on Roanoke Island in Pamlico Sound. Even though it remained there for only one year, the importance of this small beginning, this first colony, is impossible to exaggerate; for from it sprang all further determined attempts to plant a permanent colony in North America, which finally came to fruition in the time of King James I, with the direct backing and help of Francis Bacon.

(98) TITLEPAGE TO *GENERAL AND RARE MEMORIALS* (1577) by Dr. John Dee.

Appendix A

TITLEPAGE TO JOHN DEE'S *GENERAL AND RARE MEMORIALS* *pertayning to the Perfect Arte of NAVIGATION*

Written by Dr. John Dee in 1566, published in 1577, and intended to be a "Hexameron or Plat Politicall of the Brytish Monarchie."

The illustration, which is set in a square frame, depicts a sky or heaven containing:-

(a) the Source of all light, the Holy Word or Logos of God, JHVH;

(b) the Sun, emblem of Light, the *Christos,* blazing with 24 rays (in two sets of twelve, depicting the double Zodiac, "male' and 'female'), with the heavenly 'clouds' all around that veil the Light from profane eyes;

(c) the Moon and ten stars, emblem of the reflected Light, *Sophia,* received by the Mind and seen as a vision or archetype, which archetype is represented by the cabbalistic ten stars or *Sephiroth* that are reflected in the mind. *Cabbala* means 'that which is received'. Its principal archetype in the Hebraic tradition is the Tree of Life containing the ten *Sephiroth* or divine Principles (*i.e.* 'Lights'). These are arranged on the Tree in three sets of three, culminating in a single one (*i.e.* 3 + 3 + 3 + 1). This arrangement is shown here with the Tree 'descending', as it were, from the Sun towards the Moon; for the whole 'descent' or unfolding of Divinity from the Source of Light, to the radiance of Light itself, and then the reflection of that Light as the Vision in the Mind, is depicted in this illustration as proceeding from the top right-hand corner of the picture to the top left-hand corner.

The Sun, emblem of the Christos, or Love-Wisdom, that is the Great Initiator and Teacher of all, is set in the top centre of the picture. Flying directly out from the Christ-Sun is the Archangel of the Sun, *Michael* ('He who is like unto God'), correctly depicted as flying (or radiating) towards the sphere of the Moon where he will be seen as the archetypal form of Christ-light. St. Michael represents the Christ principle of inspiration, illumination and beauty, and is the Prince and central principle of all the other great Archangels. He is pre-eminently the Angel of the Heart of God, and is associated directly with the land of Britain. He carries the flaming sword of light or truth in his right hand, whilst in his left he bears the shield of protection emblazoned with the Red (or Rosy) Cross, which Cross is his particular sign and symbol. St. George is the representation of St. Michael in terms of the human soul and, like St. Michael, he is directly associated with Britain as its patron saint - for Britain is the land of the radiant heart and of the Rosy Cross. St. Michael hovers above a peninsular of Britain, guarding the land of Britain from its enemies approaching the English Channel from the continent.

Four ships filled with soldiers, representing an armada, are shown approaching the coast of Britain *via* the sea, whilst on land four Spanish soldiers represent an army marching up to the English Channel, two with swords drawn, two with muskets on their shoulders, and all four with flaming torches in their hands with which they are threatening Britain. The army is led by an officer on horseback holding a spear. The intentions of the army and armada are obvious, and hostile to Britain. The English Channel blocks the way of the army, and a ship sailing towards England from the continent depicts the ships that will transport the invading army over the protecting waters of the Channel.

But, near the foot of the illustration, a small ferry boat is shown having crossed the English Channel, and a Spanish envoy has landed on English soil and is being greeted by an English officer or nobleman, with right hand extended in a sign of friendship and peace. This scene takes place in an area of English landscape delineated by a hedge of trees. This first scene of greeting leads on to the next scene, again delineated by a band of trees, in which the Spaniard is being led through the landscape by the English officer and towards a walled city that is set in the third scene. In the third scene the area concerned is no longer bounded by trees but by the walls of the city. At the highest point of the city the dome of a temple can be seen. This dome has a pinnacle with a pennant flying from its top. There are also two further pennants shown, each flying from a pinnacle, and each pinnacle is the top of another kind of roof, giving in total three kinds of roofs (and buildings) which are positioned in a highly significant way. The two gentleman can be seen in miniature in one corner of the walled city.

The progession of this 'drama' is from left to right of the picture, paralleling the progression from Moon to Sun to Source in the sky above. It is an exact depiction of the drama of initiation as portrayed in Freemasonry, showing the first three degrees of Craft Masonry: (1) the **Making**, when the candidate for initiation is admitted and accepted into the Lodge as an Entered Apprentice; (2) the **Passing**, when the Apprentice is passed into and through the second degree as a Fellow Craft Mason, being instructed in the Liberal Arts and Sciences as he progresses through the degree; and (3) the **Raising**, in which the Fellow Craft undergoes a sacrificial death in the court of Solomon's temple in Jerusalem ('the City of Great Peace') and is then raised from the dead to become a Master Mason. The three pinnacles and their respective buildings identify the places of the three 'Lights' of the Lodge, which in the higher degrees are also represented by the three Principals of the Lodge.

That this the intended meaning of the picture is further identified by the principal symbols of the three Craft degrees which are shown above the three scenes. Above the scene of **Making** is a great gateway, signifying the entrance into the Masonic Lodge, leading into a town beyond and a rocky hill behind. Note that the gate is open, ready to receive the newly accepted candidate into the Masonic community. The town signifies the community or fellowship of the Freemasons, which fellowship or friendship (the expression of compassion and goodwill) is the first lesson that the Entered Apprentice has to learn in the First Degree.

On the top of the hill, and directly beneath the guardian Archangel Michael, stands the figure of the goddess *Fortuna* ('Fortune'). Dee associates her with the youthful god *Occasio* ('Chance' or 'Opportunity'), which he says is now manifest. This is an accurate equation, for *Occasio,* as Opportunity, used to be depicted as a youth with a lock of hair falling over his brow (representing the forelock by which the opportune time might be seized) and holding a pair of scales in his hand whilst suspended on a razor's edge. He had winged heels and his feet rested on a globe. But early on in the Renaissance this male figure of *Occasio* was superceded by the female figure of *Fortuna* ('Fortune'). The goddess Fortune is usually depicted as naked and winged, with the forelock of hair streaming out in front of her brow. Like the god Opportunity, she normally is shown standing on a globe. Instead of the pair of scales, she has a wheel - the Wheel of Fortune - which she turns, raising the humble and abasing the proud (although it is also the wheel or cycle of time which she turns, in which the life process takes place for everyone). Fortune is known as the mistress of the sea, and other attributes of hers are the rudder, a billowing sail, a shell, a dolphin, and a ship. She is also depicted occasionally with a cornucopia, a die, and a bridle, with which she rewards the worthy and curbs the wild excesses of man. As with Opportunity, the globe on which Fortune stands represents not only the world but also the seeming instability of fortune, which turns and turns about, and rolls onward. This is contrasted in the Ancient Wisdom teachings with the goddess Virtue, who stands upon a solid cube, symbol of stability. But there are more esoteric interpretations of the sphere and the cube, that relate to spiritual form and material form respectively. This latter interpretation will assist in an understanding of what Dee has shown in his picture - for the circle, triangle and square are geometric symbols respectively for spirit, soul and body.

In Dee's portrayal of *Fortuna,* she is shown pointing deliberately to her forelock with her left hand, and hence to the opportune moment that should be seized, whilst with her right hand she is holding out a wreath of laurel - the victor's crown - towards the British ship. Instead of standing on a globe, she is shown standing with her left foot on a rocky pinnacle of the hill and with her right foot on a white stone carved as a pyramid. This white stone not only depicts the foundation stone (representative of himself) that the Entered Apprentice has to place in the north-east corner of the Lodge, but it also shows that stone as being the capstone or apex of the Pyramid-Temple that is to be built masonically. In other words, this stone is the stone that is "rejected by the builders", being "neither square nor oblong". It is the equivalent to the key stone of the arch or dome used in modern Freemasonic ritual during the Mark Degree. Biblically it is referred to as the 'cornerstone', for it unites all the corners of the Pyramid-Temple and is the final stone to be placed - the temple's crowning glory. Hence *Fortuna* holds an equivalent symbol of that crown of light, to make this clear.

Furthermore, in the First Degree of Freemasonry the Entered Apprentice is given a masonic apron made of lamb's skin (or woven of lamb's wool), symbolic of innocence and purity, and of the path of love and virtue that leads to the perfect oblation as an *Agnus Dei* or spotless sacrifice. The goddess *Fortuna,* who is normally depicted entirely naked, is shown by Dee as wearing the virtuous apron, sign of true chastity. The Greek words shown on the banner outside the town - " ΙΟ ΤΗΣ ΑΣΠΗΑΛΕΙΑΣ ΠΗΡΟΥΡΙΦΜ " - clearly relate to this First Degree.

Moving further into the land of Britain we come to the primary symbol of the Second Degree, shown above the second scene of **Passing**. This is the ear of corn near a stream (here the sea) of water, symbolic of the ripening and illumination of the Fellow Craft's mind as he studies and comes to understand the Seven Liberal Sciences and their Arts - the sciences and arts that lead to liberation or freedom from the thrall of materialism and ignorance.

Kneeling by the ear of corn is the figure of "Respublica Britannica", making her supplication and holding another banner that carries an inscription pertaining to the Second Degree - " ΣΤΟΛΟΣ ΕΞΩΠΛΙΣΜ ΜΕΝΟΣ ". Although she (representing the Fellow Craft Mason) has not yet reached the level of the Arch Degree, symbolised by the ship, yet she is able to gaze towards it and beyond it, perceiving what it is and what it might mean.

Proceeding further, we come to a skull, deliberately drawn so as to show only half of it. The skull is a primary symbol of the Third Degree, which is to do with death and entombment. Not only does it signify the corpse and the dissolution of the body, but it also alludes to the manner of death which is by three blows to the skull. In further symbolism, it refers to *Golgotha,* the Place of the Skull, upon which Jesus, the Son of Man, was crucified and within the rock of which he was entombed - the particular hill representing the skull of Adam. This skull is also a symbol of the Grail Cup or Alchemical Retort, in which the water of innocence is turned into the wine of spiritual knowledge.

However, the degrees of initiation do not stop with death (for death is but a change from one state of being into another), but lead straight on to resurrection and the higher degrees of mastery. The culmination of the Third Degree is a type of resurrection, *i.e.* **Raising,** but it also involves a reburial and is but a foreshadowing of the real resurrection or rebirth that takes place in the Fourth or Arch Degree. This Arch Degree (known as the Royal Arch) is depicted by the great ship in which sails the Sovereign, Queen Elizabeth, and her three Principals of the Royal Arch.

Another word for ship is 'ark', a word associated with the Greek word *arche* from which we have our English word 'arch'. *Arche* means 'the Beginning without End', and refers to the Circle of Existence which contains all life. It is the womb or vessel of life, in which all life processes take place. The Ark of the Covenant is, in its most profound sense, this divine womb, in which is contained the Word of God. By analogy, all pure and complete life forms are microcosmic vessels of life, bearing the divine Word and Light of God, and manifesting that Word of Truth. Such perfect forms are arks or temples of Light. In terms of the human being, it refers particularly to man's soul which is, when perfected, an immaculate vessel or ark of God's holy Law and Spirit of Truth. Thus it is depicted as an actual ship or vessel - the Royal Ark - sailing the seas of time and matter, and bathed in the Light from on high that descends into matter from the heavenly Mind of God.

The perfect, immaculate vessel in which shines the Light of God is the Christ soul - the Virgin Mary who has given birth to (*i.e.* become) the Christ manifestation. The immaculate aspect of the ship that is depicted in Dee's picture is represented by the Queen, who is symbolically known as the Virgin Queen and given the titles of the Virgin Mary. Furthermore the Queen's name, 'Elizabeth', is derived from the Hebrew *Eli-Sheba* and means 'consecrated to God', which is exactly the condition of such a perfect soul. She (as a symbol) is the living, manifest expression of Divinity, sailing as the helmsman in the ship of her outermost form, and which displays the ensign of Christ at the top of its three masts - the *Chi-Rho.* The pinnacles of the masts with their *Chi-Rho* emblems are deliberately drawn as rising above the horizon of the sea and penetrating into the heavenly realm of the Christ-Sun and the Source of all Light, for such is the consciousness of the Christ soul who is born from the Virgin.

Such a ship or ark is the principal symbol of the Royal Arch Degree of Freemasonry. Here it also signifies the Arch Lodge or Supreme Lodge of Freemasonry - the Lodge of the Adepts. These are those human souls who have conquered death and gained the victor's crown of full illumination and mastery. In this Arch Lodge are the three Principals who represent the three-fold expression of the Sovereign, and who together rule the Lodge. They also represent the three further degrees - Fifth, Sixth and Seventh - which complete the Christhood or full mastery of the illumined soul.

The Supreme Lodge is at the heart of all things, for the heart is the throne or seat of the Sovereign of Light, and this secret is symbolised by the name of the vessel, *Europa.*

Europa's picture and name are clearly painted on the side of the ship, showing her riding the famous bull of the incarnate and disguised Zeus. The inner meaning of Europa's mythological story tells the relevant truth that pertains to this degree of initiation. It also reveals a further truth that relates to the role of Europe in the world, and in particular the role of Britain, which Dee certainly seems to have known and appreciated. It is not by chance that Britain is (or used to be) called the White or Pure Land - the land and dowry (or kingdom) of the Virgin Mary - and whose guardian angel is the Archangel Michael. It is *par excellence* the Land of the Rosicrucians that Francis Bacon refers to as the Island of *Bensalem,* 'the Son of Peace', in his *New Atlantis.* The true Rosicrucians are those 'knights' of the Rose Cross who have attained the Arch Degree and entered into the degrees of Christhood (*i.e.* full illumination and mastery) that begin with the resurrection. They are those who have seen into the heart of God and who consciously rest in that holy Bosom, partaking of the knowledge and bliss of Christ consciousness, and enjoying perfect fellowship.

All of this great truth, in which the armada and armies of the continent can be seen as the unresolved and aggressive passions, thoughts and deeds of the savage and ignorant man, are contained within the Masonic Square that frames the picture. This square frame has a symbolic orientation which should be noted, as it is significant. In the First Degree of initiation the foundation stone is laid in the north-east corner of the Lodge. Judging by the way that the island of Britain is drawn, and according to this esoteric symbology, the north-east corner is clearly signified by the island promontory, which corresponds to the top left-hand corner of the square frame. This is not the natural or exoteric orientation, but a reversal of the natural layout of the lands represented. This gives the esoteric orientation of the whole illustration such that East is represented by the top of the picture. Since Light symbolically comes from (or rises in) the East, Dee has rightly shown the East as being the place of the heavens, from which the light shines down upon the rest of the landscape. The West is signified by the foot of the picture, where lies the actual land or earth that receives the light. In other words, Dee has orientated his picture in the same way that Solomon's Temple was orientated, and the way in which lodges of today are still orientated. Not surprisingly, in the Greek inscription around the frame Dee proclaims that it is a 'British Hieroglyph'.

ΤΟΥ ΑΥΤΟΥ ΠΛΗΘΩΝΟΣ

ΣΥΜΒΟΥΛΕΥΤΙΚΟΣ ΠΡΟΣ ΤΟΝ ΔΕΣΠΟΤΗΝ ΘΕΟΔΩΡΟΝ, ΠΕΡΙ ΤΗΣ ΠΕΛΟΠΟΝΝΗΣΟΥ.

GEORGII GEMISTI PLE-thonis ad Principem Theodorum de Rebus Peloponnes: Oratio Posterior:

GVLIELMO CANTERO,
Interprete.

Vm in Naui *Gubernator, omnia pro suo solet arbitrio dirigere, quæ ad Vectorum Salutem pertinēt: Tùm in Exercitu, Dux, quæ ad Militum Victoriam. (Alioqui nec Vectoribus, nec Militibus, vel tantillo tempore res rectè procederent, nisi ab Vno vtræq; viro regerentur: & apertissimè, maximis in periculis, Monarchiam, simul tutissimam & vtilissimam deprehendimus:) Veruntamen, et in Naui Licere videmus * Vectorum alicui, qui consilio suo communem videatur Salutem iuuare posse: et in excercitu, Militi similiter: illi quidem apud Gubernatorem, huic autem ante Ducem, quæ in mētem venerint exponere. Illis porrò cùm audierint, vel arripere Licet consilium, vel benignè et sine acerbitate reijcere. Siquidem non iniquè ferendum est, Siquis in Periculo communi,* Salutis publicæ *curam ad se quoq; nonnihil arbitretur pertinere. Enimuerò hac quoque in Gente nostra, atque Vrbe, tibi quidem, vir Clarissime, totius est rei cura commissa: perq; multorum maiorum, Regum, Patrisq;, etiam Regis, manus, hæc ad te deuenit nostræ Regionis hæreditas: in qua tibi omnia, prout tibi ac nobis videantur profutura, Licet omniū pleno consensu disponere. Cum autem tuæ quidem sic se res habeant, nostræ autem in magno versentur discrimine: (Siquidem, vndique nobis,*

* The maner how a Subiect may dutitully say his mynde, in Matters, beside, or aboue his Charge and gouernment.

Cicero his wordes to the like intent may aptly herewith be conferred: *Te ipsum moneo Pansa (quanquam non eges consilio, quo vales plurimū: tamē etiā summi Gubernatores in magnis tempestatibus, à vectoribus admoneri solent) hunc tuum tantum apparatum, tamq; præclarum ne ad nihilum recidere patiare. Tempus habes tale, quale nemo habuit vnquam. Hac grauitate Senatus, hoc studio Equestris Ordinis, hoc ardore Pop. Ro. potes in perpetuum metu & periculo Rempub. liberare. Cicero: Philippica. 7.*

I. iij.

(99) **PAGE 69,** ***GENERAL AND RARE MEMORIALS*** (1577) by Dr. John Dee.

which , alſo , incomparably ſurpaſſeth all Humayn Policie : though the ſame be moſt carefully vſed, for matters tending greatly and chiefly to his Glorie , and Honor : as , vndoutedly , this doth . And ſo , are the Three former Douts, briefly , and ſufficiently (as this place may admit) Diſſolued , and taken away .

Therfore , Seeing No Dout , or Obiection , (worthy of Conſideration , or , of any Importance ,) is remayning , to the Contrary of that moſt zealous *WISH* , or rather , *SVPPLICATION* , which (certainly) is agreable to the Harty Deſire , of the moſt parte of all Naturall , and Faithfull Subiects of this Kingdom , at this very Preſent : (vpon our own State , both , in it ſelf, chiefly , and abſolutely : And alſo , in Reſpect of our Forrein Neighbors , their vnaſſured Frendſhip , being Conſidered) : Why ſhould not we *HOPE* , that , *RES-PVBL . BRYTANICA* , on her knees , very Humbly , and erneſtly Soliciting the moſt Excellent Royall Maieſty , of our ELIZABETH,(Sitting at the *HELM* of this Imperiall Monarchy : or , rather , at the Helm of the *IMPERIALL SHIP* , of the moſt parte of Chriſtendome : if ſo , it be her Graces Pleaſure) ſhall obteyn , (or Perfect Policie , may perſwade her Highnes ,) that , which is the Pyth , or Intent of *RES-PVBL. BRYTANICA*, Her Supplication ? Which is , That , ΣΤΟΛΟΣ ΕΞΩΠΛΙΣΜΕΝΟΣ, may helpe vs , not onely , to ΦΡΟΥΡΙΟΝ ΤΗΣ ΑΣΦΑΛΕΙΑΣ: But make vs , alſo , Partakers of Publik Commodities Innumerable , and (as yet) Incredible. Vnto which , the *HEAVENLY KING* , for theſe many yeres laſt paſt , hath , by *MANIFEST OCCASION*, moſt Graciouſly , not only inuited vs : but alſo , hath made , *EVEN NOW*, the Way and Means , moſt euident , eaſie , and Compendious : In-aſmuch as , (beſides all our own ſufficient Furniture , Hability , Induſtry , Skill , and Courage) our Freends are become ſtrong : and our Enemies , ſufficiently weake , and nothing Royally furniſhed , or of Hability , for Open Violence Vſing : Though their accuſtomed Confidence , in Treaſon, Trechery , and Diſloyall Dealings, be very great. Wherin , we beſeche our *HEAVENLY PROTECTOR*, with his *GOOD ANGELL* to Garde vs , with *SHIELD AND SWORD*, now , and euer . Amen .

ΙΕΡΟΓΛΥΦΙΚΟΝ ΒΡΥΤΑΝΙΚΟΝ

* * *

G.iij. Truly

(100) **PAGE 53,** ***GENERAL AND RARE MEMORIALS*** (1577) by Dr. John Dee.

NOTES, Part I:

Section 1: Introduction

1. "And all the days of Enoch were three hundred sixty and five years:
And Enoch walked with God: and he was not; for God took him."
(*Genesis* v. 22-24.)

"By faith Enoch was translated that he should not see death; and was not found, because God had translated him: for before his translation he had this testimony, that he pleased God." (*Hebrews* xi. 5.)

"All his action was with the Holy Ones, and with the Watchers [*i.e.* the Archangels] during his life." (*Enoch* xii. 2.)

"No one else has ever been created on earth to equal Enoch, for he was taken up from earth." (*Ecclesiasticus* xliv. 14.)

Note that Enoch was aged **365** when "God took him" - *i.e.* raised him up to the highest heaven, where he sits on the throne of the Messiah, higher than the highest angels (according to Cabbalistic tradition). He was seen as being a pattern of repentance - the example of man made perfect and glorified. 365 (the number of days in a year) is a cosmic number - a complete and perfect cycle of time in respect of our world. The numbers given in the Torah or Bible are cabbalistic numbers, or ciphers, with profound hidden meanings. With this in mind, it is significant that Enoch is said to have invented the science of writing, arithmetic and astronomy (which also included the psychological and metaphysical science that today we call astrology). He is reputed to have filled 300 books with his writings (*c.f.* 300 years is the number of years specified in which Enoch "walked with God").

Enoch is identified with Atlas, and with the whole Atlantean culture, and is the name not just of a single outstanding and perfect man, but also of a whole line of high initiates (true king-priests) who constituted the 'light' of Atlantis and the guardians of its sacred wisdom knowledges.

2. *Jude* 14.

3. *Acts* 7: 22.

4. The 'Tree of Life' is the basis of the *Cabbala,* the 'received' wisdom-knowledge passed on to the 'Elect' of Israel by Moses.

5. *Cosmos* = a total Entity or Universe whose integral parts or principles constitute an ordered and harmonious system of beauty. The Absolute Cosmos or Truth, that emanates forth from the Divine Heart in love, manifests itself in three aspects: Divine, Human and Natural (*i.e.* as Spirit, Soul and Body).

6. These are equated cabbalistically with height and depth, east and west, south and north respectively. See *Sepher Yetsirah,* the Book of Formation.

7. *Heru-pa-khret* is depicted either as a child wearing the double crown of Pharaoh, or as a youth wearing the triple *Atef* crown, symbol of liberty. As Harpocrates he is referred to as the god of silence, and is represented as holding one of his fingers to his mouth to denote silence. Sometimes he is shown clothed in a wolf's skin, bespangled with eyes and ears, to intimate the hearing and the seeing of truth, but he holds a finger to his mouth to denote caution and silence, for such truths cannot be spoken of without great care ("Cast not pearls before swine") and the deepest truth, which is beyond reason, cannot be openly talked about but only experienced inwardly. Sometimes he is shown seated upon, or emerging from a lotus, symbol of the heart. The rose is an alternative symbol for the lotus, and both the lotus and the rose signify discretion and silence.

8. *Thoth* is an ancient Greek version of the Egyptian *Tat* or *Tot,* meaning 'Under' or 'Beneath'; for the Word of Truth lies under or beneath all things, as the very foundation or source of life. As such, Thoth is the Lord of the Underworld, which is the Heart centre - the Alpha and Omega of all life.

9. The dignitaries of each temple were organised in the form of the Tree of Life, each grade or office corresponding to one of the principles of that spiritual Tree. The grade of prophet was the starting point, corresponding to the ninth principle of the Tree of Life, called 'Foundation'. The other eight principles were represented by officers ascending in order up to the highest temple office of Demiurge, who corresponded to the principle called 'The Crown'. The supreme office of Pharaoh was above all the others and summed them all up in a unity as being their source and creative principle. The offices were as follows:-

0. ***Pharaoh*** (King)	THE CREATOR or VOICE
1. ***Demiurge*** (Initiator)	CROWN
2. ***Hierophant*** (Teacher)	WISDOM
3. ***Stolista*** (Purifier and Guardian)	INTELLIGENCE
4. ***Hierogrammatista*** (Secretary)	MERCY
5. ***Thesmophores*** (Guide and Tester)	JUDGEMENT
6. ***Zacoris*** (Treasurer)	BEAUTY
7. ***Komastis*** (Overseer of Banquets)	VICTORY
8. ***Odos*** (Orator/Chanter)	GLORY
9. ***Saphenath Pancar*** (Prophet)	FOUNDATION
10. [The Temple]	[KINGDOM]

Each temple had its own set of dignitaries, but supreme over them all were the four principal temples of Egypt at Hermopolis, Memphis, Heliopolis and Thebes. These four were associated with the cabbalistic 'worlds' of Emanation, Creation, Formation and Action respectively. The Pharaoh was supreme officer and ruler of all four.

10. Thoth is the Lord of *Maât,* Lord of Truth.

11. *Hermes* is really a Greek word for the Egyptian *Takhut,* which signifies the visible appearance or manifestation of *Tat,* the Word of Truth, as the perfect human soul. The Hebrew form of the word is *Tachath.* The Roman name, *Mercurius,* is derived from the Egyptian *Maâ kHeru* ('True Words').

The Egyptian *kH* is the Latin *Ch* and the Greek *C.* Hence the *Chi-Rho* (*CR*) in Greek - written as the well-known symbol of Christ (*Cristos*), meaning 'the Anointed' and derived from *Criw* ('I anoint') - is a direct translation into Greek of the Egyptian *kHeru.*

12. Hebraism has the same rule, and no one can become recognised as a rabbi (teacher) unless he is (or has been) married and raised a family. The role of *rabbi* is played by men, as they represent the 'force' aspect of Truth (*i.e.* the teaching, initiating principle), whilst the role of the home-maker and guardian of the holy flame of the hearth is played by women, who personify the 'form' aspect of Truth.

13. Science is an understanding of certain laws or truths; art is the best possible practice of those laws according to the understanding gained. The final knowledge acquired as a result of the experience of both the science and the art is a knowledge beyond intellectual understanding. It is wisdom, or knowledge of the Lord (*i.e.* Truth), known as full illumination. Science and art are stages leading to this full illumination.

14. The Holy Spirit is essentially a term for the whole Thought of God, which consists of three integral aspects - the Will or Word of God, the Vibration or Sound of that Word, and the Vision of that Vibration. The first is the specific impulse of love or desire that gives rise to and is the basis of thought; the second is the creative force or movement of thought; and the third is the reflected thought-form or idea concerning the thought. The three can be termed Inspiration, Intuition and Intelligence. The Greeks referred to them as *Logos, Christos* and *Sophia.* In the Vedas they are called *Brahmâ, Vishnu* and *Shiva.* The Ancient Egyptians used the terms *Tat, Ptah* and *Râ.* In Hebraic cabbala they are called *Kether* ('Crown'), *Chokmah* ('Wisdom') and *Binah* ('Intelligence'). The Christian Church names them as Father, Son and Holy Ghost.

    Nowadays the tendency is to use the term 'Holy Spirit' for only the one aspect of Divine Thought, the Holy Ghost, although the Spirit of God is really the whole triune Thought of God.

15. *The Advancement of Learning,* Bk.I.

16. Such a deliberate plan was by no means unique and expressed only in the land and civilisation of Egypt. In many parts of the world, in many periods of the world's history, other nations have been guided by sages to consciously build their countries and civilisations into temples of light. But there is no doubt that Egypt was an extraordinary one in that it has left us with a great and clearly visible 'light', which has inspired and guided subsequent Western civilisations ever since. The Hebraic nation that was led by Moses from Egypt into the land of Palestine, and which adopted the Egyptian name *Is-Ra-El,* was perhaps the most obvious and well-documented initiatory experiment with a nation that was directly inspired by the Egyptian experiment.

17. A.J.Festugière, *Corpus Hermeticum,* collection Budé, Vol.II, p.326: Hermes Trimegistus speaking to Asclepius.

18. Isha Schwaller de Lubicz, *Her-Bak,* Appendix 1.

19. The symbol of *Hapi* should be born in mind when studying the history of the Pharoah Akhenaton, whose role in the master plan of Egypt was to end one cycle and give birth to the next. Akhenaton deliberately personified Hapi and the various stages of the Nile Inundation.

20. Due to the precession of the equinoxes this is no longer so. During the Age of Taurus the midsummer sun arose in the sign of Leo; during the Age of Aries it arose in the sign of Gemini; and during the Age of Pisces it arose in the sign of Cancer. Now, as we enter the Age of Aquarius, the midsummer sun will arise in the sign of Taurus, the Bull.

21. Horus is depicted with a hawk's head upon a man's body. The hawk can stay poised in the air, has penetratingly clear sight, is a formidable hunter and can swoop rapidly upon its prey. It is the natural enemy of the serpent, which it is able to overpower and slay. Thus the hawk represents the clear-sighted intellect or mind of man, that perceives truth clearly and accurately and is able to grasp or comprehend it. As the serpent-slayer, it signifies the victory or command of man's illumined thought over the 'serpentine' powers of his being signified by the god Seth.

22. *c.f.* Eros, or Cupid, the Chief of all the gods - the other gods being but aspects or principles of the One God, Love. The bow and arrow are Egyptian emblems for the heart centre (*Tefnut*) and its love-light (*Shu*).

23. *Mem-nefer* (Coptic *Menfe*) was the name of the pyramid of Pepy I (*i.e.* Pharoah *Meryre* or *Mery-Râ,* 6th Dynasty, 'Old Kingdom', *c.* 2289-2255 BC) at Saqqara, the necropolis of Memphis, which was adopted by the time of the 18th Dynasty as a name for the whole city.

24. *Chakra* is a Vedic term meaning 'wheel' - a wheel or vortice of energy focused in and spinning around a central 'hub' or axis. The axis of each chakra is located upon the central axis or spine of the whole being. From, or through, these seven main centres are controlled all the functions of the body and soul. They have their origin in the Spirit, or spiritual Archetype - a thought-form of Light that projects its pattern of seven major focal points (and many minor focal points) of energy and consciousness upon matter, and thereby builds all material living form. In the fully evolved human being this pattern of intelligent energy is understood to reach its fullest and most harmonious expression.

25. The Greeks called this principle of Emanation, *Logos* ('the Word'). The Christian Church call it the 'Father'. The Brahmins call it *Brahmâ,* the first 'person' of the triune expression of God in the aspect of *TAT,* the Son.

In the Vedic teachings, *TAT,* the Truth (the unsounded Word of God), signifies the Divine Consciousness immanent in the heart of all creation. It is comprised of *Brahmâ-Vishnu-Shiva,* which are the three *TAT* expressions of *SAT,* the Absolute Divinity. *SAT-TAT-AUM* signify the three principal aspects of Existence, and may be seen as corresponding to (or rather, giving rise to) the three phases of the life-process in man (*i.e.* desire-thought-action); for *SAT* is pure Love in which is contained everything in potential, whilst *TAT* is the Consciousness or Thought (*i.e.* Wisdom) of that Love which designs, governs and teaches all that it creates, and *AUM* is the dynamic power or energy that does the work. *Brahmâ-Vishnu-Shiva* are the three divine aspects of *SAT-TAT-AUM* as they appear in terms of *TAT,* the Son.

26. The Greeks called this principle of Creation: *Christos* ('the Voice of Truth'). The Brahmins call it *Vishnu.* The Christian Church call it the 'Son'.

27. The Greeks called this principle of Formation, *Sophia* ('the Vision of Truth', 'the Image of God'). The Brahmins call it *Shiva.* The Christian Church call it the 'Holy Ghost'.

28. *i.e.* Amen.

29. R. A. Schwaller de Lubicz, *Sacred Science,* ch. 8.

30. *Matthew* 5: 8.

31. *I Kings* 8: 61.

32. *Nut* is associated with *Nun,* and is the archetypal or spiritual form of *Nun. Nun* is the primordial Space, dark and void, from which all life emerged. *Nut* is the aspect of that Space which is differentiated as the womb of heaven - the feminine aspect of the divine Mind in which divine Thought is conceived; whilst *Geb* is the other aspect of Space (*i.e.* the material form of *Nun*) which is differentiated as earth, or matter. *An* is the masculine aspect of the heavenly Mind. *An-Nut* (*Annu*) is the combined male-female Mind, the spiritual expression of *Heh* and *Hehet* (the masculine and feminine aspects of 'Spatial Infinity' that reside in the heart of God as the principle or potential of Mind).

33. The word 'pyramid' means a temple or holy structure that is set on fire with light, normally in the shape of the sacred mountain or hill, denoting the '**A**'.

34. "Bring me my bow of burning gold,
Bring me my arrows of desire,
Bring me my spear of light untold,
Bring me my chariot of fire...."

From *'Jerusalem',* by William Blake.

35. Hymn to Osiris. See Wallis Budge, *Egyptian Religion,* ch. 2.

36. The evening and morning sun respectively.

37. 'The house of Horus'.

38. Harmachis = *Heru-em-akhet* ('Horus in the horizon'). This name was given to the Great Sphinx at Giza.

39. *i.e.* the Underworld.

40. Hymn to Ra, *c.* 1100 BC, from the papyrus of Nekht (Brit. Mus. No.10471). See Wallis Budge, *Egyptian Religion,* ch. 1.

41. *I Corinthians* 15: 35-57.

42. "For all knowledge and wonder (which is the seed of knowledge) is an impression of pleasure in itself..." - Francis Bacon, *Advancement of Learning,* Book I.

43. Francis Bacon and his Rosicrucian group utilised this mystery sign in their published literature, in the famous '**AA**' headpieces and, in inverted form, as the secret sign of Cassiopeia, '**VV**'. (See FBRT Journal I/2, *The Virgin Ideal.*)

44. The double-A (**AA**) and double-T (**TT**) were used as cabbalistic ciphers in the English Renaissance, and are signatures of the Baconian-Rosicrucian group. The double-A was used in symbolic headpieces to titlepages, dedication pages and other highly significant pages. According to research carried out by Y. Ledsem (see *Baconiana* Vol.VIII, *Third Series,* No.31), the earliest printed use of the *AA* in Renaissance headpieces appears to have been in *De Furtivis Literarum Notis Vulgo. De Ziteris* (1563), by Ioan. Baptista Porta.

Ledsem says:-

> "It is used only once - over the dedication *Ioanni Soto Philippi Regis.* There is no other head-piece in the book. John Baptist Porta was, with the exception of Trithemius, whom he quotes, the first writer on cypher. At the time at which he wrote cypher-writing was studied in every court in Europe. It is significant that this emblematic device is used in the earliest period in which headpieces were adopted, in a book which is descriptive and is in fact a text book of the art of concealment. I cannot find any impressions from this particular block elsewhere, but the exact design appears in our literature in 1590 [*i.e.* the 1590 edition of *Stratioticos* by Leonard Digges].
>
> "The earliest English printed book in which I have been able to find the light **A** and the dark **A** is *The Arte of English Poesie,* printed by Richard Field and bearing date 1589."

The double-T was normally used to sign the end of a titlepage, dedication page, letter or chapter, used either on its own as '**T.T.**' or embodied as the capital letters in a suitable name - even as the supposed printer's or publisher's name or initials. It is inscribed as a signature on the Shakespeare memorial in Westminster Abbey.

45. The Bacon brothers utilised this symbology in the use of their name, **BAC-ON**, as an esoteric cipher for Truth, as well as for their own signature (and a signature of their 'Shakespeare' fraternity).

Section 3: The Egyptian Mysteries of Initiation

1. The existing account of the *Crata Repoa* ritual appears to have been derived, at least in part, from *The Ritual of Initiations* by Humberto Malhandrini, published in Venice in 1657. It appeared for the first time in 1770, printed anonymously in German. This was republished in 1778, and then translated into French that same year. It was first translated into English at the close of the 19th century by the Masonic scholar, Dr. John Yarker, and published in *The Kneph,* a short-lived Masonic journal. It was first published in English in book form by Manley P. Hall in 1937, under the title *Freemasonry of the Ancient Egyptians.*

2. See other FBRT Journals, which deal with other traditions and their Mysteries in order to provide a suitable comparison. By such comparisons the essential truth may be glimpsed over and over again, underlying all approaches to and expressions of life.

3. The civil (or agricultural) calendar of Egypt had three seasons: *Pert* (November-February), *Shmw* (March-June), and *Akh.t* (July-October). But the temple calendar recognised the four seasons that underlie all nature, in all countries, which are related to the alchemical elements and initiation.

4. **Eve** is the mother of the solar child. She is the human consciousness that develops into understanding, born (or extracted) from the heart (rib) of Adam; Adam being the essential desire nature of man - the will. Eve thus stands for the second degree of initiation, in which the understanding of man is developed. In a greater sense Eve, perfected as **Mary** (*i.e. Ave Maria*), represents the whole of the Lesser Mysteries of initiation. The Egyptian archetype of Eve, who becomes Mary, is **Isis**. The Egyptian archetype of Adam, who develops into Joseph, is **Osiris**.

5. The axe did not in fact **physically** touch the initiate's head. The pharaoh's skill was such that it was gauged to come within a hair's breadth of the initiate's forehead. The effect on the new initiate was psychological, inducing a temporary state of shock and swoon, simulating death.

6. When a real swoon was induced, the soul of the initiate, released from the physical body, actually 'descended' into the realm of Hades and met the gods who are the true judges of the quick and dead. When he awoke from the swoon, as if from a sleep, he was able to remember his out-of-the-body experience vividly. Even if the initiate did not swoon, an aspect of his soul did encounter the same metaphysical experience, but it was not likely to have been so profound or memorable.

   Another method was to place the initiate in the sarcophocus whilst he was still awake but in a state of raised consciousness; and then, by striking the side of the sarcophocus (which was tuned like a bell to a special note), the soul of the initiate was sent out of its physical body and on its journey into other spheres. During this time the *Hierophant* and the *Thesmaphores,* using their own extended powers and awareness, would assist and guide the soul during its experience. The experience was to help and heal all those lost souls in the purgatorial regions that he could find or was led to.

7. > "By the 'term of anger' is meant the time required to master irascibilities of the animal soul. The eighteen months is a Qabbalistic allusion - the 18 is composed of 8 plus 1, or 9, the number of human nature or sublunary imagination."
   > (Manley P. Hall, *Freemasonry of the Ancient Egyptians.*)

8. > "His arms and legs are then extended in the form of the spread eagle or phoenix, the posture used in the old rites as a symbol of death and resurrection." (Manley P. Hall, *Freemasonry of the Ancient Egyptians.*)

   This 'spread eagle', often shown with two heads, is a well-known Masonic and Rosicrucian alchemical symbol. It is used with great significance on the obverse of the Great Seal of the United States of America (which has a pyramid on the reverse).

9. The knightly investments correspond to the 'armour of God' with which the saintly person is invested, which represent his endowments as a saint that he then must learn to use:-

Buckler of Isis	=	Shield of Faith
	=	human wisdom, the reflective mind, providence.
Scimitar of Anubis	=	Sword of the Spirit
	=	truth, justice, right realisation, active energy.
Buskins of Anubis	=	Shoes of the Preparation of the Gospel of Peace
	=	understanding put into practice, movement.
Mantle of Orci	=	Breastplate of Righteousness, or of Faith and Love
	=	protection, concealment, virtue, zeal.
Helmet	=	Helmet of (the Hope of) Salvation
	=	illumination, wisdom, invisibility and revelation.
Bow and arrows	=	the power of eloquent love
	=	the Word and Voice of God.
Spear	=	illuminating ray, persuasion carried afar
	=	That which destroys ignorance and vice.

10. *i.e.* the head of the Gorgon Medusa - the *Python* or feminine aspect of the great Dragon, whose male aspect is called *Typhon* (*i.e. Set* or *Seth*). The act of decapitating the woman was a contrived drama for the purpose of teaching, and the 'woman' was in fact a realistic man-made model.

11. The new **robe** signified the new or resurrected body - the celestial form of the adept. It is the 'golden fleece' of the Argonauts, and the 'spiritual body' that St. Paul writes about. It was tied about the waist by a **cord**, signifying truth, life, fellowship, love, concord. It was tied in a half-bow, representing the sign of the *Ankh,* and signifying not only the strength of brotherhood but also the delicacy of it, as the bow could be untied with one sharp pull on the end of the cord - such a 'pull' representing any desire, thought or deed that is impure or inharmonious.

12. The **owl** is an attribute of Isis, and of the Greek Athena and Roman Minerva. It symbolises the initiate of the 4th Degree. The owl is able to see in the dark, and can hunt and kill rodents and snakes with great efficiency - creatures that symbolise the lower dark nature of man which are conquered and transmuted by the owl-like higher nature. The owl is also supposed to be very wise.

13. The **palm** represents glory, triumph, victory over sin and death, or resurrection, which is the true and everlasting **fame** that any soul can achieve. The date palm is a symbol of the Tree of Life. The phoenix is often associated with the date palm, as the phoenix is a symbol of the resurrected soul of the 4th Degree. The Egyptians also used it to signify their calendar, the tree producing a new branch for each month.

14. The **olive** tree, with its branch and fruit, represents peace, concord, plenty, wisdom, illumination or knowledge of God, immortality and fruitfulness. It is a symbol of the Golden Age, for the Golden Age of each soul is attained in the resurrection of the 4th Degree - the soul having become pure gold alchemically, shining with the golden light of love and wisdom.

15. The significance of ***IOA*** is profound. It is a fundamental word of God as the Holy Spirit - the Light that incarnates into matter. The letters ***IOA*** signify the divine love, wisdom and power that generate loving, understanding and service (*i.e.* creation, formation and manifestation). The *A* in this holy Name is the final '**A**' of revelation, or completion. All three letters signify the signs or sounds of the Holy Trinity - the expression in sound and light of the Eternal One, Who is the initial (but unheard) '**A**'. The vowels were never written in the Egyptian language and therefore formed the esoteric or secret part of the language. Whereas the written consonants signified the reflected light of God, the unwritten vowels stood for the radiant sound or voice of God. The Hebrew language adopted the same rules. Only the adept was ever initiated into the correct use of the vowels and therefore into the correct pronunciation of the language. To sound a word correctly takes intuitive knowledge, and this is the mark of the adept. Correctly sounded words are a great power, both creative and destructive. The speaker must know what he is doing and be morally sound and good; but then, only a true adept will be able to know the correct sounding of a word at any time or place.

    The correct sounding of a word, inwardly and outwardly, constitutes the true secret of all initiation, and is the *only* true secret of Freemasonry and Rosicrucianism - a secret which, by its very nature, can never be divulged but only discovered through personal achievement. This is the meaning of the 'Lost Word', for the truly spoken word is the direct expression of the Word or Will of God that lies in the heart. Inwardly it is the *Christos* or Bliss Consciousness (Intuition); outwardly it is the manifestation of the creative Christ power wielded by a true adept or 'Rosicrucian' - one who has risen victorious over death as the Golden One and Red King.

    IOA is the inner sound of the God-name that is outwardly written as ***JHVH***. Uniting inner and outer, the word can be written as ***JIHOVAH***, but only an adept will be able to sound this Great Name correctly. It is represented by the **Dove**. The name **John** is derived from it.

16. **100** is a cabbalistic number, 10 x 10, signifying the Ultimate All. The conquest of *Typhon* is the demonstration of complete mastery and control of the lower self by the higher self, and thus the full and harmonious integration of the whole being (spirit, soul and body) as a perfect organ of creativity.

    The Rose Cross Knight, by virtue of slaying the dragon, becomes the controlled and perfect expression of the 100 'heads'. Thus his cipher number is **100**. Francis Bacon and his cabbalistic 'Rose-Cross' group used this cipher as a means of personal identification. FRANCIS BACON, for instance, adds in Baconian Simple Cipher to 100 (FRANCIS = 67, BACON = 33). He and his group were known as '**C**' men - **C** being the Roman numeral for 100.

    The **comma** was often employed in the printed text by the Baconian fraternity as an embellishment to give the '100' signature. It was invariably used in the 'framework' of emblems. In addition, the comma signifies the so-called Pythagorean Comma - the mysterious and inexpressible interval that completes a scale.

17. ***CHYMIA***, as the word of the 5th Degree, is of course only a blind. It is a good description of the meaning and work of the 5th Degree (*i.e.* the Philosophical or Hermetic Work), but the word itself cannot be properly written. With the word ***IOA*** of the 4th Degree, the essence of vibration had been reached. Further degrees delve deeper into this vibration and its power, discovering what it can and does do, and how it may be appropriately used. The philosopher is the 'lover of Wisdom', and his alchemical work is the **loving** of Wisdom. It is gentle loving which constitutes the alchemical fire, as well as being the heavenly water that dissolves all problems.

    > "Truth, which only doth judge itself, teacheth that the enquiry of truth, which is the love-making or wooing of it, the knowledge of truth, which is the presence of it, and the belief of truth, which is the enjoying of it, is the sovereign good of human nature." (Francis Bacon, Essay *Of Truth*.)

18. This is the same crypt that had been used in the 3rd Degree.

19. The sacred dance, or series of dances, represented the motions of the stars, constellations, sun and planets in the sky, and the various motions and cycles of time and space as they occur in nature as a result of the 'dance' of heavenly thought.

20. ***IBIS***, the crane, is a symbol of Thoth. Like the heron, it is reputed to be the king of fishers - the Rich Fisher or Grail King. For this 6th Degree its particular significance is that the ibis is known to have extraordinary eyesight which enables it to catch fish expertly. It watches from above the water that which swims below in the water. It is the symbol of vigilance and seership - the true seer watching the events of the world from the high vantage point of raised and illumined consciousness, seeing and understanding clearly the motions and cycles of life, and able to identify and fish out those souls who become ready to enter the higher degrees of initiation.

    The amusing story concerning the crane (or heron) is that it stands on one leg whilst the other leg is held cradling a stone. If it should happen to doze off, its hold on the stone would loosen and the stone would fall in the water, or onto its other foot, awakening the bird abruptly. In this way the bird keeps itself ever watchful and awake. The Fisher King is said to be lame, and perhaps this is one of the reasons!

    It is the ibis, stork or crane that symbolically carries the baby in a cradle slung from its long beak, giving rise to the story that every baby is brought into the world by the stork. This is another way of saying the same thing as the fishing story, the baby being the fish that the stork lifts out of the water of the Lesser Mysteries and into the fiery (or sunlit) air of the Greater Mysteries. Such a child is the newly-born adept soul - a resurrected soul of the 4th Degree of initiation, that the Egyptians called *Moses* ('He who is drawn out of the waters').

21. Astronomy and astrology were then one science - metaphysical causes and philosophy not then being seperated from physical results and material science, but being embraced as one universal science and art. Also considered was the fact that there is an outer astronomy and an inner astronomy, the former being concerned with what may be seen of the heavenly bodies with the outer eyes (*i.e.* the study of nature and its laws, which express divinity), and the latter being concerned with the heavenly spheres of the mind, that can only be observed with the inner eye of the mind and which constitute the spiritual causes of natural effects.

22. *Pamylach,* or *Oris circumcisio* - circumcision of the tongue. Manley Hall remarks that this would seem to be a figurative expression, by which the Egyptians wished to say that the new prophet had acquired all knowledge which they could give him: his tongue was acute, and he was permitted to speak of all knowledge. (See *Freemasonry of the Ancient Egyptians.*) The symbolic meaning of circumcision is purification by the cutting away of all impurities, as an act of dedication or covenant. Hence circumcision of the tongue signifies that purification of the whole creative self (or initiated self) has taken place, as compared with the circumcision of the sexual organ which represents the purification of the lower self that has taken place as a prelude to initiation.

    There may also have been an actual circumcision of the tongue, cutting the skin beneath the tongue that holds the tongue to the base of the mouth, as is done by some yogis. This frees the tongue so that it can be rolled back into the throat, thereby cutting off the passage of air to the lungs. This is a highly skilled yogic practice in which the yogi demonstrates that he is master of all his bodily functions and can exist without breathing. The practice is usually used by the yogi when he wishes to operate in his soul body, free of the physical body, for a long period of time, and in this way he leaves his physical body in a state of suspended animation. This suspended animation of the physical body has been known to last several decades or even centuries in some cases, but for this to happen further important precautions must also be taken. This is possibly one of the keys to the ancient use of certain pyramid chambers.

23. *Maneras* = the houses of the *Manes* (*i.e.* the souls of the departed - in particular the celestial forms and consciousness of the high initiates). Whereas the procession represented the supraphysical ascent to the kingdom of the gods, the entry into the houses of the *Manes* signified the return to the physical body made by the prophet after his initiation.

24. *Oimellas,* or *oinomeli,* is composed of wine and honey (*i.e.* equivalent to mead), and was known as the Cup of Sweetness, the ambrosia of the gods (as compared to the Cup of Bitterness of the 4th Degree). This is the true Grail - the Cup of Pure Knowledge or Remembrance (*i.e.* divine realisation or illumination). This Grail represents the high initiate himself - the prophet who is become all sweetness in his nature, highly illumined and beautiful in body and soul. It also signifies that Pure Light of Divine Knowledge which the prophet is able to perceive and enjoy in a state of constant and holy communion.

25. The *Ankh* or *Crux Ansata* is the Cross of Life, the proper symbol of the high initiate who had become a perfect and full life-giving spiritual source for others. Its philosophical meaning is summed up in the words "the life bestowing." St. Paul refers to this state of the arisen and glorified human soul as "a life-giving spirit." The prophet was to wear this golden symbol of the perfect man constantly upon his heart.

26. The *Etangi* was a full, white, striped robe. The fullness signified the ampleness of the perfected nature and the free-flowing life forces that pour from the prophet and of which he imbibes. The whiteness symbolised the purity of the soul and the perfect state of illumination (*i.e.* of pure White Light) in which he dwelt. The stripes represented the perfect discipline of the initiate's life and control over his body and nature.

27. The square tonsure represented the full opening and use of the crown chakra:-

> "The master also received the tonsure, a ritual that has been preserved in most mystical orders from the most remote times. The part of his hair not shaved was arranged in a square coiffure, and he most frequently wore a beard. The ritual of the tonsure, or the head shaving, was based upon the releasing of the spiritual consciousness through the parietal foramina, two small openings in the crown of the skull. This area was called the Gate of the Gods, and the tonsure came to represent released consciousness or the use of the third eye."
>
> (*Freemasonry of the Ancient Egyptians.*)

The circular tonsure on the crown of the head, coupled with the square coiffure of the hair, symbolised:-

(1) the squaring of the circle, or the incarnation of spirit into matter and the subsequent raising of matter to the spirit to such an extent that they became fused in marriage, 'in heaven' (*i.e.* in the state of bliss consciousness), as the celestial soul.

(2) the intuitive light of wisdom (the open circle) conceived and perceived as an idea within the intelligence (the square coiffure) of the human mind.

(3) the conceptual powers of imagination (encircling) being given rational sense by the analytical powers of the intellect (squaring).

28. ***Adon*** is the root of *Adonis,* meaning 'Lord' as a title of God. The ***A*** signifies the heart expression of divine love, as also being the sign of full realisation, whilst the ***Don*** means 'Lord', the master or ruler who has real power and authority. Hence the full interpretation of *Adon* is 'Lord of Love' and 'Lord of Light'. It is used as an intimate title of God, and also as the title of all God-realised and perfected souls - *i.e.* the *Adonai* or 'lords from heaven', who 'speak with authority' and are 'anointed' with the Love-Light of God.

(See *I Corinthians* 15: 47, and *Matthew* 7: 29.)

29. *i.e.* The Demiurge was the equivalent of the Lord Chancellor.

30. The *Primum Mobile* ('First Swirlings') is the first movement of love as e-motion, the life-force. It is life; it is an impulse of love; it is the **will** of God. It is the birth of divinity, and is the power that brings all things to birth (*i.e.* the supreme initiating power, the beginning of all things).

31. The feet symbolise the understanding (*i.e.* standing under) - the intelligence that puts all things into action. It is the support of the holy wisdom (signified by the crown), and the receiver of its light so that it may be 'earthed'. For this reason it should be noted that Sir Francis Bacon, for instance, is sometimes shown with large roses on his shoes, as well as wearing a high black hat.

32. Manley Hall's translation of the *Crata Repoa* names this office as *hierostolista* instead of *hierogrammatista*. *Hierogrammatista* is the title of this office as given by Clemens Alexandrinus ("our sole ancient authority" according to *Encyclopedia Britannica*), and is certainly more suitable. *Hierostolista* seems not only inappropriate but incorrect, for the *hiero-* prefix should place this office above that of *stolista* rather than beneath it, which is not the case.

33. The *thesmophores* is called *horoscopus* by Clemens Alexandrinus.

NOTES, Part II:

Chapter 1: Return to England

1. Francis Bacon, *Sylva Sylvarum,* cent. X, exp. 986. See also F.B.R.T. Journal I/4, *The Great Vision.*
2. See F.B.R.T. Journal I/4, *The Great Vision.*
3. Ibid.
4. *i.e.* melancholic depression.
5. Note found amongst Anthony Bacon's private papers, 'The British Museum', vol I, page 14.
6. Letter written by Lord Burghley to La-Motte Fenelon, dated Dec.1579.
7. Francis was left a small amount of property and an income of £300 a year. See Joel J. Epstein, *Francis Bacon: A Political Biography* (Ohio, 1977), ch.2, p.24.
8. Rawley's 'Life of Bacon', *The Advancement of Learning* (1640).
9. Robert Dudley first became the patron of a company of players in June 1559. Players were treated as vagabonds in Elizabethan England, and either imprisoned or run out of towns, unless they were adopted into the retinue of some great lord of the realm and thus given his protection. Robert Dudley was one of the first noblemen to adopt a company and have it known by his name.

   James Burbage (1531-1597) started life as a joiner, but by 1572 had settled in Shoreditch and become a leading actor in Leicester's Men. In 1576 he built the first known permanent theatre, called 'The Theatre', in Halliwell (Holywell) Street, Shoreditch, which was the prototype of the theatres that followed - *viz.* the Curtain, the Rose (1587), the Swan (1595) and the Globe (1599). He was referred to as "the first builder of playhouses" by his elder son, Cuthbert (born 1567). In 1596 James bought the old hall in Blackfriars Monastery, that had been used for rehearsals and performances of the boys' Choirs of St. Paul's and the Chapels Royal (until suppressed in 1589 for scurrility), intending to create an indoor theatre there; but he was never granted permission. He died soon afterwards. James' younger son was Richard (born 1568). Soon after their father's death, on 28th December 1598 Cuthbert and Richard organised the complete pulling down of The Theatre and the transportation of the materials and properties across the river Thames, to rebuild it as The Globe on Bankside.
10. Spenser's name was first used when *The Faerie Queene,* Part I, was published in 1590 bearing the name 'Ed. Spenser', on the titlepage.
11. See F.B.R.T. Journal I/4, *The Great Vision,* p.225.

12. Elizabeth I and Robert Dudley were married in secret in the latter part of 1560, the wedding taking place at Baynard's Castle in front of a few witnesses, Sir Nicholas Bacon performing the ceremony in his capacity of Lord Keeper of the Great Seal. Elizabeth was then pregnant with the child who was to be born on 22nd January 1560/1, named Francis and discreetly adopted by Sir Nicholas and Lady Anne Bacon. (See F.B.R.T. Journal I/3, *Dedication to the Light.*)

13. The Dudleys are the same as the Dudderleys, Dudderleighs, Didderleys, Tiderleys, Tuderleys. Tidirs or Tudors. Even the seemingly different names of Tite, Tyte, Tyt, are variations of the Tudor family name, due to the enormous variations in local dialects. Phonetically 'Tudor' is the same as 'Judah', and in this we can begin to see the significance of the family name. Moreover, the word 'tutor' is the same name; and the Tudors were indeed the tutors or teachers of their kingdom, who carried the 'grail' or tradition of wisdom teaching (as also the Jews claimed to do). The Tudors are, in terms of bloodline, the ancient Grail Line - the bearers of the Holy Grail, lineal descendants of the royal line of the Cymry, the 'oldest race on earth'. Related to this, in terms of symbolic meaning, is the fact that 'tite' means 'red-haired' (alluding to the symbolic importance of red or fiery hair, emblem of illumination, as evinced in the story of St. Bride or Bridget).

    'Tite' is also a term for the 'bacon fat' which was used to light beacons (pronounced 'bacons').

    Not until after the Elizabethan period did English spelling become more standardised, once a national language had been truly born through the deliberate efforts of John Dee, Francis Bacon and a few others.

    For some of this research into the origins of our written language and these new cipher discoveries, I am grateful to David Millward who has been mainly responsible for making these recent discoveries.

14. See F.B.R.T. Journal I/4, *The Great Vision,* p. 108.

15. See F.B.R.T. Journal I/4, *The Great Vision,* pp. 35, 41.

16. See note 13. It is also interesting that the word 'bard' means not only a minstrel-poet, but also a thin slice of bacon used to cover meat. It is unlikely that this allusion to the veiling of truth was missed by the Bacon family. ("My meat is to do the will of Him that sent me, and to finish His work." *John* iv. 34.)

17. The "esoteric currents of the Renaissance" might be labelled 'Hermetic-Cabalistic', amongst other things. It was a fusion of Hebraic-Christian Cabbala, Egyptian Hermeticism, Orphic (or Platonic) Gnosis, and (in the case of Tudor England) Druidic Wisdom, with imput from Middle Eastern and Far Eastern sources. Its flow and development through Europe can be traced in the lives of such men as Ramon Lull, Marsilio Ficino, Pico della Mirandola, Johannes Reuchlin, Francesco Giorgi, Albrecht Dürer, and Henry Cornelius Agrippa. (See Frances A. Yates, *The Occult Philosophy in the Elizabethan Age.*)

18. F.B.R.T. Journal I/3, *Dedication to the Light.*

19. F.B.R.T. Journal I/4, *The Great Vision.*

20. *i.e.* in previous F.B.R.T. Journals.

Chapter 2: John Dee: the Magus

1. John Dee's father was Rowland Dee, a gentleman server to Henry VIII. His mother was Johanna Wild. John Dee drew up a family tree of his notable pedigree (British Museum, Cotton Charter XIV, I), indicating that his family was of the greatest antiquity and highest respectability, and was traceable to Roderick the Great, the ancient Prince of all Wales. (See Peter French, *John Dee, the World of an Elizabethan Magus,* ch.2, p.20.)

2. For this, I am grateful to David Millward, whose cipher discoveries are making it possible to see some of the occulted picture more clearly.

3. Peter J. French, *John Dee, the World of an Elizabethan Magus,* ch.2, p.34.

4. *Liber de Ludo Alae.*

5. Noel Fermor, 'Two Renaissance Magi', *Baconiana,* vol.LXV, no.182, pp.12-13.

6. Gabriel Naudé, 'Preface', *Vita propria* (1643).

7. Ewen MacDuff has perhaps done the greatest amount of work on discovering and deciphering the Baconian ciphers which employ their modified form of the Cardano grille method. Ewen MacDuff considers that Anthony Bacon was the one who principally learnt the method and adapted it for Francis' use in the Baconian publications. The adaptation was specifically to provide a means in the printed text by which the keys to making and using the 'grille' could be found by future readers of the books.

8. John Dee, *The Philosophicall and Poeticall Originall occasions of the Configurations, and names of the heavenly Asterisms,* and *The true cause, and account (not vulgar) of Floods and Ebbs* (1553).

9. Peter J. French, *John Dee, the World of an Elizabethan Magus,* ch.3, p.41. Dee's original *Supplication* is in the British Museum, Cotton MS. Vitellius. C. VII, art, 6.

10. *De superficierum divisionibus* was a work by Euclid translated into Arabic. See Raymond Clare Archibald, Euclid's Book on Divisions of Figures (Cambridge, 1919). See also Peter J. French, *John Dee, the World of an Elizabethan Magus,* ch.2, p.37.)

11. Dee, *Rehearsal,* p. 10.

12. Dee, *Rehearsal,* p. 10. (See Peter J. French, *John Dee, the World of an Elizabethan Magus,* ch.2, p.39.)

13. Dee, *Rehearsal,* p.21.

14. Dee, *Rehearsal,* p. 27. (See Peter J. French, *John Dee, the World of an Elizabethan Magus,* ch.3, p.43.)

15. *John* i. 1.

16. The key to Truth is perfect love, intuitively and perceptively recognised. Love is the Word, the Christ - Who is both the Truth and the Teacher of Truth.

    First have perfect love. Intuition is the ability to be aware of and listen to that perfect love. Perception is the ability to see in vision what one is aware of intuitively, and to understand it. The gold and silver keys that St. Peter (or Janus) carries, represent the intuition (gold) and perception (silver) of divine love (the key).

17. *Luke* viii, 8-10.

18. *Luke* x, 23-24.

19. See F.B.R.T. Journal I/2, *The Virgin Ideal.*

20. *Luke* xxiii, 34.

21. This final illumination is also known as 'Joy'. It manifests as a light not only in the mind but also irradiating the whole face, causing what is referred to as 'the countenance of the Lord' or 'face of God'. This light is perceived as a halo or corona ('crown of light') shining from the head of the illumined person, who has reached an ecstacy of perfect knowledge concerning some aspect of Truth.

22. "*Tam Marti quam Mercurio*" - "As much by Mars as by Mercury". (See 'Hemetes and the New Rosicrucian Birth', F.B.R.T. Journal I/4, *The Great Vision.*)

23. Peter J. French, John Dee: *the World of an Elizabethan Magus,* ch.3, p.57.

24. *Ibid.*

25. Dee, 'Mathematicall Preface' to Henry Billingsley's translation of *Euclid* (1570).

26. One of the reasons, which cannot be discussed here, but which can nevertheless be hinted at, is Mortlake's position in terms of the landscape temple of *Caer Troia* (of which London forms the head and the landscape zodiac centred on Kingston forms the heart), and its relationship to at least one of the networks of 'ley' energies - these networks or patterns of energy forming the energy body of the world's intelligence or mind.

Chapter 3: The English Areopagus and Arcadia

1. See F.B.R.T. Journal I/4, *The Great Vision.*

2. *Ibid.*

3. Biliteral cipher from the Letter by E.K., *Shepheard's Kalendar* (1579). See Elizabeth Wells Gallup, *The Bi-literal Cypher of Sir Francis Bacon* (1901).

4. Biliteral cipher from *Love's Labor's Lost,* Shakespeare Folio (1623). See Elizabeth Wells Gallup, *The Bi-literal Cypher of Sir Francis Bacon* (1901).

5. Biliteral cipher from *Midsummer Night's Dream,* Shakespeare Folio (1623). See Elizabeth Wells Gallup, *The Bi-literal Cypher of Sir Francis Bacon* (1901).

6. In enumerating a list of Francis Bacon's genuine works, Bacon's chaplain, Dr. William Rawley, states that in the last five years of his life Francis Bacon composed the greatest part of his books and writings both in English and Latin, and proceeds to give them in the following order:-

   > "*The History of the Reign of King Henry the Seventh; Abcedarium Naturae, or a Metaphysical Piece,* which is lost...."

   Basil Montague, in his *Life and Works of Lord Bacon,* vol.III, p.830, records it as follows:-

   > "The *Abcedarium Naturae* (or Alphabet of Nature), a fragment of a book written by Lord Verulam, and entitled *The Alphabet of Nature.* "

   The *Abcedarium Naturae* was written in Latin with the use of certain Greek letters, and was referenced to a numerical system. What has been published is noted to be but a "fragment" of the whole book; but this fragment is all that is needed by a careful thinker in order to reconstruct the original scheme, and is clearly intended by Bacon to be presented in this way. It was first published in 1679 by Archbishop Thomas Tenison, one of Rawley's successors in the guarding and release of Baconian documents, who added his own English translation. Tennison cryptically entitled this 'fragment' as *The Lord Bacon's Physiological Remains.*

7. Henry Seymour, 'The History or Mystery of Edmund Spenser', *Baconiana vol. xxii, no. 87.*

8. *'Three proper and wittie familiar Letters: lately passed between two Universitie men: touching the Earthquake in Aprill last, and our English refourmed Versifying. - With the Preface of a wellwiller to them both. - Imprinted at London by H. Bynnemann, dwelling in Thames streate, neare unto Baynardes Castell, Anno Domini, 1580, Cum gratia et privilegio Regiæ Majestatis.'*

   'Two other very commendable Letters, of the same mens writing: both touch the foresaid Artificial Versifying, and certain other Particulars, - More lately delivered by the printer. - Imprinted at London by H. Bynnemann, dwelling in Thames streate, neare unto Baynardes Castell, Anno Domini, 1580, Cum gratia et privilegio Regiæ Majestatis.'

9. John Still (1543-1608), who was Master of St. John's College, Cambridge, and author of *Gammar Gurton's Needle* (an early comedy), may be the "Maister Still" referred to.

10. From this one might possibly infer that the English Areopagites numbered either seven or eight. **7** is the number of the Seven Spirits, the seven colours of the spectrum, the seven major notes of the octave, and the seven stars of the *Pleiades, Ursa Major, Ursa Minor, etc.,* Certainly the Harvey-Immerito letters refer by name to a group of seven scholar-poets. But **8** (or 7 + 1) is a special number that Francis Bacon seems to have worked with also. **8** is the number of the Holy Ghost or Holy Breath, and refers to the Origin out of which the seven Spirits of God (the *Logoi* and Rays of Light) emerge, and to which the seven manifestations of life return. It is the cipher for Knowledge. In another sense, it is the cabbalistic number for *Hod,* 'Glory', which refers to the 'mercurial' capacity and illumination of the mind, and the ability of the mind to both learn and communicate its ideas.

    In Freemasonry, seven brethren perfect a lodge. The 'eighth' person can be understood as being the overseer or Archmason who inspires and teaches the others. Francis also used this pattern of 7 + 1 in his cabbalistic scheme of the 'Seven Wise Men of the West' - seven masks representing or standing in for one hidden author. A similar working of this scheme of ancient usage is that of the French Pléiades, who were a group of seven scholar-poets who were deliberately brought together in order to create a French Renaissance of language and culture. (See F.B.R.T Journal I/4, *The Great Vision.*) The Pléiades greatly inspired Francis and his friends.

    It should be noted that the group of seven French poets derived their name from the group of seven bright stars in the constellation of *Taurus.* This group of stars, called *Pleiades* ('the Assembly of the Judges'), marks the throat chakra of the celestial Bull. The Bull itself is associated with the outpouring of light into the minds of mankind, and is directly related to the *Beltaine* festival of May, and to its corresponding Eastern festival of *Wessak* - the festival of the Buddha.

    But *Ursa Major* and *Ursa Minor* are also composed of seven bright stars each. The English Areopagus, because of their identification with Arcadia, clearly set themselves up to represent *Ursa Minor,* the Little Bear, which contains the present North Pole star, called *Al Ruccaba* ('the Axis') or *Arcas* ('the Stronghold of the Saved'). The star *Arcas* is so named after Arcas, the son of Zeus and Callisto. The whole constellation, of which *Arcas* is the first star, is also called *Arcas,* and this is the celestial Arcadia - the 'Stronghold of the Saved' or 'Place of the Elect'.

11. The laws are inscribed on the twin pillars, *Boaz* and *Jachin,* in Freemasonry.

12. Lempriere's *Classical Dictionary.*

13. See F.B.R.T. Journal I/4, 'Hemetes and the New Rosicrucian Birth', *The Great Vision.*

14. Lempriere's *Classical Dictionary.*

15. In other versions of the legend, Arcas is equated with the constellation *Bootes,* or Arcturus.

16. The name 'Phoenician' is derived from:-

    (1) *Phoenix,* which (a) describes the royal colour, purple-red; (b) is the name of the mythical bird, of gorgeous plumage, which continually renews itself by dying in its self-made pyre of incense and then resurrecting itself from the flames rejuvenated, to live another cycle of five to six hundred years; (c) is the name of the date palm, symbol of the Tree of Life and of immortality, achieved through death and resurrection. The name thus means 'Victory', 'Fame', 'Exultation', 'Joy'. It is the Rosicrucian *Fama.* The Phoenicians are the 'Master Race' or 'Chosen People' in the true sense of the term, which has a historical connotation as well as spiritual. They are 'the Elect'.

    (2) *Phone,* meaning 'Voice' or 'Sound' - a direct allusion to the Voice of Truth that sounds the Word of God. The Voice of God is the one and only Teacher of the Universe, that is also known as the Christ Spirit or the Light - the universal vibration of Love-Wisdom. The Phoenicians are representatives of this Voice, and the elected teachers of mankind. This is the Tudor secret; but it is only earned through sacrificial death and resurrection (*i.e.* devoted, selfless service, loving and wise, followed by rebirth into a higher consciousness or knowledge of Truth).

17. Hence 'Arcas' and 'Arcadia'. The word relates to the Arc or Arch of the heavens, the Rainbow - sign of the Covenant, and the Ark of the Covenant, which also has a geometric representation in sacred architecture that is related to the Golden Mean. The Arcadians are 'the People of the Covenant' - 'the Arch People'. Noah's Ark should be seen in this context. The Great Bear is also known as 'St. Peter's Ship', and St. Peter holds the keys to the Gates of heaven, as the Hierophant or 'Revealer' of the Mysteries.

18. Ægaea had previously been a large country with lakes in its centre and surrounded by a circle of mountain ranges. It was known as a golden land, or 'Summerland', and a golden age reigned there for a time. When the Mediteranean area was flooded by water from the Black Sea bursting through the straits of the Bosphorus, Ægaea was flooded and its central low-lying part became a sea. This was the time when Albion led refugees from Ægaea back to the ancestral land that then was given the name of Albion, or Alba, 'the Pure'.

Chapter 4: Immerito

1. See F.B.R.T. Journal I/4, *The Great Vision,* pp. 233-234.

2. See Alfred Dodd, *Francis Bacon's Personal Life Story,* ch.5, 'The Court of France'.

3. This rough draft in Francis Bacon's handwriting is to be found without heading, docket or date amongst the Lambeth collection of Bacon's manuscripts.

4. See F.B.R.T. Journal I/4, *The Great Vision,* p.126.

5. Jan van der Noot was a refugee from the Duke of Alva's persecutions in the Netherlands. He later became a member of the 'Family of Love', of whom it is thought Sir Nicholas Bacon might have been another member. (See F.B.R.T. Journal I/3, *Dedication to the Light.*)

6. Frances A. Yates, 'The Triumph of Chastity', *Astraea.*

7. *Ibid.*

8. Frances A. Yates, 'John Dee and The Faerie Queene', *The Occult Philosophy in the Elizabethan Age.*

   Dee put forward his proposals for reforming the Julian calendar in 1583; but, although his work was widely admired, and although the Queen went so far as to approve the draft of a proclamation implementing Dee's suggested reforms, the Anglican bishops objected for religious reasons and the changes were not made.

9. The *Dreames* had, according to 'Immerito' himself, glosses "full as great as my *Calendar.*"

10. *i.e.* 'The origins or genealogy of the Dudleys'. This was obviously an important subject for Francis!

11. According to Harvey, the *Nine English Comedies* were given the names of the nine Muses, of whom Pallas Athena is the tenth and chief Muse, in imitation of Herodotus.

12. *i.e.* 'the Nuptial Song of the Thames'.

13. Parker Woodward, in his *Tudor Problems,* observes that "Francis had too much confidence in himself to waste anything good," and suggests that *Dreames,* with its glosse, was published as the *Passionate Century of Love* (1582) under the pseudonym of Watson; *The English Poet* was published as *Discourse of English Poesie* (1586); and *Court of Cupide* was used for the Masque of Cupid in *The Faerie Queene* (1590).

14. Gabriel Harvey writes that 'Nash', 'Lyly' and 'Greene' were three faces in one hood (*Pierce's Supererogation*). He was also fond of associating 'Euphues' as well as 'Lyly' with 'Greene'. (See Parker Woodward, *Tudor Problems.*)

15. *John* x, 1-18. (See F.B.R.T. Journal I/4, pp. 41-42.)

16. *The Shepheard's Calender,* Æclogue IX, 'September'.

17. *The Shepheard's Calender,* Æclogue I, 'January'.

18. Letter to Gabriel Harvey, signed 'E.K.' and dated 10th April 1579, published in '*Two other very commendable Letters, of the same mens writing: both touch the foresaid Artificial Versifying, and certain other Particulars, - More lately delivered by the printer. - Imprinted at London by H. Bynnemann, dwelling in Thames streate, neare unto Baynardes Castell, Anno Domini, 1580, Cum gratia et privilegio Regiæ Majestatis.*'

19. According to the original monument erected to Edmund Spenser in Westminster Abbey, Spenser was born in 1510. In 1778 the original monument was restored and the birthdate altered to read 1553. Either dates would mean that Spenser was certainly not a youth at the time in question, as a 'youth' referred to a male under the age of 21 (*i.e.* who has not yet reached his majority). According to the alternative birthdates, Spenser was either 69 years old, or 26 years old, in 1579.

    The 1510 birthdate is recorded in the 1679 edition of the Spenser Folio, and in a Latin-English version of the *Shepheard's Calender* made by Theodore Bathurst and first published in 1653. Both refer to the epitaph on the original monument to Spenser. That the original monument and its epitaph is identical with the record of it made in the 1679 Folio is confirmed by a survey of the Abbey Church of St. Peter's, Westminster, made in 1723 by John Dart and published by him soon after in two beautifully illustrated volumes.

20. *The Shepheard's Calender,* Æclogue IV, 'April'.

21. *The Shepheard's Calender,* Æclogue IX, 'September'.

22. *The Shepheard's Calender,* Æclogue X, 'October'.

23. Cognisance means 'badge', and refers to the "white beare" and the "stake" in the poem, which compose the heraldic 'Bear and Ragged Staff' emblem of the Dudleys.

24. All the Dudleys were identified by the heraldic badge, 'the Bear and Ragged Staff'. Francis himself had a right to this badge.

25. *The Shepheard's Calender,* Æclogue X, 'October'.

26. *The Shepheard's Calender,* Æclogue I, 'January'.

27. *Ibid.*

28. *The Shepheard's Calender,* Æclogue IV, 'April'.

29. *Ibid.*

30. Marginal note in George Whetstone's *Sir Philip Sydney, his Life, his valiant Death, and true Virtues* (1587).

31. William Webbe, *Discourse of English Poetrie* (1586).

    William Webbe may be another mask for Francis Bacon, as the *Discourse of English Poetry* seems to be in Francis' hand and it displays an extraordinary knowledge of the Areopagus and of Immerito's as yet unpublished and unannounced works (*e.g.* "his *Poemes*....his Dreames, his Legends, his Court of *Cupid,* his English Poet, with other....").

32. Chapter XXXI., Book I., *Arte of English Poesie* (1589).

33. *Psalm 57,* v.1:-

    > "Be merciful unto me, O God, be merciful unto me, for my soul trusteth in Thee:
    > And under the shadow of Thy wings shall be my refuge, till this tyranny be overpassed."

    Psalm 91, vv.1-2:-

    > "He that dwelleth in the secret place of the Most High
    > Shall abide under the shadow of the Almighty.
    > I will say of the Lord, 'He is my refuge and my fortress:
    > My God: in Him will I trust."

34. The Rosicrucian motto "***Sub Umbra Alarum Tuarum Jehovah***" contains a code in its five initial letters - **S.U.A.T.J.** Making use of the 22-letter form of the Roman alphabet (known as the Simple Latin Cabbala), as given in Silenus' *Cryptographia* - in which I and J are identical, U and V are identical, W is written as VV (*i.e.* double-U) and therefore does not exist as a separate letter, and Y does not exist except as an I - place the five initials on the cipher wheel as described by J. B. Porta. The five initials give the key, suggesting that the wheel should be turned five places to the right. The result is that S becomes A, T becomes B, V becomes C, A becomes F, and I becomes O, yielding **A.B.C.F.O.** Rearranged, these letters form **F. BACO**, the Latin signature of **F. Bacon**, corresponding in style to the Latin words of the motto.

    This cipher signature was discovered by the mathematician and cipherist, Dr. Speckmann of Holland. (See Lewis Biddulph, 'The Rosicrucian Three Treasures', *Baconiana* vol.xxxii, nos. 126-128.)

35. John Wilkins, Bishop of Chester, first secretary and one of the founders of the Royal Society, writes on page 237 of his *Mathematical Magick*: "Such a lamp is likewise related to be seen in the sepulchre of Francis Rosicross as is more largely expressed in the confessions of that Fraternity."

36. Extract from 'E.K.'s Epistle to Gabriel Harvey that prefaces *The Shepheards Calender.*

37. *Ibid.*

38. 'Letter of the Author's intent', written to Sir Walter Raleigh and prefacing *The Faerie Queene.*

Chapter 5: Spenser

1. The monument was erected by Anne Clifford in 1619, twenty years after Spenser's death.

2. Sonnet 60 is a deeply mystical and symbolical poem. In it there is a suggestion that the poet has lived forty years. Most commentators take this literally and seem to miss the symbolic meaning of the number 40, which equates with the symbolism of the rest of the poem. This normal kind of scholarship compares with the scholarship of those who take the forty years wandering in the wilderness by the tribes of Israel, or the forty days spent fasting in the wilderness by Jesus, at face value; whereas the numerology involved is entirely cabbalistic and conveys deep inner meanings.

3. A sizar was a serving-scholar: that is to say, they received free education, board and lodging in return for their services to the masters and wealthier students.

4. Sir Henry Sydney (1529-1586) was first appointed Lord Deputy of Ireland in 1565. He found the country in a more impoverished and turbulent state than when he had been there before (1556-1559). He returned to England in the autumn of 1567 for ten months. Returning to Ireland he stayed there until 1571, then returned home. In 1575 he returned again to Ireland with additional powers given him by the Queen. He was recalled in September 1578, living mainly at Ludlow Castle for the remainder of his life and performing his duties as President of the Welsh Marches. He died there on May 5th, 1586.

5. If Philip Sydney was in Ireland in 1576, which is not definite, he certainly did not stay long For in 1577 he was sent from England on an embassy to the Emperor Rudolf II in Prague On his return he wrote a defence of his father's policy in Ireland (*Discourse on Iris Affairs*, 1577).

6. Letter from Sir Amyas Paulet to Sir Nicholas Bacon, dated September 1577 from Poitiers.
7. F.B.R.T. Journal I/4, *The Great Vision,* pp.191-192.
8. Ludovick Briskett was an Italian who had travelled abroad as courier or companion to the Earl of Leicester's nephew, Sir Philip Sydney, in 1572-1575. He had obtained the Clerkship to the Council of Munster when Sydney's father was Lord Deputy of Ireland.
9. E. G. Harman, *Edmund Spenser and the Impersonations of Francis Bacon,* p.39.
10. On February 25th, 1591, Spenser was granted an annual pension of 50, payable half-yearly every Midsummer and Christmas. It was granted to Edmund Spenser and his assigns during his natural life, to be paid at the office of the Exchequer at Westminster "by the hands of our Treasurer and Chamberlain." (English State Papers.) The Issue Rolls for 1598-1599 are missing, but Spenser certainly drew his half-year pension payment (his last) in January 1599, as this is recorded being paid by the hand of Thomas Walker.
11. *View of the Present State of Ireland* was not actually printed until 1633. Harman considers this long prose tract was written in London during 1596. (See E. G. Harman, *Edmund Spenser and the Impersonations of Francis Bacon,* p.45.)
12. The Simple English Cabbala is the most basic cabbalistic numerology of the 24-letter Elizabethan alphabet, *viz.:-*

A	B	C	D	E	F	G	H	I	K	L	M	N	O	P	Q	R	S	T	V	W	X	Y	Z
1	2	3	4	5	6	7	8	9	10	11	12	13	14	15	16	17	18	19	20	21	22	23	24

    The numerical value of words is derived by the addition of the letter values. Word values can then be used separately or added together again.

13. Preface to *The First Booke of the Faerie Queene: containing the legend of the Knight of the Red Cross, or Of Holinesse.* (1617 folio, *The Faerie Queen,* by 'Edm. Spenser".)

Chapter 6: Euphues

1. Mary, Countess of Pembroke, was a remarkable lady. She was born at Tickenhill Palace, near Bewdley in Worcestershire on 27th October 1561 - the same year of birth as Francis Tudor (Bacon). She was the third daughter of Sir Henry Sidney, her mother being Mary, daughter of John Dudley, Duke of Northumberland. Robert Dudley, Earl of Leicester, and Ambrose Dudley, Earl of Warwick, were uncles to Mary. Philip Sydney was her brother. Although Philip was seven years her senior, nevertheless during their childhoods they were constant and devoted companions. Their childhood was spent partly at Ludlow Castle in Wales, which was their father's residence as President of Wales, and partly at Penshurst in Kent, where the family had a lovely house (and was where Philip was born). They were not wealthy enough to accept a peerage, when it was offered, but they gave a first class education to all their children. Mary became a scholar in Latin, Greek and Hebrew, and later became interested in chemistry.

    When Mary was fourteen, her sister Ambrosia died, occasioning great sorrow. The Queen wrote a letter full of sympathy and understanding to the grieving parents, and invited them to send their remaining daughter Mary to court, out of the 'unpleasant air (if it be so) into better in these parts" to be cared for by the Queen in healthier and more privileged surroundings. Mary thus became an attendant on the Queen - one of the ladies-in-waiting at court, where her beauty and grace of manner soon established her position. Soon a marriage was arranged between her and the wealthy Henry Herbert, second Earl of Pembroke, which took place on 21st April 1577. He was then forty and twice-married previously, she sixteen. Her uncle, Leicester, supplied Mary with a dowry.

The Pembrokes' country house was Wilton, near Salisbury in Wiltshire. Nearby, on the river Avon at Ivy Church, was a smaller house of theirs which Mary was especially fond of. It was here, in 1580, at Mary's suggestion, that Philip Sydney began to write his pastoral romance, *Arcadia* - probably to entertain his sister whilst she was nursing her newly-born son. They spent much of their time in literary studies, and were visited by many of the 'new set' - poets, scholars, artists, scientists and, of course, the Areopagus. Among them were Nicholas Breton, William Browne, Samuel Daniel and John Davies of Hereford. Mary herself was an acknowledged poet as well as a writer and translator of no mean accomplishment. She made ready for publishing her brother's major literary works after his death in 1586, *viz. The Countess of Pembroke's Arcadia, Astrophel and Stella* and the *Apologie for Poetrie,* translated the *Psalms* and *Antonius* into English, and produced a dialogue called *Astraea* for a proposed visit of the Queen to Wilton (which never took place). She was considered, after the Queen, to be the most admired of the Elizabethan 'femmes savantes'.

Many writers paid tribute to her, both during her life and afterwards, including Michael Drayton, John Davies, John Donne, Sir John Harrington, Thomas Nash, Ben Jonson, Samuel Daniel, Abraham France and Sir Nicholas Bretton. Francis Bacon, as 'Immerito', dedicated to her his *Ruines of Time,* which was written in memory of his cousin, Philip Sidney, and describes her, under the name of 'Urania' in his *Colin Clouts Come Home Again,* as "the well of bountie and brave mind, excelling most in glorie and great light.......the ornament of womankind."

2. John Aubrey, 'Mary Herbert, Countess of Pembroke', *Brief Lives.*

3. 'John Lyly', *Eupheus and his England* (1580).

4. The 2nd edition of *Euphues Anatomy of Wit,* corrected, was published in 1581 and titlepaged to 'John Lyly, Master of Art.' Whenever the name is subsequently used, it is followed by a like description. Never is 'Lyly' referred to as M.A. of Oxford.

5. John Lyly (?1554-1606).

6. This was first proposed by Wood who, in compiling his *Ath. Oxon* in 1691, found the record of this John Lylie and proposed that he was one and the same with the author of *Euphues Anatomy of Wit, etc..*

7. Parker Woodward, *Tudor Problems.*

8. The old monastery was protected by walls and four gates. It was situated just inside the walls of the City of London, on the corner between the river Fleet and the banks of the Thames. But it was treated as being situate within the verge of the Court, and was under the jurisdiction of the Lord Steward of the Royal Household. The monastery was dissolved in Henry VIII's time and the Friars suppressed, but, by the time of Edward VI, the buildings were in the charge of the Master of Revels. The apparel and furniture for the Court masques and revels were kept there, which, during the reign of Elizabeth I, grew to some considerable size and worth. It was, of course, an excellent place to plan and rehearse productions.

9. The lease may, in fact, have been taken by Francis Bacon, using Lyly's name.

10. Both 'Lyly' and 'Nash' are one person, Francis Bacon, masquerading under two pseudonyms. In fact the 'Lyly' pamphlet was never signed as such, and is only attributed to 'Lyly' by its style. It could just as well be attributed to 'Nash', or 'Bacon'; for they are one and the same. The pamphlets in question were written by Francis, in various guises and styles, in order to help his friend and old tutor, John Whitgift, the Archbishop of Canterbury, in counteracting the effect of the schismic Martin Marprelate pamphlets.

11. Parker Woodward, *Tudor Problems,* p.84.

12. Parker Woodward, *Tudor Problems,* pp.84-85.

13. Parker Woodward, *Tudor Problems,* p.90.

14. Parker Woodward, *Tudor Problems,* p.92.

15. 1590 is the usual date given for the first publication of *Rosalynde,* but 1588 is the more likely date - unless, as often happened, the book was first given out in manuscript form to those it was intended for in the first place. No copies of a 1588 edition are known of, and only one good copy of the 1590 edition exists.

16. *i.e.* Hellebor, a herb of Saturn, whose roots are effectual against all melancholic diseases. (See Culpeper's *Complete Herbal*). The following verse is taken from *The Anatomy of Melancholy* by 'Democritus Junior' (first published in 1621), which was preceded by an earlier version entitled *A Treatise of Melancholy* by 'T. Bright' (published in 1586). Both versions are by Francis Bacon, whose favourite classical sage and hero was Democritus.

    "Borage and Hellebor fill two scenes,
    Sovereign plants to purge the veins
    Of melancholy, and cheer the heart,
    Of those black fumes which make it smart;
    To clear the brain of misty fogs,
    Which dull our senses, and Soul clogs.
    The best medicine that e'er God made
    For this malady, if well assay'd."

17. "Dying to live" is an explicit statement of the 3rd Degree of Initiation - the sacrificial death (*i.e.* perfect service) in order to be reborn into eternal (or real) life. Also, in characteristic fashion, Francis uses an enigmatic phrase with a double meaning, for 'dying' can be interpreted as 'longing' (which gives in the end the same meaning as the mystical interpretation).

18. This would imply that when *Rosalynde* and its prefatory letter were first written and published, the Earl of Leicester was not yet dead. Leicester died on the 4th September 1588.

19. "I rang the bell that called the wits together." (Francis Bacon)

20. Robert Greene (1558-1592), son of a Norwich saddler, was at one time a player in the Earl of Leicester's Company. Like Gosson, he had been in his youth one of the Queen's Chapel choristers. When too old for this role, he was sent to St. John's College, Cambridge, as a poor boy or 'sizar' (1575-1578). He then led a dissolute life in Italy and Spain before returning to Clare Hall, Cambridge, as a reformed character to take an M.A. in 1583 and to marry. But soon he relapsed into vice once again, leaving his wife and child before 1586, and living dissolutely in London amongst villains of all sorts. His mistress was the sister of a notorious thief, 'Cutting Ball', who was later hanged at Tyburn. He was at one time a sub-dean of the Chapel Royal, but was commonly known as 'the red-nosed minister" and as "Robert the Parson". A surfeit of wine and pickled herrings ruined his health and he died in September 1592 in squalor, after a month's illness. (see Alan and Veronica Palmer, *Who's Who in Shakespeare's England.*)

21. Parker Woodward, *Towards a More Correct Biography of Francis Bacon.* (*Baconiana,* vol. xxxix, No,152.)

Chapter 7: Athena's Veil

1. *Arte of English Poesie* (1589) - anonymous. Published by Richard Field, with a dedication to Lord Burleigh, dated "May 28th." A curious manuscript by 'Edmund Bolton', entitled *Hypercritica,* probably written in 1620, was printed in 1722 for the first time. In it the authorship is ascribed to one of Queen Elizabeth's gentleman pensioners called Puttenham, "as the fame is." But there is no Puttenham recorded in the lists of the Queen's pensioners.

   '**Puttenham**' may be another jesting pseudonym of Francis Bacon - *i.e.* Putten = Put on = 'to assume'; Ham = 'Bacon'. There are many indications within the text to point to a Baconian authorship.

   The work was probably written about 1585, with amendments being made in 1589, just before publication. It carried a significant emblem or trademark bearing the motto '*Anchora Spei*'. It is of some significance then that in 1584 Thomas Vautrollier, the London printer, had published *A Treatise of the Airt of Scottis Poesie* for King James VI of Scotland, that this work falsely stated that it was "Imprinted at Edinburgh" when it had been published in London, and that it also bore the '*Anchora Spei*' emblem. It might be argued that the repeated use of the emblem was because Richard Field was Vautrollier's successor; but it is surely noteworthy that Francis Bacon, writing years later to King James, refers him to "your Majesty's Royal promise (which to me is *Anchora Spei*)."

   It seems quite likely that Francis Bacon was asked, first by King James of Scotland and then by Queen Elizabeth, to put together suitable books on poetry in which each sovereign and their respective talented subjects could have some of their verses and recollections published anonymously. In the *Arte of English Poesie* Francis not only had the opportunity to expound the rules of poetry which had been such a burning issue with the Areopagus, to put in his own observations about poetry and poets, and to tender some of his personal recollections such as private conversations with Sir Nicholas Bacon at his home, but also to give advice to the Queen as to the best way to treat the Dutch - a continuation of the statesmanlike *Letter of Advice to the Queen* which Francis presented to her Majesty in 1584. (see Spedding, *Lord Bacon's Letters and Life,* vol.I, ch.3; Walter Begley, *Bacon's Nova Resuscitatio,* vol.I, ch.1 & 2; Parker Woodward, *Tudor Problems,* ch.XIX.)

2. Francis Bacon, *Of the Advancement of Learning* (1640), p. 315.

3. Francis Bacon, *Of the Advancement of Learning* (1640), pp.115-116.

4. Francis Bacon, Preface, *Of the Advancement of Learning* (1640).

5. Francis Bacon, *Of the Advancement of Learning* (1640), p. 45.

6. Francis Bacon, Preface, *Novum Organum* (1620).

7. Epistle Dedicatory, *The French Academy* (published in French in 1578, first published in English in 1586). Published anonymously. The first English version was printed as having been "translated" by 'T.B.' Francis Bacon was the author. (See F.B.R.T. Journal I/4, *The Great Vision,* pp. 195-204.)

   '**T.B.**' is 'T. Bright' or 'Timothy Bright', who masked several of Francis' works, in particular *A Treatise of Melancholy* (1586), the original of *Anatomy of Melancholy* (1621) titlepaged to 'Robert Burton'. '**T. Bright**' conveys the idea of the Rosy and Golden Cross, which is a Tau Cross of golden Light. (*Tau* means 'Truth'.)

8. Thomas Powell, Dedication to *Attourney's Academy* (1630).

9. William T. Smedley, 'The Mystery of Francis Bacon', *Baconiana,* vol. ix, no.36.) *Ad* is the Latin for 'To', and Owen uses the abbreviation '*D*' for *Dominus* elsewhere in the text. Thus it is a reasonable suggestion that *AD. D.B.* means *Ad Dominum Baconum,* or 'To Lord Bacon'.

10. *Hermes Trismegistus* is a Greek name that primarily relates to the Word of God - Truth itself. The Ancient Egyptian name for this was Thoth, or *Tat.* But it was also used as a generic term or title of the greatest sages of Ancient Egypt in each generation - *i.e.* the Pharoah in particular, and the High Council of Egypt in general. In an even more special sense, though, the Egyptians and Greeks applied the term to a few extraordinary men who appeared now and again at critical moments in history, who were seen and acknowledged as godlike people with seemingly super-human genius and gifts, and who appeared to be the embodiment of Truth. One of these was Amenophis-son-of-Hapu, who lived in the time of Pharoah Amenophis III and his son, Pharoah Akhenaton (Amenophis IV), and who was minister and councillor to both kings.

11. Handwritten sonnet placed amongst Anthony Bacon's correspondence and preserved in the Lambeth Archiepiscopal Library.

12. Alfred Dodd, *Francis Bacon's Personal Life Story,* pp. 100-101.

13. Together with the Introduction to the thirty-two elegies by Dr. William Rawley, the *Manes Verulamiani* total **33**, which is the cipher in Simple Cabbala of 'BACON'.

14. *Manes Verulamiani,* edited by W.G.C. Gundry (F.B.S.,1950).

15. *Manes Verulamiani,* Elegy 16, anonymous.

16. *Manes Verulamiani,* Elegy 9, by R.C.,T.C.

17. *Manes Verulamiani,* Elegy 11, by T.P.

18. *Manes Verulamiani,* Elegy 17, by E.F. Regal, King's College. *Fandi fictor* is translated as 'concealed poet''. *Fictor* means more precisely a creator, contriver, maker, of whom a poet is one, an architect another, a dramatist, *etc.,* further ones.

19. *Manes Verulamiani,* Elegy 4, by R.P.

20. *Manes Verulamiani,* Elegy 32, by Thomas Randolf, Trinity College.

21. *Manes Verulamiani,* Elegy 5, by William Boswell.

22. *Manes Verulamiani,* Elegy 1, by S. Collins, R.C.P.

23. *Manes Verulamiani,* Elegy 20, author unknown.

24. *Manes Verulamiani,* Elegy 21, by William Loe, Trinity College.

25. *Manes Verulamiani,* Elegy 18, author unknown.

26. *Manes Verulamiani,* Elegy 12, by Williams.

27. The structure of this Council is based on cabbalistic principles, representing the Tree of Life. It is the ancient form of all High Councils, as seen in Ancient Egypt and elsewhere, having three Principals and seven Officers. This is the composition of a Rosicrucian Council. Royal Arch Masonry is also ordered according to this revealed law.

28. Sir Francis Bacon was created Baron Verulam of Verulam on May 12th, 1618. Almost three years later, James I created Lord Verulam as Viscount St. Alban, on January 22nd, 1621 (Francis' 60th birthday).

29. Postscript to letter written by Tobie Matthew to Francis Bacon in 1621.

30. Ben Jonson, *Discoveries.*

31. Ben Jonson, 'To the memory of my beloved, The AUTHOR', prefatory verses to the 1st Shakespeare Folio (1623).

32. That is to say, to be born with a poetic gift is not enough to make a good poet. A good poet is also **made**: that is to say, he has to acquire the experience, knowledge and learning that forms the background to and meat of his poetry, plus the acquisition of skills in the actual form or making of poetry itself. In the case of 'Shakespeare', the knowledge and skills that are to be found in the plays are considerable, and far beyond the capacity of any ordinary scholar-poet, let alone a virtually uneducated and inexperienced man such as Will Shaxpere from Stratford. Only one man of those times was known to have all the qualities, knowledges, experience and skills to have written the Shakespeare Plays: that man was Francis Bacon.

33. "The father's face lives in his issue........" points out that the quality and character of the writings themselves speak of and reveal the quality and character of the mind that composed them. To match the 'face' of Will Shaxpere to the writings commonly ascribed to him is no match at all; but in the case of Francis Bacon the 'face' and the works match admirably, for the one is the expression of the other.

34. These two lines describe why the name 'Shakespeare' is used, and summarise the purpose of the work.

35. See Mrs Henry Pott, *Francis Bacon and his Secret Society, ch.XI.*

36. Francis St. Alban, *De Augmentis.*

37. Mrs Elizabeth Wells Gallup, *The Biliteral Cipher of Francis Bacon,* p.314. (Cipher found in *Natural History,* 1635.)

38. *Ibid,* p.358.

39. Francis Bacon, *De Augmentis.*

40. *Ibid.*

41. Francis Bacon, *Advancement of Learning.*
Reppresentarione is Italian for a stage play. The English word has this meaning.

42. Francis Bacon, *Advancement of Learning.*

43. Francis Bacon, 'Orpheus', *De Sapientia Veterum* (1609).

44. *The Arte of English Poesie* (1589).

45. Shakespeare, *As You Like It,* ii, 7.

Chapter 8: The 'AA' Signature of Light

1. Andreas Alciatus, or Alciat, was born in Milan. He was described as: "Not only was he the most noble jurisconsult, but in all liberal learning and especially poetry, so experienced that he could vie with the very highest geniuses." (De Bry, *Icones Virorum Illustrium,* Pt.II, p.134.) The emblems were collected by Alciat from various sources, the first of the four "Fountains' being gathered in Milan in 1522, but not then printed. Emblem No.XLV of the 1546 'Fountain' edition was, apparently, also entirely new at that time, never having appeared before. (See W. Lansdown Goldsworthy, *Shake-speare's Heraldic Emblems; their origin and meaning,* ch.1 & 2 - H. F. & G. Witheby, London, 1928.)

   Christopher Plantin was a printer in Antwerp, and known as one of the most ardent members of the 'Family [or House] of Love'. He was responsible for printing, for instance, the works of the Familist prophets, Henry Niclaes ('H.N.') and Henry Berefelt (pseudonymously known as 'Hiel', meaning 'Light of God').

2. The motto, "*Moniti meliora*", can be translated as "Instruction bringeth improvement". This motto was painted on the wall over the dining table in the hall of Sir Nicholas Bacon's house at Gorhambury, together with a mural of Ceres teaching the sowing of the golden seeds of corn. It was a key reference to the Orphic Mysteries, to which Gorhambury was dedicated (see FBRT Journal I/3, *Dedication to the Light*). It is recorded that the young Francis Bacon used to continually ponder on the meaning of this picture and its motto, and kept inquiring about it.

3. Virgil, *Aeneid,* iii, 188.

4. The boar, which is Adonis or Mercury, signifies the human manifestation of Pure Wisdom or Light, which is Truth. The sow, which is the Bride or Venus, signifies the human manifestation of Wisdom-Knowledge, which is also known as Cabbala ("That which is received").

5. The prophecies of Paracelsus are extraordinarily exact. This great prophet foretold the supernova explosion in Cassiopeia, which occurred in 1572-3 at the time that Francis Bacon received his great vision of his life's mission, and referred to the revolution in thought which Francis was to usher in, in which all things would become revealed. Referring to the supernova, Paracelsus referred to it as

   > "....the Sign and Harbinger of the approaching revolution: there is nothing concealed which shall not be revealed, for which cause a marvellous Being [Elias the Artist] shall come after me, who as yet lives not, and who shall reveal many things."

   This great Being, called 'Elias the Artist' by Paracelsus, is directly associated in the minds of many well-known 'Rosicrucians' as being one and the same as the mystery mage, C.R.C. (sometimes translated as meaning 'Christian Rose Cross'), the supreme head of the Invisible Order. For instance, the *Order Kabbalistique de la Rose Croix,* issued in 1891, states:-

   > "Elias Artista! Genius Director of the Rose Cross, symbolical personification of the Order, Ambassador of St. Paraclet! Paracelsus the Great has predicted thy coming, O collective breath of generous vindications, Spirit of Liberty, of Science and Love which must regenerate the World."

   The regeneration and enlightenment of the whole wide world through the renewal or reformation of all arts and sciences was the stated aim of Francis Bacon, the great poet-philosopher and 'lover (or bride) of wisdom'. This is the same aim proclaimed in the early 17th century Rosicrucian manifestos, which first publicised the Rosicrucian Order and C.R.C. openly: naturally, for C.R.C. and Francis Bacon are one and the same.

In the *Anatomy of Melancholy,* first published in 1621 under the mask of Robert Burton (but based on an earlier version, *A Treatise of Melancholy,* published in 1568 under the pseudonym of T. Bright), the author, in referring to the Rosicrucian Brotherhood in the opening address 'to the Reader', associates C.R.C. (described in the Rosicrucian manifestos as the Theophrastian Master) with Elias the Artist and Francis Bacon:-

> "....that omniscious, only wise fraternity of the Rosicrucians, those great theologues, politicians, philosophers, physicians, philologers, artists, &c. of whom s. Briget, Albas Joacchimus, Leicenbergius, and such divine spirits have prophesied, and made promise to the world, if at least there be any such (Hen. Neuhusius makes a doubt of it, Valentinus Andreas and others) or an Elias artifex their Theophrastian master; whom though Libavius and many deride and carp at, yet some will have to be 'the renewer of all arts and sciences, reformer of the world, and now living', for so Johannes Montanus Strigoniensis, that great patron of Paracelsus, contends, and certainly avers 'a most divine man', and the quintessence of wisdom wheresoever he is; for he, his fraternity, friends, &c. are all ' betrothed to wisdom', if we may believe their disciples and followers."

To make it clearer, Tycho Brahe wrote that for him the supernova in Cassiopeia marked the entrance of the world into the 7th revolution (*i.e.* the seventh and last part of the Piscean Age) that would inaugurate the Golden Age (*i.e.* the Aquarian Age):-

> "Some great Light is now at hand which shall enlighten and by degrees expel the former darkness."

He went on to say that the supernova would be followed by a major conjunction of planets when the actual fulfilment of the prophecies of the supernova would be realised. This major conjunction occurred in 1603-4, together with the appearance of another supernova (this time in Ophiuchus), which marked the death of Queen Elizabeth I, the enthronement of James VI of Scotland as King of England, and the start of Francis Bacon's programme to reveal openly to the world for the first time what his great 'Rosicrucian' scheme was: *i.e.* the Great Instauration (or Restoration) - the world-wide Reformation of all Arts and Sciences, *etc.*. This programme Francis discussed and proposed openly in his so-called philosophical works that bear the name of 'Bacon'.

Robert Fludd also helped to elucidate what was going on. In his *Tractatis Apologetica* defending the Rosicrucians, he explained that the 1603-4 planetary conjunctions were a sign to the Rosicrucian Brotherhood to expand their membership and begin the great Restoration (*i.e.* Instauration) of the World, emerging into the open after having prepared their work during a period of secrecy that had begun at the time of the supernova in Cassiopeia.

6. Shakespeare, *The Tempest,* V, i.

7. The choice of "Hamilcar's son" in the emblem on page 33 of Peacham's *Minerva Britanna* - the emblem which shows the invisible knight brandishing his spear - is noteworthy. 'Hamilcar' and 'Amilchar' are very close, if not identical, in their sounding.

8. The slicing off of the crown of the head is an allusion adopted by the Mystery Schools of old to imply the full opening of the crown chakra in universal consciousness or enlightenment. The effect of this occurring is a feeling as if the top of one's head really is not there anymore, and one's consciousness is simply no longer limited by any personal form whatsoever. To reach this type of consciousness requires loving effort and self-sacrifice, hence its association with 'death'.

9. U is alternately drawn as the zodiacal sign of the Ram, ♈, which rules the head of man. The ram particularly signifies the *Agnus Dei* ('Lamb of God') which is slain as the immaculate sacrifice. It is the sign or symbol of the Master. Similarly, in the Druidic Alphabet, **Y** was the symbol for the perfect or holy man. The sign is made by a man standing upright, with his hands raised up in adoration (or reception) of the divine Light from above, thus making the sign of the **Y** - the sign of the Grail.

10. Ovid, *Metamophoses,* X, 196.

11. Hence also the real reason for the so-called 'blue blood' of royalty and nobility.

12. Shakespeare, *Winter's Tale,* IV, iv. Apollo was said to take the form either of a swineherd or a shepherd - a country swain with stick or crook.

13. *Bacon's Works* (1740), Vol,IV, p.487.

14. The date is printed as "February 1577", but this is 1588 by our modern reckoning.

15. See F.B.R.T. Journal I/4, *The Great Vision,* pp.195-204, 'The French Academy'.

16. It is of course possible that the actual number of students given was not the factual number, but that Francis adopted the four **A**'s as a purely symbolic portrayal of truth for the purposes of his book. The argument could be taken even further, and it could be said that the so-called 'French Academy' never actually existed in a historic sense as described, but was purely an emblematic story invented and written by Francis Bacon so as to convey his ideas and intentions; but my own personal feelings are that the Academy did exist as such, even if Francis did use a purely symbolic rather than factual number of students for his literary purposes.

17. Letter from Sir Ayias Paulet to Sir Nicholas Bacon (Sept. 1577). See F.B.R.T. Journal I/4, *The Great Vision,* pp.191-192, 'Return to France'.

18. See *ante* 'Spenser', p.130; and F.B.R.T. Journal I/4, *The Great Vision,* p.192.

19. See F.B.R.T. Journal I/4, *The Great Vision,* pp.180-182, 'Return to France'.

20. Francis Bacon was 16 years old on 22nd January 1577.

21. F.B.R.T. Journal I/3, pp. 32-33, and FBRT Journal I/4, pp.118, 174-177.

22. See F.B.R.T. Journal I/4, *The Great Vision.*

23. Frances A. Yates, 'Religious Processions (1583-4): The Family of Love', *Astraea* (Routledge & Kegan Paul, 1975). See also Wallace Kirsop, *The Family of Love in France* (Sydney University Press, III, 1964-5).

    Why "ten languages"? The number **10** brings to mind the ten principles and keys of the cabbalistic Tree of Life.

24. L. Voet, *The Golden Compasses* (Amsterdam, 1969), ch.I, p.91. See also Frances A. Yates, 'Religious Processions (1583-4): The Family of Love', *Astraea* (Routledge & Kegan Paul, 1975).

25. *Venus and Adonis* renders the '**AA**' in the alternative form of '**VA**' - the two triangles, which, when united, give the six-pointed Christ Star, signet of the Master. Francis Bacon used this form later on, under the mask of **V**alentine **A**ndreas, to publish the Rosicrucian manifestoes of the great work of which he was the Architect.

26. See *Tudor Problems* by Parker Woodward, pp.94-95, 178-179.

27. Francis Bacon, *De Interpretatione Naturæ.* (Works, 1740 ed. Vol.II, p.263.)

Chapter 9: A Rare and Unaccustomed Suite

1. This is the earliest dated letter written by Francis Bacon that Francis' biographer, James Spedding, was able to find. See James Spedding, *The Letters and the Life of Francis Bacon,* vol.I, ch.II.1.

2. See F.B.R.T. Journal I/4, *The Great Vision,* pp.223-240, 'Anthony Bacon, Secret Agent'.

3. This indicates that Anthony Bacon was in all probability in England at least by July 1580, in order to make report to Walsingham, Burghley and the Queen.

4. See W. Holden, 'Francis Bacon's Association with Gray's Inn', *Baconiana,* vol. XXXV, no.141.

5. Francis Bacon was admitted to the Grand Company of Ancients of Grays Inn, by Order of Pension dated 21st November 1576, along with all of Sir Nicholas Bacon's sons. (See F.B.R.T. Journal I/4, *The Great Vision,* pp.184-5.)

6. James Spedding, *The Letters and the Life of Francis Bacon,* vol.I, ch.II.2.

7. *i.e.* Saturn.

8. James Spedding, *The Letters and the Life of Francis Bacon,* vol.I, ch.V.2.

9. Six letters in all are extant concerning Francis' rare and unaccustomed suit to the Queen, all written from Gray's Inn:-

   (i) to Lady Burghley, 16/9/1580;
   (ii) to Lord Burghley, 16/9/1580;
   (iii) to Lord Burghley, 18/10/1580;
   (iv) to Sir Francis Walsingham, 25/8/1585;
   (v) to Lord Burghley, 6/5/1586;
   (vi) to Lord Burghley, undated, but written in 1592.

   A good commentary on these letters can be found in William. T. Smedley's article, 'The Mystery of Francis Bacon', Part I, *Baconiana,* vol.IX, no.33, pp.9-22.

10. Mrs. Elizabeth Wells Gallup, *The Bi-Literal Cipher of Sir Francis Bacon* (Howard Publishing Co., Detroit, 3rd ed., 1901), pp.361-362.

11. Mrs. Elizabeth Wells Gallup, *The Bi-Literal Cipher of Sir Francis Bacon* (Howard Publishing Co., Detroit, 3rd ed., 1901), pp.28-29.

12. Mrs. Elizabeth Wells Gallup, *The Bi-Literal Cipher of Sir Francis Bacon* (Howard Publishing Co., Detroit, 3rd ed., 1901), p. 46.

13. Mrs. Elizabeth Wells Gallup, *The Bi-Literal Cipher of Sir Francis Bacon* (Howard Publishing Co., Detroit, 3rd ed., 1901), p. 91.

14. *i.e.* in the Word Cipher - the principal cipher in which Francis Bacon wrote his history of his times in dramatic and emotional detail, and to which the Biliteral Cipher is but an introduction and précis.

15. Mrs. Elizabeth Wells Gallup, *The Bi-Literal Cipher of Sir Francis Bacon* (Howard Publishing Co., Detroit, 3rd ed., 1901), p.353.

16. Mrs. Elizabeth Wells Gallup, *The Bi-Literal Cipher of Sir Francis Bacon* (Howard Publishing Co., Detroit, 3rd ed., 1901), pp.353-354.

17. I think I should emphasise here that I do not think the motives of the two Cecils - Lord Burghley and Robert Cecil - were the same by any means. Robert Cecil acted through pure malice, and revenge for the misfortune of being born a hunchback, with an unlikeable nature that brought forth reactions from others that he did not like. He particularly vowed to actively bring about the downfall and humiliation of both Tudor princes - Francis [Bacon] and Robert [Devereux] - simply because of who they were; and also Anthony Bacon, because Anthony was so closely involved with and loved by both of the princes. Robert Cecil used all his wiles to bring this about, as well as to further himself so that he realised, ultimately, a position of power that was, in effect, stronger than that of the Queen herself, whom he detested. In the end he achieved his greatest aim - the domination, humiliation and tragic death of the Queen herself - the only 'medicine' to satisfy his crazed soul.

    Robert's father, Sir William Cecil (Lord Burghley), was not by any means of the same character as his hunchback son. Burghley was devoted to serving the Queen and, more particularly, his country. He was worldly-wise, although perhaps never with the same depth, height and range of wisdom as his one-time friend, Sir Nicholas Bacon. He was a peace-lover, and was always cautious, even fearful, of any scheme or person that might jeopardise the delicate status-quo. He became increasingly anxious about and even actively opposed to the Earl of Leicester, whom he viewed as one of the greatest threats to the stability of the throne and country. Francis, Leicester's son, was thus thrown into this category, particularly as a real (and rightful) contender for the throne. I think there is no doubt that Burghley considered it quite possible that Francis and Leicester might act as Francis' grandfather, John Dudley, then Duke of Northumberland, had once acted. And where would he (Burghley) be then? Burghley obviously enjoyed his own position, and he worked hard, and slyly, to obtain the same position for Robert, his son. The Cecils were not a wholly altruistic family by any means; in fact, they were (with the exception of Robert) quite a 'normal' family!

Chapter 10: Secret Intelligence and the Grand Tour

1. See F.B.R.T. Journal I/4, *The Great Vision,* pp.223-240, 'Anthony Bacon, Secret Agent'. The four missions under the alias of 'Edward Burnham' took place at the following dates:-

    (i) February-March 1578, to Picardy and the Low Countries;
    (ii) June 1578, to Champagne and the Low Countries (*via* Paris);
    (iii) April-May 1579, to Artois, Hainault and the Low Countries;
    (iv) February-April 1580, to Portugal (*via* Paris).

2. Blaise de Vigenère, *Traicts des Chiffres* (1586), dedicated to Monsieur Antoine Seguier.

3. See Edith Durning-Lawrence, 'A French Cipher Book', *Baconiana,* vol.XLIII, no.160.

4. See Pierre Amboise's *Life* of Francis Bacon, quoted.

5. See Bodley's letter, quoted.

6. Joel J. Epstein, *Francis Bacon: A Political Biography* (Ohio University Press, 1977), ch.II, p.25.

7. Speech delivered in Parliament by Francis Bacon, 28th November 1601, in which he stated that he had "been a member of the House these seven Parliaments" and that he was present when one Arthur Hall was committed to the Tower (1581). See Heywood Townsend, *Historical Collections* (London, 1680). See also the *Journals of the House of Commons, 14-18th February 1580.*

Francis Bacon also referred to specific activities which occurred in the House of Commons in "a °23 Eliz." (*i.e.* February *1580,* or 1581 modern dating). See *Parliamentary Debates in 1610,* ed. Samuel Gardiner (London, 1862), p.38.

8. Nicholas Faunt, who left Paris for England on 22nd March 1582, referred to having seen Lady Anne Bacon and Francis Bacon in a letter written by him on 15th April 1582. (Birch's *Memorials,* vol.I, p.22.)

9. The letter could not have been written sooner because (a) Bodley himself had only returned to England in 1580 after spending four years abroad in Italy, France and Germany (as stated by Sir Thomas Bodley in his *Life*), and (b) Francis Bacon, when he first went abroad in 1575. was only 14 years old, under banishment and in the charge of Sir Amyas Paulett at the French Court, and not undertaking a journey through many lands to ascertain the condition of both Church and State in each country visited (as the Bodley letter states). Neither could the letter have been written later than April 1582, as Francis Bacon's life can be accounted for in other ways after that date. The period of March 1581 to 15th April 1582 is the only unaccounted-for period of Francis' life in which he might have undertaken such a journey.

10. Pierre Amboise, *Discourse on the Life of M. Francis Bacon, Chancellor of England,* (Paris, 1631).

11. *Reliquiæ Bodleianæ,* no.ccxxxii. (See 'An Interesting Letter', *Baconiana,* vol.VI, no.21.)

12. Nicholas Faunt commenced his close acquaintance, correspondence and friendship with Anthony Bacon when both were in Paris in 1580. In the beginning of 1581 Faunt travelled into Germany, where he spent three-and-a-half months; after which he journeyed south to Italy. The next six or seven months he spent travelling between Geneva and the north of Italy, during which time he must have met Anthony Bacon in Geneva frequently (if Anthony, as it seems, was principally resident there between June-November 1581). From Geneva Faunt went to Paris, being there at least from February-March 1582. He returned to London at the beginning of April 1582.

13. Montaigne left his home in Bordeaux in the summer of 1580 and travelled north to Paris and beyond, to meet up at Beuamont-sur-Oise (north of Paris) with a young French nobleman by the name of d'Esstissac on the 5th September 1580. This young nobleman carried letters of introduction to the d'Estes and Medicis from the French Court at Paris. The enlarged party then journeyed south-eastwards, through Lorraine and Alsace to Basel, arriving there late September 1580. From thence the little party continued eastwards, following the Rhine up to Constance, then north-eastwards to Augsburg and Munich, in Bavaria, visiting many places and castles *en route.* Turning south they travelled into Tyrol, to Innsbruck, and from thence along the great trading route to Venice, *via* Trent, Verona and Padua, arriving in Venice on the 4th November 1580. From Venice they travelled to Florence *via* Padua, Ferrara and Bologna, arriving in Florence on the 22nd November 1580, where lived the Grand Duke of Tuscany, Francesco de Medici. After two days in Florence the party journeyed to Sienna, and from thence made their way south to Rome where they stayed for four-and-a-half months (December 1580 to mid-April 1581). Montaigne left Rome on the 19th April 1581 (probably about the same time that Francis set sail from England on his grand tour), and returned to Florence *via* Ancona, Pisa and Lucca, arriving in Florence on the 22nd June 1581. After returning to Lucca and Sienna again, then Rome and Florence, Montaigne eventually began his journey home. He finally left Rome on the 15th October 1581, travelling to Milan *via* Piacenza and Pavia. Milan was left on the 28th October and, after reaching the Alps *via* Leghorn and Turin, he crossed the mountains into Provence and hence home to Bordeaux.

14. See Frances A. Yates, 'The Joyeuse Magnificences', *Astraea: the Imperial Theme in the Sixteenth Century,* for a good account of these entertainments.

Chapter 11: The Greatest Birth of Time

1. In describing the list of contents of his manuscript entitled *Temporis Partum Masculum,* Bacon states:-

   > "The first chapter of a book of the same argument [*i.e.* of the *Valerius Terminus*], written in Latin, and destined to be separate and not public."

2. James Spedding, *The Works of Francis Bacon* (London, 1857), vol.III, p.524.
3. Letter written by Francis Bacon to Father Fulgentio describing his Great Instauration. The letter is undated, but probably belongs to the period 1622-3. It was printed in *Baconiana* (London, 1679), p.200.
4. Alfred Dodd, *Francis Bacon's Personal Life Story* (London, 1910 & 1986), p.126.
5. Alfred Dodd, as a Freemason, may have had access to as yet unpublished manuscripts that form the remainder of Francis Bacon's *Temporis Partus Masculus* - documents which I have not seen and can therefore not comment upon one way or the other.
6. See Francis Bacon's essay, 'Cupid, or an Atom', in his *The Wisdom of the Ancients,* for a fuller explanation of Cupid.
7. For the significance of the 'W' or 'Double-V' and Cassiopeia, see F.B.R.T. Journal I/2, *The Virgin Ideal.*
8. See Francis Bacon's essay, 'Pan, or Nature', in his *The Wisdom of the Ancients,* for a fuller explanation of Pan.
9. There is cipher evidence pointing to the possible authorship (or principal authorship) of *Philostrate* as being by Francis Bacon, with Blaise de Vigenère acting as a mask and 'good pen' for Francis Bacon. See the work done by Arthur B. Cornwall, Brooklyn, New York - in mss, 'A Study of Anagrammatic Decipherments in the 1637 Paris edition of *Philostrate* by Blaise de Vigenère'. If this is so, it explains many things neatly - not least the truth concerning the true inventor of the Biliteral cipher - and suggests also that the principal author of *Traicts des Chiffres* may have been Francis Bacon, with Vigenère assisting and masking Bacon.
10. James Spedding, *The Works of Francis Bacon* (London, 1857), vol.III, p.525.
11. This nature shows itself strongly in Francis' own cipher records of his early history, particularly in his Word Cipher which is quite open and honest in its record of his own thoughts and passions as they moved and turned from day to day.
12. *Shiva* is the Third Person of the Hindu Trinity of *Brahma-Vishnu-Shiva.* Shiva corresponds to the Holy Ghost of the Christian Trinity, and is cabbalistically known as the Holy Intelligence - the 'feminine' aspect of Truth, which the Greeks personified as Sophia and as Pallas Athena. This is the primary impulse of the Baconian work, which is to do with the raising and illumination of the intelligence or consciousness of mankind. It is the civilising impulse; and Francis Bacon, whose soul is sometimes referred to as the Master R, is (as the Master R) known esoterically as the Lord of Civilisation for our present world epoch. Francis Bacon was the primary incarnation of the Master in respect of laying the foundations for his work in the Age of Aquarius. Hence he referred to himself as the Herald of the New Age.

13. Interestingly, from November 1581, when the first' Grand' Accession Day Tournament was held, to November 1602, when the final one was held (Queen Elizabeth died in March 1603), twenty-one Tournaments were held. **21** is particularly the number of the Virgin Bride, or World (*i.e.* the Soul) - the 'age' of maturity of a soul.

14. See 'Early Elizabethan Chivalry and Pageantry', F.B.R.T. Journal I/4, *The Great Vision.*

15. See 'Early Elizabethan Chivalry and Pageantry', F.B.R.T. Journal I/4, *The Great Vision.*

16. Roy Strong, *The Cult of Elizabeth* (Thames & Hudson, 1977), ch.IV, p.135.

17. At Whitehall it was possible for the Tournament to be opened to the general public on a large scale. Special stands were erected around the lists, allowing several thousands of spectators to see the grand spectacle.

18. In a letter to Anthony Bacon (25th January 1595) Francis states that the Queen, antagonized by his opposition to a policy of hers in Parliament in 1595, angrily referred to having "pulled" him "over the bar." See Joel J. Epstein, *Francis Bacon: A Political Biography* (Ohio, 1977), ch.2, p.25.

19. See Frances A. Yates, *Giordano Bruno and the Hermetic Tradition* (London, 1964), ch.XI, p.204.

20. See Frances A. Yates, *Giordano Bruno and the Hermetic Tradition* (London, 1964), ch.XII, p.206.

21. This science has also been used to great effect in Christianity - in Gregorian chant, for instance.

22. Philip Sidney, *Apology.*

23. Philip Sidney, *Apology.*

24. Philip Sidney, *Apology.* Compare with:-

    "Ignorance is the curse of God; Knowledge the wing wherewith we fly to heaven." (Shakespeare, II *Henry VI,* iv, 7.)

25. Adrian Gilbert was the brother of Sir Humphrey Gilbert and half-brother of Sir Walter Raleigh. Adrian was the chemist whom Lady Pembroke (*i.e.* Mary Sidney) employed at Wilton as her 'laborator'. Both Adrian Gilbert and John Davis were involved with Dee in his magical exercises, which included 'Angel summoning.' John Davis was the last seaman whom Dee ever taught.

26. Peter J. French, *John Dee: the World of an Elizabethan Magus* (London, 1972), ch.7, p.177.

27. It is interesting to note that, in 1616, Edmund Bolton, a historian, poet and friend of Francis Bacon, interested King James I in the forming of a Society for the Advancement of Learning, on Baconian and Freemasonic lines, called "King James, His Academe, or College of Honour." There were three classes of membership, and all members were to "Love, Honour and Serve one another according to the Spirit of St. John." The symbol of the Society was a green ribbon with the letters 'J.R.F.C.' ('Jacobus Rex, Fundator Collegii') beneath the imperial crown inscribed on it in gold.

28. See E. G. Harman, *Edmund Spenser and the Impersonations of Francis Bacon,* ch.XII, p.305.

29. See Peter J. French, *John Dee: the World of an Elizabethan Magus* (London, 1972), ch.6, pp154 - ch.7, p.181.

30. Hakluyt, *Principal Navigations* (Everyman edition), ch.VI, pp.3-5.

Chapter 12: The Catholic Menace

1. In the *Histoire Naturelle de Mre. Francois Bacon* - the French edition of Bacon's *Sylva Sylvarum* produced by Pierre Amboise and published in France in 1631 - are several sections of writings by Francis Bacon that are not to be found in other editions. In chapter 5, for instance, Francis, in discussing whether a man can still move once his head has been cut off, adds the following piece of historical information regarding his own life and experience:-

   > "I have nevertheless seen in Scotland the body of a gentleman of high rank and influence from which the head had just been severed, and I have seen this same headless body, when placed without delay in a wooden coffin, actually burst the coffin with great force. But I am unable to give any reason for it."

   This incident in all likelihood relates to what occurred in Edinburgh in March or April 1584, when Malcolm Douglas of Mains and John Cunningham of Drumquhassel were brought to trial for having conspired to seize the person of the king whilst on a hunting match. Both were condemned to death and, "with a speed which argued terror in the government, were executed in the public street of Edinburgh, before the sun had set, on the day of their trial." (Sir Walter Scott, *History of Scotland,* vol. II, ch.XXIII, p.203.) Shortly afterwards the Earl of Arran was arrested and executed as being the official responsible for the plot to seize the king. With his fall the reign of terror apparently ceased, and Sir Walter Scott records no more of these political executions of gentlemen of high rank and influence, as he would undoubtedly have done had any such taken place in Scotland afterwards.

   These executions took place whilst Sir Edward Wotton was in Edinburgh visiting James VI, and would probably have been witnessed by him and his train. The speed that the executions were carried out could have meant that the coffins were, because of the haste, badly made; and one of the coffins might conceivably have been burst apart by a *rigor mortis* of one of the bodies, seeing as they were placed quickly into the coffins after their heads had been lopped off. The same may have been true of the Earl of Arran. (See W. L. Goldsworthy, *Shake-Speare's Heraldic Emblems,* ch.II, pp.50-52.)

2. Thomas Vautrollier's printing business was in Blackfriars, London. He was a Huguenot. Richard Field became apprenticed to him in the 1580's and, when Vautrollier died, he took over the printing business.

3. Edmund Bolton (1575?-1633?), a historian and poet, was brought up a Roman Catholic and educated at Trinity College, Cambridge. Afterwards he resided at the Inner Temple, London. He became a retainer of the Duke of Buckingham, and through the latter's influence he was given a position in the Court of James I. Bolton was a contributer to *England's Helicon* (1600), and wrote numerous works, such as *Hypercritica* (1620?), *The Roman Histories of Lucius Iulius Florus* (transl. 1618), and *Nero Caesar, or Monarchie Depraved* (1624).

   It was Edmund Bolton who, in 1616, approached James I about the scheme to establish an 'Academie or College of Honour' in London, based on Baconian principles for the advancement of learning. It is thought that the location of this Academy was at Canonbury in Islington, and, since the building in question was erected by Prior Bolton in the reign of Henry VII, it might be another example of Bacon's humour to choose Edmund Bolton to be publicly associated with the request to the king for the founding of the Academy: for, without doubt, Francis Bacon was himself, in 1616, in a far better position to have made the request to James than Bolton was, and to have seen it carried out.

4. Simon Studion, *Naometria* (an unpublished manuscript dedicated to the Duke of Würtemburg, 1605, now in the Landesbibliothek at Stuttgart), folio 35 and folio 122 of the dedication.

5. See Alfred Dodd, *Francis Bacon's Personal Life-Story* (London, 1910 & 1986), ch.VI, p.123.

6. Hakluyt, *Principal Navigations.* See A. L. Rowse, *The Expansion of Elizabethan England* (London, 1955), ch.VI.

SELECT BIBLIOGRAPHY

The following publications are amongst those consulted and generally recommended, and are included in the following list as a possible selection of reading and reference matter for those who wish to study further. Bibliographies and reference material are included in many of these books listed, and the student will be able to follow his own trail once started.

This book list is additional to and forms a continuation of the Bibliography given in FBRT Journals I/3, *Dedication to the Light,* and I/4, *The Great Vision.* For a more complete bibliography the student should refer to the Select Bibliographies in these previous journals.

> "Some books are to be tasted, others to be swallowed, and some few to be chewed and digested."

Anonymous, *The Arte of English Poesie* (1589).

Cyril Aldred, *Akhenaten, Pharoah of Egypt - a new study* (Thames & Hudson, 1968).

Alice Bailey, *The Soul and its Mechanism: The Problem of Psychology* (Lucis Press, London, 1976).

Francis Bacon, *Sylva Sylvarum* (1626).

Francis Bacon, *Novum Organum; or a True Guide to the Interpretation of Nature* (orig,.publ. in Latin,1620), translated by the Rev. G. W. Kitchen (Oxford University Press, 1855).

Francis Bacon, *The Essays,* edited with an introduction by John Pitcher, and an appendix containing a selection from *The Wisdom of the Ancients* (Penguin Books, 1985).

Francis Bacon, ('Democritus Junior' *alias* Robert Burton), *The Anatomy of Melancholy* (1621).

Francis Bacon, ('T. Bright, *alias* Timothy Bright). *A Treatise of Melancholy* (1586).

Rev. Walter Begley, *Bacon's Nova Resuscitatio, or The Unveiling of his Concealed Works and Travels* (Gay & Bird, London, 1905).

Edwin Bormann, *Francis Bacon's Cryptic Rhymes, and the truth they reveal* (Siegle, Hill & Co., London, 1906).

Robin Brumby, *Doctor John Dee, or the Original 007* (Dacorum College, Hemel Hempstead, 1977).

Sir E. A. Wallis Budge, *Egyptian Religion: Egyptian Ideas of the Future Life* (Routledge & Kegan Paul, London, 1972).

Titus Burckhardt, *Alchemy: Science of the Cosmos, Science of the Soul,* transl. by William Stoddart (Stuart & Watkins, London, 1967).

Thomas Swinburne Carr, *A New Classical Lexicon of Biography, Mythology and Geography* (Simpkin, Marshall & Co., London, 1858).

David Cecil, *The Cecils of Hatfield House* (Constable & Co., 1973; Sphere Books, 1975).

Rodman R. Clayson, *Egypt's Ancient Heritage,* Rosicrucian Library Vol. XXXI (AMORC, San Jose, California, 1971).

Leorhard Cottrell, *The Lost Pharoahs* (Pan Books, London, 1956).

Peter Dawkins, *The Great Vision,* FBRT Journal I/4 (FBRT, 1985).

Peter Dawkins, *Dedication to the Light,* FBRT Journal I/3 (FBRT, 1984).

Dr. John Dee, *General and Rare Memorials pertayning to the Perfect Arte of Navigation* (1577).

Dr. John Dee, *The Heptarchia Mystica of John Dee,* transcribed, introduced and annotated by Robert Turner (Aquarian Press, Wellingborough,1986).

Dr. John Dee, *The Rosicrucian Secrets: Their Excellent Method of making Medicines of Metals also their Lawes and Mysteries,* edited with a preface and introduction, and critical and explanatory notes, by E. J. Langford Garstin (Aquarian Press, Wellingborough, 1985).

Dr. John Dee's *The Hieroglyphic Monad,* transl. and with a commentary by J. W. Hamilton-Jones (Watkins, London, 1947).

Nicholas Culpeper, *Culpeper's Complete Herbal, and English Physician* (facsimile of 1826 edition publ. by Gareth Powell, Hong Kong, 1979). The original edition of Culpeper's *The English Physician Enlarged* was published in 1653.

Savitri Devi, *A Son of God: The Life and Philosophy of Akhenaton, King of Egypt* (The Philosophical Publishing House, London, 1946).

Alfred Dodd, *Francis Bacon's Personal Life Story,* Vol.I: *The Age of Elizabeth* and Vol. II: *The Age of James* (Rider & Co., London, 1986).

Sir Edwin Durning-Lawrence, Bt., *Bacon is Shakespeare,* together with a reprint of Francis Bacon's *Promus of Formularies and Elegancies* (Gay & Hancock, London, 1910).

Joel J. Epstein, *A Political Biography* (Ohio University Press, Ohio, 1977).

Johannes Fabricius, *Alchemy: The Medieval Alchemists and their Royal Art* (Rosenkilde & Bagger, Copenhagen, 1976).

Winston Graham, *The Spanish Armadas* (Doubleday & Co., New York, 1972).

W. Lansdown Goldworthy, *Shake-speare's Heraldic Emblems; their origin and meaning* (Witherby, London, 1928).

Manly P. Hall, *Freemasonry of the Ancient Egyptians* (Philosophical Research Society, Los Angeles, 1973).

Edward George Harman, *Edmund Spenser and the Impersonations of Francis Bacon* (Constable & Co., London, 1914).

Douglas Hill, *The Illustrated Faerie Queene: A Modern Prose Adaptation* (Newsweek Books, London, 1980).

Martin Holmes, *Shakespeare and Burbage* (Phillimore, London, 1978).

John H. Ingram, *Christopher Marlowe and his Associates* (Grant Richards, London, 1904).

John Ivimy, *The Sphinx and the Megaliths* (Abacus, London, 1956).

Lucie Lamy, *Egyptian Mysteries: New Light on Ancient Knowledge* (Thames & Hudson, London, 1981).

J. Lemprière, *Lemprière's Classical Dictionary* (Bracken Books, London, 1984).

Isha Schwaller de Lubicz, *Her-Bak: The Living Face of Ancient Egypt,* transl. by Charles Edgar Sprague (Inner Traditions International, New York, 1978).

Isha Schwaller de Lubicz, *Her-Bak: Egyptian Initiate,* transl. by Ronald Fraser (Inner Traditions International, New York, 1978).

Isha Schwaller de Lubicz, *The Opening of the Way: A Practical Guide to the Wisdom of Ancient Egypt,* transl. by Rupert Gleadow (Inner Traditions International, New York, 1981).

R. A Schwaller de Lubicz, *The Egyptian Miracle: An Introduction to the Wisdom of the Temple,* transl. by André and Goldian VandenBroeck (Inner Traditions International, New York, 1985).

R. A Schwaller de Lubicz, *The Temple in Man: The Secrets of Ancient Egypt,* transl. by Robert and Deborah Lawlor (Autumn Press, Brookline, Massachusetts, 1977).

R. A Schwaller de Lubicz, *Sacred Science: The King of Phaaonic Theocracy,* transl. by André and Goldian VandenBroeck (Inner Traditions International, New York, 1982).

R. A Schwaller de Lubicz, *Symbol and the Symbolic: Ancient Egypt, Science and the Evolution of Consciousness,* transl. by Robert and Deborah Lawlor (Inner Traditions International, New York, 1978).

Manfred Lurker, *The Gods and Symbols of Ancient Egypt: An Illustrated Dictionary* (Thames & Hudson, London, 1980).

John Lyly, *Euphues and his England* (1580).

John Lyly, *Euphues Anatomy of Wit* (1581).

John Baines ad Jaromir Málek, *Atlas of Ancient Egypt* (Phaidon, Oxford, 1980).

Christopher Marlowe, *The Works of Christopher Marlowe,* edited with notes and introduction by Francis Cunningham (Chatto & Windus, London, 1884).

Kurt Mendelssohn, *The Riddle of the Pyramids* (Thames & Hudson, London, 1974).

Michel de Montaigne, *Essays,* transl. and with an introduction by J. M. Cohen (Penguine Books, 1958).

Charles Nicholl, *The Chemical Theatre* (Routledge & Kegan Paul, London, 1980).

Jill Purce, *The Mystic Spiral: Journey of the Soul* (Thames & Hudson, London, 1974).

Stanislas Klossowski de Rola, *Alchemy: The Secret Art* (Thames & Hudson, London, 1973).

A. L. Rowse, *The Expansion of Elizabethan England* (Macmillan & Co., London, 1955; Papermac, 1981).

Edouard Schuré, *The Mysteries of Ancient Egypt* (Multimedia Publ. Corp., New York, 1971).

Alan W. Shorter, *The Egyptian Gods:* A Handbook (Routledge & Kegan Paul, London, 1937).

Dr. Herbert Silberer, *Hidden Symbolism of Alchemy and the Occult Arts,* transl. by Smith Ely Jelliffe (Dover Publications, New York, 1971).

William Smith, *Classical Dictionary of Biography, Mythology and Geography (Murray, London, 1886).*

William Smith, *Dictionary of Greek and Roman Antiquities* (Murray, London, 1884).

Lewis Spence, *The Mysteries of Egypt, or The Secret Rites and Traditions of the Nile* (Rudolf Steiner Publ., New York, 1972).

Edmund Spenser, *Complete Works of Edmund Spenser,* Globe Edition, edited by R. Morris (Macmillan & Co., London, 1879).

Edmund Spenser, *The Faerie Queene, Shepheards Calender and Poems, etc.* (1st Folio, 1611; 2nd Folio, 1617.)William Webbe, Discourse of English Poetrie (1586).

G. Walter Steeves, *Francis Bacon: A Sketch of his Life, Works and Literary Friends* (Methuen & Co., London, 1910).

Sir Philip Sydney, *The Countess of Pembroke's Arcadia,* edited and with an introduction and notes by Maurice Evans (Penguine Books, 1977).

Peter Tompkins, *The Magic of Obelisks* (Harper & Row, New York, 1981).

Peter Tompkins, *Secrets of the Great Pyramid* (Allen Lane, London, 1973).

William Webbe, *Discourse of English Poetrie* (1586).

John Anthony West, *Serpent in the Sky: The High Wisdom of Ancient Egypt* (Wildwood House, London, 1979).

George Whetstone, *Sir Philip Sydney, His Life, His Valiant Death, and true Virtues* (1587).

Neville Williams, *Francis Drake* (Weidenfeld & Nicolson, London, 1973).

Alchemy: A Comprehensive Bibliography of the Manly P. Hall Collection of Books and Manuscripts, including related material on Rosicrucianism and the writings of Jacob Boehme, edited by Ron. Charles Hogart (Philosophical Research Society, Los Angeles, 1986).

Baconiana, the Journal of the Francis Bacon Society.

The Book of the Dead: Famous Egyptian Papyri (Papyri of Ani, Hunefer, Anhai), with commentaries by Evelyn Rossiter (Regent Books, London, 1984).

The Hermetic Journal, edited by Adam McLean (publ. quarterly by The Hermetic Research Trust).

The Kybalion: A Study of the Hermetic Philosophy of Ancient Egypt and Greece, by Three Initiates (Yogi Publication Society, Chicago, 1922).

Manes Verulamiani, a collection of Latin elegies first written and published in 1626 in tribute to Francis Bacon, herein collected and translated with notes, edited by W. G. C. Gundry (Francis Bacon Society, London, 1950).